Sustainable Development Using Geospatial Techniques

Scrivener Publishing
100 Cummings Center, Suite 541J
Beverly, MA 01915-6106

Publishers at Scrivener
Martin Scrivener (martin@scrivenerpublishing.com)
Phillip Carmical (pcarmical@scrivenerpublishing.com)

Sustainable Development Using Geospatial Techniques

Edited by

Disha Thakur

Sanjay Kumar

Har Amrit Singh Sandhu

and

Chander Prakash

Scrivener Publishing

WILEY

This edition first published 2024 by John Wiley & Sons, Inc., 111 River Street, Hoboken, NJ 07030, USA and Scrivener Publishing LLC, 100 Cummings Center, Suite 541J, Beverly, MA 01915, USA

For more information about Scrivener publications please visit www.scrivenerpublishing.com.

Wiley Global Headquarters
111 River Street, Hoboken, NJ 07030, USA

For details of our global editorial offices, customer services, and more information about Wiley products visit us at www.wiley.com.

Library of Congress Cataloging-in-Publication Data

ISBN 978-1-394-21434-1

Front cover images supplied by Adobe Firefly
Cover design by Russell Richardson

Set in size of 11pt and Minion Pro by Manila Typesetting Company, Makati, Philippines

Printed in the USA

10 9 8 7 6 5 4 3 2 1

Contents

Preface

Sustainable development addresses pressing global challenges such as climate change, resource depletion, and social inequalities by promoting responsible consumption, renewable energy, and inclusive policies so that the needs of the present generation are met without compromising the ability of future generations to meet their own needs. By embracing such practices, societies can achieve long-term prosperity, resilience, and equity, fostering a healthier planet and equitable communities. The rapid pace of technological advancement has opened up several avenues for tackling the pressing issues associated with sustainable development. In this context, geospatial techniques have emerged as a powerful tool, offering unprecedented capabilities to monitor, analyze, and manage our natural and built environment. They enable precise mapping and analysis of climate, vegetation, water resources, and urban growth, facilitating informed decision-making for sustainable practices.

This book is a collaborative effort stemming from our diverse and vast experience that addresses the critical need for innovative solutions in the pursuit of sustainability. As environmental degradation, climate change, and resource management challenges escalate globally, the demand for effective strategies to mitigate these issues becomes increasingly urgent. The book provides a comprehensive overview of how geospatial techniques can be harnessed to achieve sustainable development goals, bridging the gap between technical know-how and practical applications. Combining theoretical foundations with real-world case studies equips practitioners, policymakers, and scholars with the necessary knowledge and tools to implement sustainable practices effectively. In the present book, readers will find detailed discussions on the various applications of geospatial techniques in sustainability initiatives. From integrating remote sensing data in agricultural applications to using Geographic Information Systems (GIS) in urban planning and resource management, each section delves into specific methodologies and their impact on promoting sustainability. The case studies included not only highlight successful implementations but also address the challenges and limitations encountered, providing a balanced perspective on the potential and constraints of these technologies.

Furthermore, the book emphasizes the importance of interdisciplinary collaboration in advancing sustainable development. Drawing on expertise from fields such as environmental science, engineering, and urban planning showcases how a holistic approach can lead to more effective and resilient solutions. This integrative perspective is crucial for addressing the complex and interconnected nature of global sustainability challenges.

Ultimately, this book aims to inspire and inform a wide range of stakeholders, from academics and researchers to practitioners and policymakers, about the transformative potential of geospatial techniques. We hope to contribute to the ongoing efforts to create a more sustainable and equitable world by fostering a deeper understanding of these tools and their applications.

Dr. Disha Thakur
Dr. Sanjay Kumar
Dr. Har Amrit Singh Sandhu
Prof. Chander Prakash

1

Development of a Two-Layer Meta-Classifier–Based Drought Stress Detection System for Wheat Crop Sustainability

Ankita Gupta[1]*, Lakhwinder Kaur[2] and Gurmeet Kaur[3]

[1]*Department of Computer Science and Engineering, CT Group of Institutions, Shahpur Campus, Jalandhar, Punjab, India*

[2]*Department of Computer Science and Engineering, Punjabi University, Patiala, Punjab, India*

[3]*Department of Electronics and Communication Engineering, Punjabi University, Patiala, Punjab, India*

Abstract

Focusing on the Raj 3765 wheat variety's chlorophyll fluorescence–based canopy level images, this research aims to formulate an ultraprecise, bilayered meta-classifier to curtail economic adversity within wheat farming. Preprocessing of images, utilizing the total variation - L1 (TV-L1) with a primal-dual algorithm, succeeded by contrast stretching (Min-Max), emerged as paramount. This pipeline's choice hinged on mean squared error (MSE), peak signal-to-noise ratio (PSNR), and structural similarity index measure (SSIM) average values. Regarding feature selection and optimization, 23 gray-level co-occurrence matrix (GLCM) features, 9*255 color band features, 10 landmarks, and 20 wheat canopy geometric features proved significant. These features substantially enhanced machine learning algorithms' efficacy. Most prominent classifiers are included in this research, *viz*, random forests (RF), naive Bayes (NB), K-nearest neighbors (KNN), and support vector machines (SVM), with RF achieving an accuracy of 93.06%, followed by NB and KNN. A blend of these classifiers showed RF and NB flourishing in the initial layer and KNN in the succeeding layer, exhibiting numerical constancy relative to the difference between training and testing accuracy. Our custom meta-classifier

**Corresponding author*: gupta89ankita@gmail.com

Disha Thakur, Sanjay Kumar, Har Amrit Singh Sandhu and Chander Prakash (eds.) Sustainable Development Using Geospatial Techniques, (1–40) © 2024 Scrivener Publishing LLC

accomplished an apex accuracy of 95.8%. The research concludes with a case study showcasing the potential of such advancements in mitigating farmers' losses.

Keywords: Wheat, water-stress, machine learning, phenotyping, chlorophyll fluorescence, image processing, meta-learning, sustainability

1.1 Introduction

Wheat and barley are two main crops on which consistent research has been done since the independence of India [1]. Wheat is a nutrient-rich, inexpensive source of energy that is highly consumed by all sections of society throughout the world [2]. Due to an accelerated increase in the population, the country (India) will not be able to feed its population due to lower yields in these crops [3]. The Indian government has been making sincere efforts to overcome this challenge by introducing new high-yielding and stress-tolerant wheat varieties [4]. This can be accomplished by increasing the area under irrigation and technological regimes and developing sustainable agriculture practices [5].

Techniques proved to be most fruitful in differentiating stress-tolerant and stress-susceptible genotypes utilizing physiological tools like canopy morphological analysis [6]. Most of the authors are using the crop water stress index (CWI) as a key metric for the identification of water stress by measuring canopy temperature [5, 7]. But still, there is a need to develop a technique that is purely nonintrusive in a cost-effective manner. It has been observed that an aggregate of 60% of global agriculture uses rain-fed farming practices, which are under threat due to unprecedented rains [8]. This underlines the need for the development of an affordable phenotyping platform for wheat crops under drought and controlled conditions for stress identification. The authors can identify the most dominant factors that are influenced by drought stress nonintrusively by analyzing color spectrum, texture, and statistical features [9–11]. The yield losses can be significantly affected by the water stress caused by the deficiency in soil moisture levels. This leads to an urge to analyze the soil and its characteristics for the development of efficient drought control systems.

There are standard protocols (gravimetric method) for conducting soil sampling, and the analysis of the water content in the soils is done in the laboratory using instrumentation [12, 13]. The method requires a lot of fieldwork and does not exactly give the status of plant conditions in a non-invasive manner. Another approach is the analysis of the subsurface of the soil using multiple instruments and sensors. This method requires field work and the installation of moisture probes along with other peripheral equipment that can be connected to computer networks [14, 15]. The neutron probes use radioactive material to analyze the water content in the soil. Multiple instruments with similar capacity and functionality use principles of resistance, dielectric permittivity, thermal conductivity, and thermalized neutrons [16–18]. Through the soil-water retention or soil-water characteristic curve [19], tensiometers relate the relationship between pressure potential and soil water tension to soil water content [20]. This relationship is unique to each soil or porous medium.

Furthermore, indicators known as "organ level stress indicators" can also be used to identify water stress in wheat [21]. In this study, various parts of the wheat plant, including the canopy, roots, pollens, grains, development stages, hormonal regulation, and leaves, have been analyzed to gain insights into the signals emitted by these components in response to water stress. The objective was to understand how each part of the plant indicates the presence of water stress. This objective can be accomplished by employing high-throughput imaging technologies, such as hyperspectral imaging, thermal imaging, and chlorophyll fluorescence (CHF) imaging. These advanced imaging techniques enable the computation of spectral vegetation indices and xanthophyll indices using satellite imagery and cameras, which capture and quantify the "photosynthetic activity" within a given frame. The utilization of these imaging technologies proves to be highly valuable in assessing and monitoring the impact of water stress on different parts of the wheat plant [22].

1.2 Literature Review

Water stress in plants is a complex phenomenon that has significant implications for their growth and development. As researchers have explored this subject, they have observed a range of changes in plants exposed to dehydration stress, including reductions in the number of leaves, leaf size, leaf area, and leaf longevity [15, 23–26]. These alterations ultimately lead to a decrease in the overall canopy size and shape of the plants. Furthermore, water stress also affects the root-to-shoot ratio and the biomass of the plants, as they struggle to absorb water in different parts of their body under drought conditions [27].

To gain a deeper understanding of water stress in wheat, researchers have conducted experiments using field setups to impose water stress at various stages of crop growth and development [28, 29]. They have focused on parameters such as canopy biomass and root-shoot size to study drought tolerance in wheat crops [30, 31]. It has been observed that during extended dry periods, crops extend their roots deeper into the soil, altering their root-shoot ratio and canopy morphology [31–33].

In addition to the noninvasive techniques mentioned earlier, other technical methods have proven valuable in detecting water stress in plants. Thermal imaging has been widely used to assess changes in plant temperature distribution, providing insights into their water status and stress levels [6, 34]. Hyperspectral imaging, on the other hand, analyzes the interaction of plants with different wavelengths of light, allowing researchers to identify specific spectral signatures associated with water stress [35–38]. These techniques, along with advancements in digital imaging instrumentation and high-throughput phenotyping capabilities of cameras, have transformed precision agriculture. Moreover, the use of 3D imaging and LiDAR has gained prominence, enabling detailed analysis of plant architecture and structure, as well as precise measurement of plant height, canopy structure, and biomass [39].

Traditional plant phenotyping methods, involving manual processing of agricultural samples, are time-consuming and inefficient. However, the advent of data collection tools like cameras and unmanned aerial vehicles (UAVs) has revolutionized this field [10, 15, 40]. CHF imaging using remote sensing is a valuable tool for detecting and assessing drought and water stress conditions in vegetation. It measures the light emitted by chlorophyll during photosynthesis and provides indicators of stress and changes in photosynthetic efficiency. By quantifying fluorescence signals, remote sensing enables the early detection of water stress, estimation of water use efficiency, and large-scale monitoring of vegetation health. This information aids in timely intervention and resource allocation, supporting effective drought management and sustainable agricultural practices.

Machine learning approaches have significantly improved the accuracy of CHF-based systems by leveraging time series data for trend analysis. By automating the process of feature extraction, machine learning pipelines facilitate the development of water stress detection systems [41–43]. In the field of wheat research, drought stress detection plays a crucial role in sustainable crop management. While deep learning models have shown promise, the utilization of meta-classifiers remains largely unexplored. Deep learning techniques can also be utilized when technical specialists are not available [44, 45]. However, statistically, it can also be observed that the researchers are directly using deep learning algorithms that have high overload and low interpretability.

Furthermore, the availability of datasets for drought water stress detection, particularly at the canopy level utilizing CHF imaging, is a challenge, especially for Indian wheat varieties. Such datasets are rarely available and pose a limitation in conducting comprehensive studies on drought stress in wheat crops. To address this limitation, it is crucial to prioritize the collection and sharing of canopy-level CHF datasets for Indian wheat varieties.

Analyzing existing literature reveals several challenges in drought and water stress detection. Issues include limited data availability and quality, achieving adequate spatial and temporal resolutions, complexities in remote sensing interpretation, calibration, and validation difficulties, integration of multisource data, and scaling and transferability concerns. Overcoming these challenges necessitates interdisciplinary collaboration, data sharing, methodological advancements, and continued research to develop robust and reliable approaches for effective drought/water stress detection.

Meta-classifiers can contribute to overcoming limitations in drought and water stress detection while promoting cost-effectiveness and sustainability [46]. By utilizing ensemble learning techniques, meta-classifiers can harness the collective knowledge of multiple base classifiers, enhancing the accuracy and robustness of predictions [47]. This can address data availability and quality challenges by leveraging diverse data sources and improving model performance. Meta-classifiers also offer the potential to optimize computational resources, reducing the need for high-resolution imagery and enabling efficient processing of large datasets. Through feature selection and dimensionality reduction, meta-classifiers can improve interpretability and mitigate complexities in remote sensing interpretation. Moreover, the ensemble approach of meta-classifiers provides an opportunity for model calibration and validation through cross-validation and model averaging. By leveraging the strengths of various base classifiers, meta-classifiers can enhance the integration of multisource data and enable the transferability of models across different regions and ecosystems. This cost-effective and sustainable approach can contribute to advancing drought and water stress detection while optimizing resources and promoting long-term environmental sustainability. This chapter highlights the potential of meta-classifiers in enhancing the effectiveness of drought stress detection in wheat crops, leading to sustainable agricultural development.

1.3 Background

In this background section, we will provide an overview of the concepts, terms, and important definitions that are necessary to understand the topic "Development of a two-layer meta-classifier–based drought stress detection system for wheat crop sustainability with the help of CHF image analysis.

a. Two-layer meta-classifier: A machine learning model that uses multiple algorithms to make predictions. In this case, it is used to detect drought stress in wheat crops. Meta-classifiers are machine learning algorithms that combine the predictions of several base classifiers to improve the overall performance. They are designed to overcome the limitations of single classifiers, such as bias and overfitting. In the context of drought stress detection, a two-layer meta-classifier–based system can be used to classify wheat water stress utilizing CHF signals embodied in wheat canopy images, with high accuracy and robustness [48].
b. Drought stress: A condition where a plant is exposed to prolonged periods of water shortage, leading to reduced growth and yield. Drought stress is a major abiotic stress factor that affects the growth and productivity of crops, including wheat. It is a complex phenomenon caused by a lack of water supply and often leads to irreversible damage to the crop. Therefore, timely detection and monitoring of drought stress are essential for effective management and sustainable crop production [49].
c. Sustainability: The ability of a system to maintain productivity and function over time without degrading the natural resources it depends on.
d. Wheat crop: A cereal grain plant widely cultivated for its seed, which is a staple food in many parts of the world. (Variety = **RAJ 3765**) [50].
e. Chlorophyll fluorescence (CHF): The emission of light by chlorophyll molecules in plants when excited by light. It can provide information about plant photosynthetic performance and stress. Fluorescence is the emission of light by a substance that has absorbed light or other electromagnetic radiation. In plants, CHF signals provide a noninvasive and sensitive tool for monitoring photosynthetic performance and stress responses. The fluorescence signals are generated when chlorophyll molecules in photosystem II (PSII) absorb light energy and become excited. The excited electrons then emit fluorescence as they return to their ground state. The intensity and shape of the fluorescence signals can be used

to quantify various photosynthetic parameters, such as the efficiency of light energy conversion and the quantum yield of photosystem II [51–53].

f. Chlorophyll fluorescence analysis: CHF analysis involves the examination of CHF signals captured within the wheat canopy. By utilizing image processing operations, this analysis can provide valuable insights into the crop's stress status in a nonintrusive manner.
g. Drought/water stress in wheat crop [54]: A condition where the amount of water available to a wheat plant is insufficient to meet its physiological needs, leading to reduced growth and yield.

The incorporation of several different ideas is required for the development of a two-layer meta-classifier–based drought stress detection system for wheat crop sustainability with the assistance of CHF signals analysis embodied in CHF image [23, 35, 55].

1.4 Problem Formulation

The study's problem is to develop an innovative algorithm that can detect drought/water stress in Indian wheat crop varieties **RAJ 3765** in a noninvasive manner while conserving water resources. The objective is to solve the problem of excessive use of water and pesticides due to unscientific agricultural practices and a lack of crop stress detection.

The problem mentioned above can be effectively addressed by analyzing the chlorophyll signals present inside the wheat plant. By examining these signals, valuable insights can be gained, leading to a successful solution for the given problem. The CHF analysis in the wheat crop involves examining the fluorescence effect captured in the wheat image using image processing techniques. This analysis encompasses various mathematical operations, such as color channel analysis, texture analysis, landmark analysis, and wheat canopy component analysis. Additionally, a two-layer meta-classifier system is utilized for the detection of water stress using CHF images. By applying these methods, a comprehensive understanding of CHF can be obtained, allowing for effective analysis and interpretation of the water stress in wheat crops.

The design of the meta-classifiers is rooted in the hypothesis that if the accuracy of previous machine learning algorithms, tested for their effectiveness in identifying water stress, falls below 95% even after hyper-parameter tuning, there arises a necessity to develop a meta-classifier for wheat water stress detection. This approach aims to address the limitations of existing base automation models and enhance their performance in accurately detecting drought/water stress in wheat plants.

1.5 Methodology for Drought/Water Stress Detection

In this section, an explanation of the steps taken to solve the problem stated in the problem formulation section is given. The first step of the model-building process is to select of appropriate wheat variety CHF image dataset, followed by CHF image analysis of the selected wheat variety, utilizing four key features, *viz*, color features, GLCM texture features, landmark features, and wheat canopy components. Furthermore, picking the right automation procedures that are capable of reliably predicting the presence of stress in the chosen wheat variety is important.

1.5.1 Wheat Variety (V)

The wheat variety selected for research, RAJ 3765, was developed by the Rajasthan Agriculture Research Institute (RARI) in Durgapur. This particular variety originates from Uttar Pradesh, a state that contributes the highest percentage (35.03%) to the national wheat production. Additionally, RAJ 3765 is cultivated in other regions, including Punjab, Haryana, Delhi, western Uttar Pradesh, Rajasthan (except Kota and Udaipur divisions), the foothills of Himachal Pradesh, Jammu and Kashmir, and Uttarakhand. In our research context, we have chosen this improved wheat variety from western Uttar Pradesh, which requires irrigation and is sown later in the season. The dataset comprises CHF images of the "**RAJ 3765**" variety of wheat. The experiments were conducted for 135 days in a controlled environment. However, the images of 60 days are considered for this research work as they represent the days of growth of the wheat canopy.

1.5.2 Chlorophyll Fluorescence Image Analysis (CFIA)

The CHF image modality was considered for the design of the water stress detection system in this research. According to the given treatment, the CHF image dataset contains two groups of images: group 'X' well-watered (controlled) and group 'Y' drought (water deficit). The plants were cultivated in 12-inch-diameter pots. For each of the 60 days of these investigations, 24 images are captured daily. The image dataset contains a compilation of (1440 x 2) RGB images with 72 dots per inch resolution taken from a public repository [56]. The dataset and code for the model development are openly available for replication [57]. It was found from the literature that the thermal activity of the photons during the fluorescence signal's emission introduces "random noise" in these CHF images. So, the image dataset is simulated with different noise levels to analyze the effectiveness of applied denoising filters. In this context, we evaluated the denoising methods at three distinct noise levels: 0.27, 0.5, and 0.8. The rationale behind choosing these specific noise levels was likely based on the consideration of state-of-the-art techniques in image denoising and their applicability to real-world scenarios, to cover a broad spectrum of noise intensities, and it was empirically observed that the noise level with a mean of 0 and a standard deviation of 0.27 yielded the highest values of the evaluated performance parameter, indicating better performance. Furthermore, an interesting observation was made regarding the tendency of the fluorescence effect to exhibit a trend from minimum to maximum fluorescence, which introduces constraints on contrast. This observation stems from the recognition that images captured at the initial stages of fluorescence activity tend to be darker compared to those approaching maximum fluorescence. As a result, there is a desire to enhance the contrast of the images to better capture and differentiate the varying fluorescence levels throughout the process. Therefore, it is important to preprocess these images using appropriate noise removal and contrast enhancement approaches. The primary element that is most likely to be disturbed whenever there is stress is texture. As a result, procedures for noise reduction and contrast amplification that can preserve an image's texture are carefully chosen. The wheat canopy images are preprocessed for enhancement in three ways: a) individual image denoising, b) individual contrast enhancement, and c) cascading of a) and b). The comparison of the applied preprocessing techniques is done using six potential pipelines where denoising is followed by enhancement procedures as shown in Table 1.1.

This leads to significant improvement in the quality of wheat canopy images. The six selected pipelines were developed using two denoising filters, *viz*, non-local means (NLM) denoising and total variation L1 denoising with primal-dual algorithm (TV-L1 with a PDA), and three contrast enhancement approaches, *viz*, contrast stretching (Min-Max) [CS(Min-Max)], contrast limited adaptive histogram equalization (CLAHE), and histogram equalization (HE). It can be concluded from the visual inspection of the results shown in Table 1.1 that TV-L1 with a PDA followed by CS (Min-Max) is highly effective in producing high-quality images in comparison to other cascaded approaches. Quantitative justification of

Table 1.1 Comparison of six different preprocessing pipelines.

Pipeline	Workflow	Noise removal	Contrast stretching	Preprocessed image
1	NLM + CS (Min-Max)	 NLM	 CS (Min-Max)	
2	TV-L1 with a PDA + CS (Min-Max)	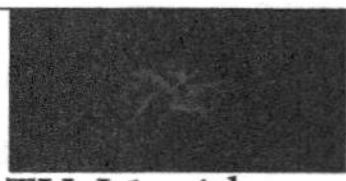 TV-L1 with a PDA	 CS (Min-Max)	
3	NLM + CLAHE	 NLM	 CLAHE	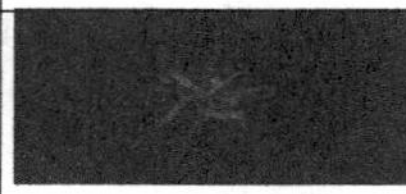
4	TV-L1 with a PDA + CLAHE	 TV-L1 with a PDA	 CLAHE	
5	TV-L1 with a PDA + CLAHE	 NLM	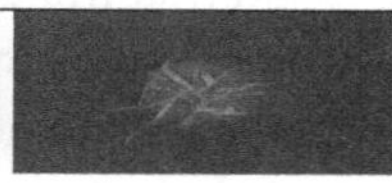 HE	
6	TV-L1 with a PDA	 TV-L1 with a PDA	 HE	

this can be achieved through the use of three performance assessment metrics: average mean squared error (MSE), average peak signal-to-noise ratio (PSNR), and average structural similarity index (SSIM). The empirical analysis involved a total of 2880 image pairs, including the ground truth images (original noise-free images) and the reconstructed images obtained after applying the denoising methods followed by enhancement approaches. Since TV-L1 with a PDA followed by CS(Min-Max) yields the higher values for all three metrics, *viz*, average MSE = 0.127, average PSNR = 57.092, and average SSIM = 0.891 proved to be highly effective in producing high-quality images [58], evidenced in Table 1.2.

After the preprocessing step has been completed successfully, images are subjected to a combinational segmentation technique known as the "CFit k-means" algorithm [59]. According to statistics from a related paper [9], there was a remarkable improvement in the segmentation accuracy of curve fit-based K-means (CFit K-means algorithm) by approximately

Table 1.2 Performance comparison of six different cascaded pipelines.

Performance parameters	NLM + CS (Min-Max)	TV-L1 with a PDA+ CS (Min-Max)	NLM + CLAHE	TV-L1 with a PDA + CLAHE	NLM + HE	TV-L1 with a PDA + HE
Average MSE	0.582	0.127	0.938	0.734	0.714	0.329
Average PSNR	50.481	57.092	48.408	49.473	49.593	52.958
Average SSIM	0.789	0.891	0.614	0.866	0.761	0.871

36%, after the application of preprocessing operations [24]. Following the successful completion of the segmentation phase, images are subjected to a data augmentation procedure including horizontal/vertical flip and rotation at four different angles (45°, 90°, 135°, and 180°), to increase the dataset size by a factor of 10 and improve automation process accuracy. The dataset and code for the development of an automatic water stress detection system are publicly available [57].

1.5.3 Feature Extraction and Selection

The segmented CHF image after the completion of the augmentation process is subjected to a feature engineering process to extract drought-related key features. Features were categorized into five different categories, *viz*, color percentage of various nine bands (blue, green, red, lightness, green-magenta, blue-yellow, hue, saturation, and value), photosystem II (PSII) based features, GLCM texture features, wheat canopy components, and landmark features, all these features are rigorously tested for their suitability following the proper selection procedure.

These attributes are extracted from the plant images and compared to build an autonomous ML-based identification model. Each of these features was evaluated in the comparative study using a backward elimination technique for feature selection. Performance evaluation for features is done on the basis that. If a subgroup of features provided a global diagnostic accuracy greater than 80% (designated as baseline accuracy) throughout the testing phase for top-performing classification algorithms, *viz*, logistic regression (LR), linear discriminant analysis (LDA), K-nearest neighbors (KNN), decision tree (DT), naïve Bayes (NB), random forest (RF), support vector machine (SVM), extra trees (ETC), and gradient boosting (GBC), that subgroup was taken into account in the wheat stress detection model for the further research on its performance. The machine learning

Table 1.3 AUC test score of RF algorithm on two different feature combinations.

Feature set	Description of feature combination	AUC test score
1	**23 GLCM features and color features from 9 bands total (9*255) features**	**91.164%**
2	**23 GLCM features and color features from 9 bands total of (9*255) features, 20 wheat canopy components (10 path length, 10 path angle), and 10 landmark features**	**93.064%**

model's accuracy measure was used as "removal criteria" to eliminate the overloaded features. As a consequence, a specific combination based on the outcome of baseline accuracy was selected, and the remaining features were eliminated.

The comparative accuracy performance evaluation process concerning each tested feature, both individually and in combination, serves to validate the results of the feature selection procedure. The results of the removal criteria indicate that only two combinations can achieve an accuracy greater than baseline accuracy, *viz*, **combination 1:** comprises two feature sets [23 GLCM and (9*255) color features (blue, green, red, lightness, green-magenta, blue-yellow, hue, saturation, and value) obtained from nine bands]; **combination 2:** comprises a set of four feature sets [23 GLCM and (9*255) color features (blue, green, red, lightness, green-magenta, blue-yellow, hue, saturation, and value) obtained from nine bands], top 10 landmark features, and 20 wheat canopy components (10 path lengths and 10 path angles). Among the two combinations obtained, **combination 2** demonstrated the highest diagnostic accuracy (AUC score) for the top-performing RF classifier, as indicated in Table 1.3. Analyzing the testing AUC score of the best-performing algorithm, i.e., RF depicts that its accuracy in terms of AUC score gets hiked by almost 2%. Details of the AUC score improvement of best-performing algorithm RF are given in Table 1.3 [60].

Consequently, **combination 2** was chosen and the remaining characteristics were eliminated. This combination considerably enhances the accuracy of classification algorithms. The details of outcomes produced by image-based features for water stress detection in wheat crops is given below.

1.5.3.1 Gray-Level Co-Occurrence Matrix Texture Features

Segmented CHF images underwent texture analysis to extract 23 gray-level co-occurrence matrix (GLCM) texture variables. Following successful extraction, correlation analysis was conducted using the Pearson correlation coefficient on these 23 GLCM variables. This analysis aimed to gain valuable insights into the effects of water stress on the wheat canopy by analyzing the changes observed in the values of these variables, as depicted in Figure 1.1. The GLCM metrics comprise 23 texture-based features, *viz*, autocorrelation (autoc), contrast (contr), correlation (corrm), correlation (corrp), cluster prominence (cprom), cluster shade (cshad), dissimilarity (dissi), energy (energ), entropy (entro), homogeneity (homom), homogeneity

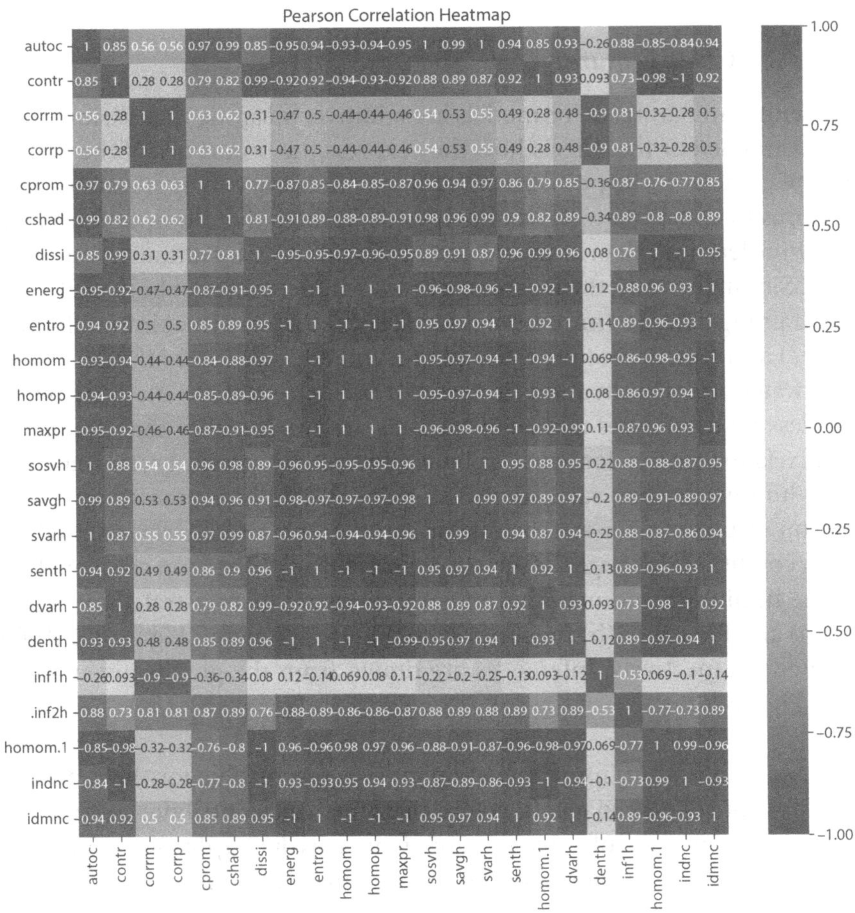

Figure 1.1 Pearson correlation analysis for 23 GLCM texture variables.

(homop), maximum probability (maxpr), sum of squares (sosvh), sum average (savgh), sum variance (svarh), sum entropy (senth), difference variance (dvarh), difference entropy (denth), information measure of correlation1 (inf1h), information measure of correlation2 (inf2h), inverse difference (INV) is homom (homom), inverse difference normalized (INN) (indnc), and inverse difference moment normalized (IDMNC). The outcomes of the Pearson correlation analysis indicate that there exists a significant degree of correlation among all the variables. This finding suggests that considering all the variables collectively would be advantageous in constructing a water stress detection system, results of correlation-based analysis are given in Figure 1.1. This analysis provides insights into the changes in the wheat canopy caused by water stress, and the GLCM feature analysis provided valuable information about the "change" caused by water stress in the wheat canopy.

1.5.3.2 Color Features

Generally, in nature, color models are additive, meaning that various hues are combined or that this is how they express the properties of the matter. In the context of the problem at hand, it is necessary to aggregate various features to construct a stack of pertinent features that do not burden the machine learning algorithm and do not provide it with redundant information. Simultaneously, the machine learning model must not be overfitted to the training data. Fo (minimum fluorescence) to Fmax (maximum fluorescence) as a function of time and photosynthetic stage in both the control and water-stressed situations is depicted in Figure 1.2(a to d), using image processing techniques and color analysis, we can quantify the proportion of pixels for each color, such as blue, blue-yellow, green, green-magenta, hue, luminance, red, saturation, and value. Using histograms, the distribution of pixel values for each color can then be visually represented for both conditions (drought/water stress and control). The crop color analysis at the minimum and maximum fluorescence are demonstrated by Figure 1.2(a) and Figure 1.2(b) for control and Figure 1.2(c) and Figure 1.2(d) for drought, respectively. Inferences are drawn from the graphs given above (see Figure 1.2):

a. Anthocyanins, which are pigments responsible for the red, purple, or magenta colors in plants can contribute to the proportion of magenta pixels in images during drought conditions. Anthocyanin production in plants can be induced by various environmental factors, including drought stress. During drought stress, plants may produce higher levels of

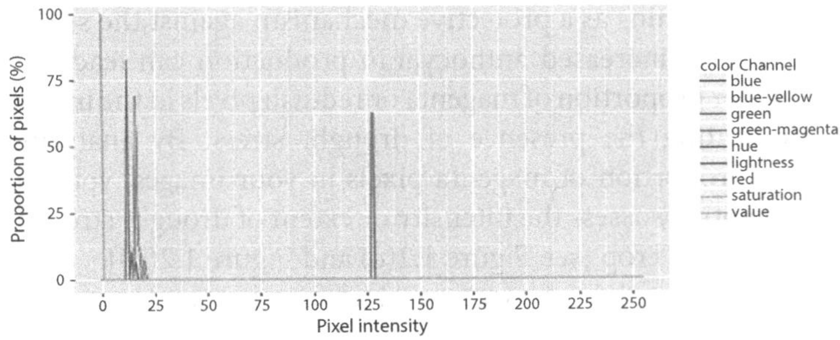

(**a**) Color feature Analysis histogram for control image at minimum fluorescence

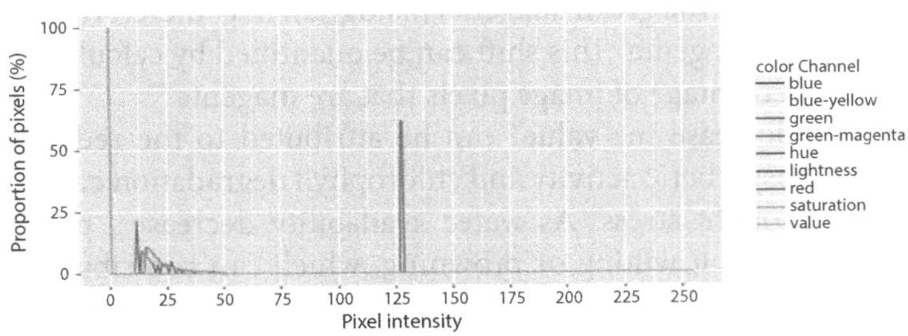

(**b**) Color feature Analysis histogram for control image at maximum fluorescence

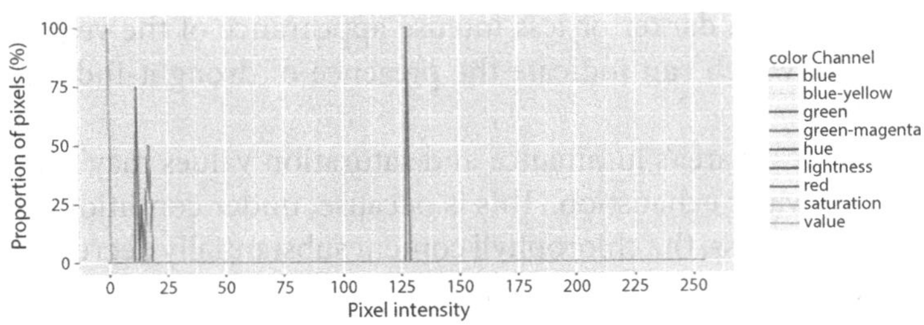

(**c**) Color feature Analysis histogram for water-stressed image at minimum fluorescence

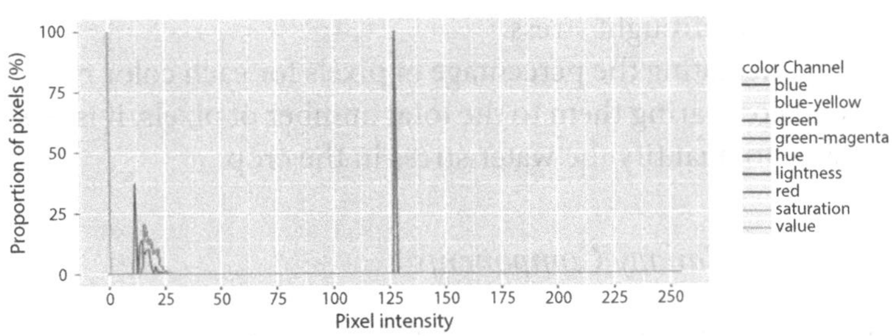

(**d**) Color feature Analysis histogram for water-stressed image at maximum fluorescence

Figure 1.2 Graphical representation of Fo (minimum fluorescence) to Fmax (maximum fluorescence) as a function of time and photosynthetic stage in both the control and water-stressed images.

anthocyanins as a protective mechanism against the stressors. This increased anthocyanin production can lead to a higher proportion of magenta or reddish pixels in the image, indicating the presence of drought stress. By analyzing the proportion of magenta pixels in your images, you can potentially assess the intensity or extent of drought stress in the wheat crop [see Figure 1.2(c) and Figure 1.2(d)].

b. A large number of magenta pixels on the graph could be indicative of water stress. Water stress can reduce a plant's chlorophyll content, altering the green-to-magenta ratio of an image. The green-magenta histogram may therefore shift toward magenta. This shift can be quantified by calculating the percentage of image pixels that are magenta.

c. The decrease in "value" can be attributed to the reduced photosynthetic activity and chlorophyll degradation caused by drought stress. As water availability decreases, plants may exhibit wilting or browning, which can contribute to lower overall brightness. By analyzing the value component in images, changes in brightness associated with drought stress can be assessed. A decrease in the value component suggests a darker or less intense appearance of the vegetation, which can indicate the presence of drought-induced stress.

d. The leafy area's luminance and saturation values may indicate water exhaustion. This is because, under conditions of water stress, the chlorophyll content substantially decreases, causing the color and texture of the wheat canopy to deteriorate. The decrease in saturation can be attributed to the fading or desaturation of colors in the vegetation due to the effects of drought stress.

e. By calculating the percentage of pixels for each color model and comparing them to the total number of pixels, it is possible to quantify the water stress in the crop.

1.5.3.3 Wheat Canopy Components

The canopy components, such as foliage, branches, and branch angles, can provide extremely useful information for determining whether or not wheat crops are experiencing water stress. Image analysis techniques can be used to capture and quantify morphological and spatial changes induced by water stress in these components. For instance, a decrease in leaf area, leaf angle, and leaf chlorophyll content can be indicative of water stress, whereas an increase in branch angle can be indicative of a response to water stress. The leaves wilting is an additional sign of water scarcity. By evaluating these components using imaging, we can develop algorithms that accurately detect and quantify water stress in wheat fields. This will enable more efficient agricultural management, resulting in increased crop yields. The extraction of wheat canopy components from the segmented wheat crop image is required for plant isolation and gaining detailed insights about each section of the plant body, a "PlantCV" [26, 61] library is required for morphological functional analysis to understand the details and alterations in each component, details of morphological operations are given in Figure 1.3.

These morphological functions provide information regarding the length, angle, and curvature of leaves. After reading the segmented image, it is binarized, and the first operation on that image is to "zoom in" on the fine details of each connected component of that image using Zhang's

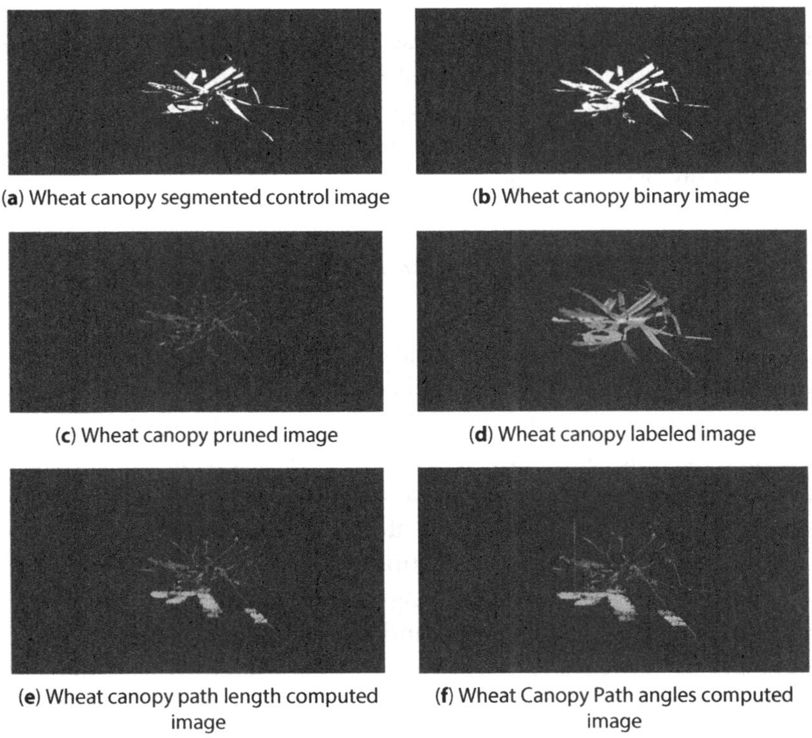

(**a**) Wheat canopy segmented control image
(**b**) Wheat canopy binary image
(**c**) Wheat canopy pruned image
(**d**) Wheat canopy labeled image
(**e**) Wheat canopy path length computed image
(**f**) Wheat Canopy Path angles computed image

Figure 1.3 Wheat canopy component analysis for segmented control image.

algorithm (2D)-based skeletonization. The execution of the skeletonization algorithm requires one special parameter called "pixel wide parameter," which defines the thickness of the constructed skeleton and is set to 1 in this case. After skeletonization, images contain barbs that may impede the calculation of essential features that must be eliminated using a pruning function. After skeletonizing and pruning the desired image, branch, and tip nodes are identified and labeled. For further analysis, the image has been divided predominantly into three major segments. Specifically, stem, branches, and leaf objects. By measuring the length of the branch (path length) of each segment, the Euclidean distance of each segment, the curvature of each segment, the angles between each segment, the tangent of each segment, and the insertion angles of each segment useful insights about canopy can be made. Following skeletonization and pruning, the identified segments are subjected to the curve fitting process [linear regression (LR) procedure known as least-squares $y = mx + c$, where m represents slope and c = intercept) to determine the angles between them. The final extracted numerical feature is the geodesic distance (the length of the shortest path between two segments). In this investigation, we utilize two canopy characteristics: path length and path angles. As canopy components, the top 10 path angles for each image and the top 10 path lengths for each image are computed. The output of various morphological operations on control segmented images is given in Figure 1.3(a to f) to maintain the brevity of the work only control output is given similar operations are performed on drought images too.

1.5.3.4 Wheat Canopy Landmark Features

Examining landmark points within an image of a wheat canopy can aid in the detection of water stress in the crop by providing crucial information about the canopy's structure and characteristics. Landmark points are distinct locations within an image that can be used to identify and track particular elements over time. In a variety of imaging applications, landmark points can be utilized. These landmark data points can be retrieved and analyzed with the aid of the Oriented FAST and Rotated BRIEF (ORB) algorithms. This will enable the identification of canopy structural changes that may indicate water stress. Changes in the density or distribution of landmark points, for instance, may indicate a change in the plant's overall health, which may be a result of water stress. Therefore, it is possible to determine whether wheat is experiencing water stress by analyzing images of the entire wheat canopy. The top 10 landmark data points were extracted as they represent the shape of the canopy better as compared to points

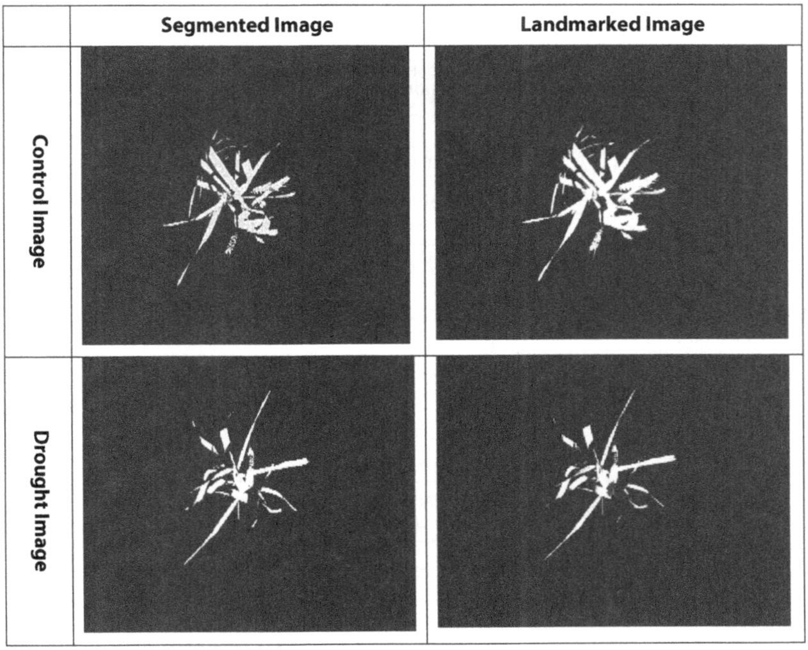

Figure 1.4 Landmark features on segmented wheat canopy.

that are far from the center of the image and used as input to the machine learning algorithms. The output of the landmark features obtained for both control and stressed crops is demonstrated in Figure 1.4.

The classification process considers the combination of these four feature sets as input. The detailed flow of the automatic water stress detection process is illustrated in Figure 1.5.

1.5.4 Water Stress Detection in Wheat Crops

This section explains the construction of the two-layer meta-classifier that can be utilized for detecting drought/water stress in the Indian wheat variety RAJ 3765. To date, few studies have created classification models for various CHF image datasets using a layered (stacked) generalization. Consequently, this method of developing a custom algorithm has been adopted. In addition, the issue of limited dataset size has been resolved through the use of data augmentation strategies. As mentioned previously, dataset augmentation helps in creating a highly generalized machine model. A detailed design of experiments (DoE) was conducted to identify the most important aspects, and it was discovered that only four of the five

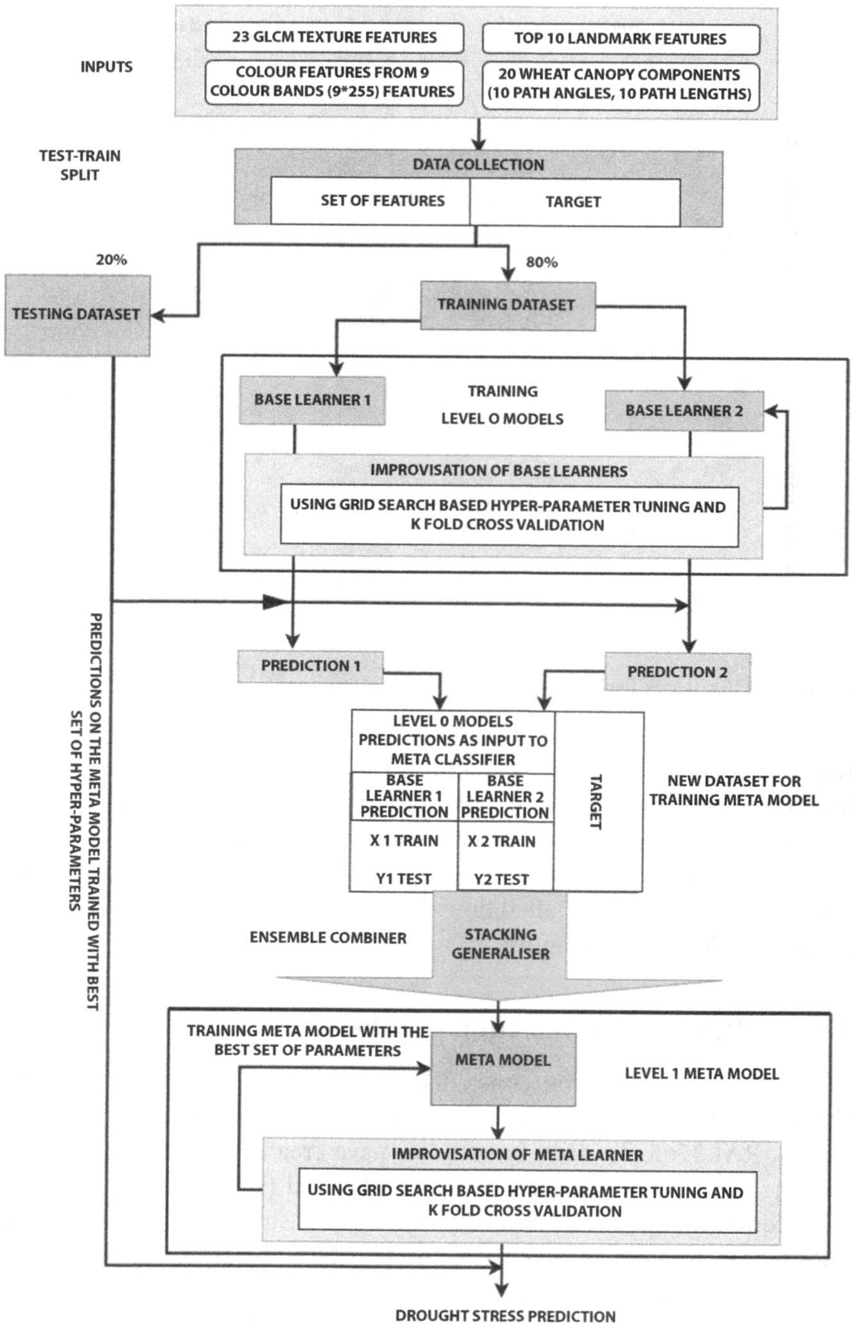

Figure 1.5 Flow of meta-learning stacking ensemble for drought stress detection.

analyzed features, *viz*, 23 GLCM features, 9*255 color features, 20 wheat canopy components (10 path lengths and 10 path angles), and landmarks features, were essential for the successful working of the automation models. The detailed comparison of all tested base learners is given in Figure 1.6. For details of the respective AUC scores of the learners (see Table 1.4).

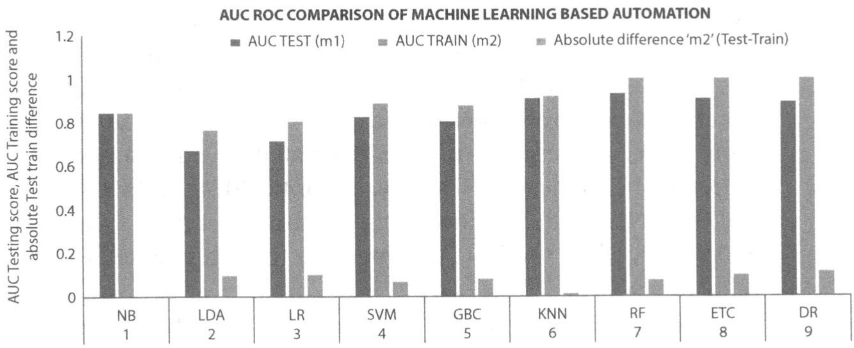

Figure 1.6 AUC ROC score comparison of different classification algorithms.

Table 1.4 AUC ROC performance evaluation of nine base learners/classifiers.

S.no	Algorithms	AUC test (m1)	AUC train (m2)	Absolute difference 'm3' (test-train)
1	NB	0.84597268	0.84749576	0.00152308
2	LDA	0.66880706	0.76467457	0.09586751
3	LR	0.71098665	0.8053969	0.09441025
4	SVM	0.82517629	0.88908612	0.0639098
5	GBC	0.8051109	0.881208612	0.0760977
6	KNN	0.91075046	0.91987461	0.00912415
7	RF	0.93064728	1	0.06935272
8	ETC	0.90773481	1	0.09226519
9	DT	0.8923432	1	0.1076568

Inferences that can be drawn from Table 1.4 are given below:

- The comparison among nine high-performing classifiers, *viz*, LR, LDA, KNN, DT, Gaussian NB, RF, SVM, ETC, and GBC, was thoroughly analyzed before the construction of a meta-classifier.
- Intensive research on this aspect has resulted in the identification of the highest achievable diagnostic accuracy range, which is 93.064%, for the current dataset when utilizing the aforementioned four feature sets in combination.
- The efficacy of nine classifiers was compared using metrics such as the AUC test score, the AUC train score, and the absolute difference between the AUC test and the AUC train score. Among the models evaluated, the RF model achieved the highest AUC test value of 0.93064. Following that the KNN model attained the second-highest AUC value of 0.91075, while the ETC obtained the third-highest AUC value of 0.90773.
- Notable is the fact that the DT algorithm has the largest absolute difference between its test and train scores, *viz*, (0.1076568), which may be indicative of overfitting. The classifiers with the smallest absolute difference, namely, NB (0.00152308), KNN (0.00912415), SVM (0.0639098), and RF (0.06935272), are less susceptible to overfitting.
- According to the analysis of the area under the curve (AUC) analysis, the RF, KNN, and ETC perform well on this specific dataset. In contrast, the LR (0.09441025), LDA (0.0958675), ETC (0.09226519), and DT (0.1076568) models demonstrate an enhanced propensity for overfitting.
- After conducting an extensive exploration of various feature combinations, it has been established that the RF algorithm attains the highest achievable accuracy of 93.064%. Furthermore, based on collective performance measures, the RF model can serve as a representative model for all tree-based high-performing models. Additionally, the RF algorithm can handle multicollinearity, making it a versatile and effective choice for classification tasks.
- This rigorous investigation ensures that the chosen features are optimized for accuracy in the classification process. However, even after employing a grid search-based algorithm to tune the hyper-parameters, it has been established

that there is still room for improvement in accuracy. To surpass the specified accuracy threshold of 95%, a custom meta-classifier has been implemented, which integrates six logical sequences given below:

Step 0: Test train split: For a given dataset 'D' of m x n dimensions, with the test-train split of 20% to 80%, design a custom two-layer stacking algorithm.
Step 1: Analysis: Analysis of each existing machine learning algorithm, A = {A1, A2, A3, ... An} has been performed on the same dataset 'D'.
Step 2: Selection: The selection of the top three performing algorithms based on a metric ('m' = {mv1, mv2, mv3}), where m1 = AUC test score, m2 = AUC train score, and m3 = absolute difference between the AUC training and AUC testing score.
Step 3: Hierarchy of model construction: The machine learning ensemble model is composed of two levels, models in level 1 are called base learners, and the model in level 2, which is the final estimator called meta-learner. Different combinations of base learners and meta-learners were explored to construct the two-layer stacking-based meta-classifier for drought stress detection. The goal was to find the optimal configuration that yields the highest performance in detecting drought stress. This involved assessing the mean accuracy and standard deviation (std) of different ensemble combinations and selecting the most effective combination based on the evaluation results. Hence, in step 4, an explanation of how the meta-classifier was constructed is discussed in detail.
Step 4: Logic: The logical sequence of steps that were used to construct the meta-classifier can be understood from the pseudo logic given below:

- Initialize base1, base2, meta
- Initialize prediction scores list
- For base1 in [RF, NB, KNN, SVM]:
- For base2 in [RF, NB, KNN, SVM]:
- If base1 != base2

  - o Set layer1 = base1 + base2
  - o For meta IN [RF, NB, KNN, SVM]:
  - o Set layer2 = meta
  - o Set model = layer1 + layer2
  - o Evaluate the model using cross-validation
  - o Add score to scores list

- Sort the scores list in descending order
- Print the best combination and its score

As per the pseudo logic, the process is as follows.

a. The base estimators are trained on the training data using the feature matrix.
b. The predictions from the base estimators are combined and used as features for the meta-estimator.
c. The meta-estimator is trained on the combined features and predicts the target variable.
d. Combination generation: To generate all possible combinations of base and meta-estimators, we use the itertools. The product function from the itertools library generates all possible cartesian products of the two lists. The mathematical formula used is the Cartesian product formula, which returns the set of all possible ordered pairs (x, y) of elements x from the first set and y from the second set.

Step 5: Validation: The final architecture and the implementation of the custom meta-classifier were based on the cross-validation process that evaluated the performance of the stacking ensemble; we used fivefold cross-validation.

1.5.4.1 Results and Discussions

The effectiveness of the stacking ensemble algorithm was evaluated through the utilization of appropriate performance metrics demonstrated in Figure 1.7. Table 1.5 presents an assessment of the algorithm's performance, showcasing the diverse algorithmic combinations to develop a meta-classifier. The combination that yielded an accuracy score surpassing the 95% threshold is the final selection sequence for the two-layer meta-classifier.

a. According to the provided performance metric, the top four algorithms are RF, SVM, NB, and KNN. For the best performance, these algorithms were jointly tested utilizing various experimentation sequences.
b. After performing significant experimentation for the creation of two-layered stacked meta-classifiers, the four pipeline combinations were created given below:

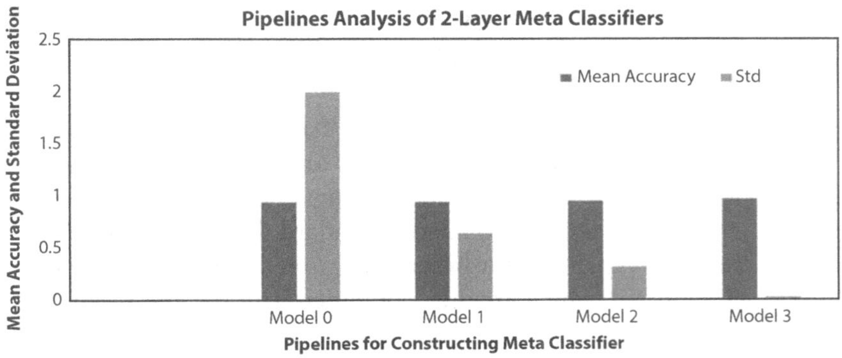

Figure 1.7 Two-layer meta-classifiers 1 to 4 comparison graph.

Table 1.5 Pipeline of the models evaluated to finalize the meta-classifier.

Model no	Meta-model	Base or initial model(s)	Final model	Mean accuracy	Std
0	RF, NB, KNN, and SVM	RF (estimator = 10, depth = 5), NB, KNN (neighbors = 5)	SVM (kernel = radial basis function)	0.931	1.993
1	RF, SVM, and KNN	RF (estimator =10, depth = 5), SVM (Kernel = Radial basis function)	KNN (neighbors = 5)	0.932	0.636
2	RF and KNN	RF (estimator = 10, depth = 5)	KNN (neighbors = 5)	0.942	0.313
3	RF, NB, and KNN	RF (estimator = 10, depth = 5), NB	KNN (neighbors = 5)	0.958	0.025

Models = {(algo 0 = [RF, NB, KNN, SVM]), (algo1 = [RF, SVM, KNN]), (algo2 = [RF, KNN]), (algo3 = [RF, NB, KNN])}

c. From Table 1.5, it can be inferred that Model 3, developed using RF and NB as base learners and KNN as the meta-learner, is the optimal model among all the models, with a mean accuracy of 0.958 and a standard deviation of 0.025, the third pipeline is the best option. Initial models consist of RF and NB, which are then combined with the end model, KNN using a stacking ensemble approach. Additionally, the combination of RF and NB as the initial models contributes to the improved accuracy of the pipeline. Based on these findings, it can be concluded that the stacking ensemble technique can be used to derive more precise meta-classifiers from multiple base models. Moreover, Model 3 follows the best-discovered pipeline for the development of a two-layer stacked meta-classifier because it achieved the highest mean accuracy with a relatively low standard deviation, indicating that it is a stable and reliable model.
d. However, in Model 0 and Model 1, the standard deviation is too high in comparison to Model 2 and Model 3, indicating that the performance of the models is not consistent. This inconsistency can be attributed to the complexity of the models, the type of models used, and the order in which the models are stacked to construct the meta-model. Consequently, we are compelled to drop these models and the SVM algorithm.

Based on the analysis of mean accuracy, standard deviation, and model complexity, it can be concluded that Model No. 3 is the optimal pipeline.

1.6 Application of Meta-Classifier–Based Detection for the Development of the Sustainable Model for Drought Detection

Meta-classifiers can be effectively utilized in the application for the development of sustainable agricultural practices to mitigate the losses incurred due to water stress, which significantly impacts the yield of wheat crops cultivated per unit area of land. By incorporating meta-classifiers into the decision-making process, agricultural stakeholders can make informed choices and implement adaptive measures to optimize water usage, enhance crop productivity, and ensure the sustainable growth of wheat farming practices.

1.6.1 Meta-Machine Learning for Overcoming Agricultural Losses

Agricultural practices in India are facing numerous challenges, including water stress, excessive water and pesticide use, and unscientific agricultural practices, leading to significant losses in crop yield. The potential losses due to these factors can be represented by the variable "L". In this research, we propose an **additive model** to estimate the potential reduction in losses due to the implementation of water conservation measures (wcm), scientific and precise agriculture (spa), and appropriate technology (at) [62]. We also propose a cost-effectiveness analysis of these measures to identify the most sustainable and effective agricultural practices. The proposed model is based on the latest research papers [63]; reports [64] and the work done in this direction will be used to address the challenges facing agriculture in India. The different associated variables required to build the model are detailed below:

The primary variable associated with the proposed foundation model has been detailed below.

- V: Indian wheat crop variety(s)
- W: Water stress detection in Indian wheat crops [healthy and water-stressed]
- P: Excessive use of water and pesticides due to unscientific agricultural practices
- C: Lack of detection of crop stress
- F: Chlorophyll fluorescence imaging analysis
- T: Two-layer meta-classifier system
- E: Effectiveness of different classifiers in identifying water stress
- L: Total losses caused due to (W) water stress, excessive use of water, and pesticides due to unscientific agricultural practices (P) in India, where L_{ws} is the loss due to water stress, L_{eu} is the loss due to excessive use of water and pesticides, and L_{up} is the loss due to unscientific agricultural practices.

- R: Potential reduction in losses, caused due to water stress, excessive use of water, and unscientific agricultural practices in India, where R_{wcm} is reduction due to the implementation of water conservation measures, R_{spa} is reduction due to scientific and precise agriculture, and R_{at} is reduction by the use of appropriate technology.
- p_{wcm}: It is proportional reduction due to the implementation of water conservation measures, p_{spa}: It is reduction due to scientific and precise agriculture, and p_{at} is reduction using appropriate technology.
- C_{wcm}: It is the cost per unit area of implementing water conservation measures.
- C_{spa}: It is the cost per unit area of implementing scientific and precise agriculture.
- C_{at}: It is the cost per unit area of implementing appropriate technology.
- EC_{wcm}: The effective cost of implementing water conservation measures.
- EC_{spa}: The effective cost of implementing scientific and precise agriculture.
- EC_{at}: The effective cost of implementing appropriate technology.
- A: Represents the total area of agricultural land in India.

The created meta-classifier can be used to noninvasively identify water stress in Indian wheat crop varieties RAJ 3765 while preserving water resources. The goal is to detect water stress by addressing the losses caused due to the excessive use of water and pesticides, caused by illogical agricultural practices and a dearth of crop stress detection.

Therefore, the study aims to develop an algorithm that can identify water stress (W) in Indian wheat crop varieties (V) using CHF imaging analysis (F), which includes color channel analysis, texture analysis, landmark analysis, and wheat canopy component analysis. The algorithm employs a two-layer meta-classifier system (T) for the detection, based on the results of a series of studies that tested the effectiveness of different classifiers (E) in identifying water stress. The objective is to solve the problem of excessive use of water and pesticides (P), due to unscientific agricultural practices and a lack of detection of crop stress (C) while conserving water resources through the promotion of sustainable farming techniques. If the accuracy

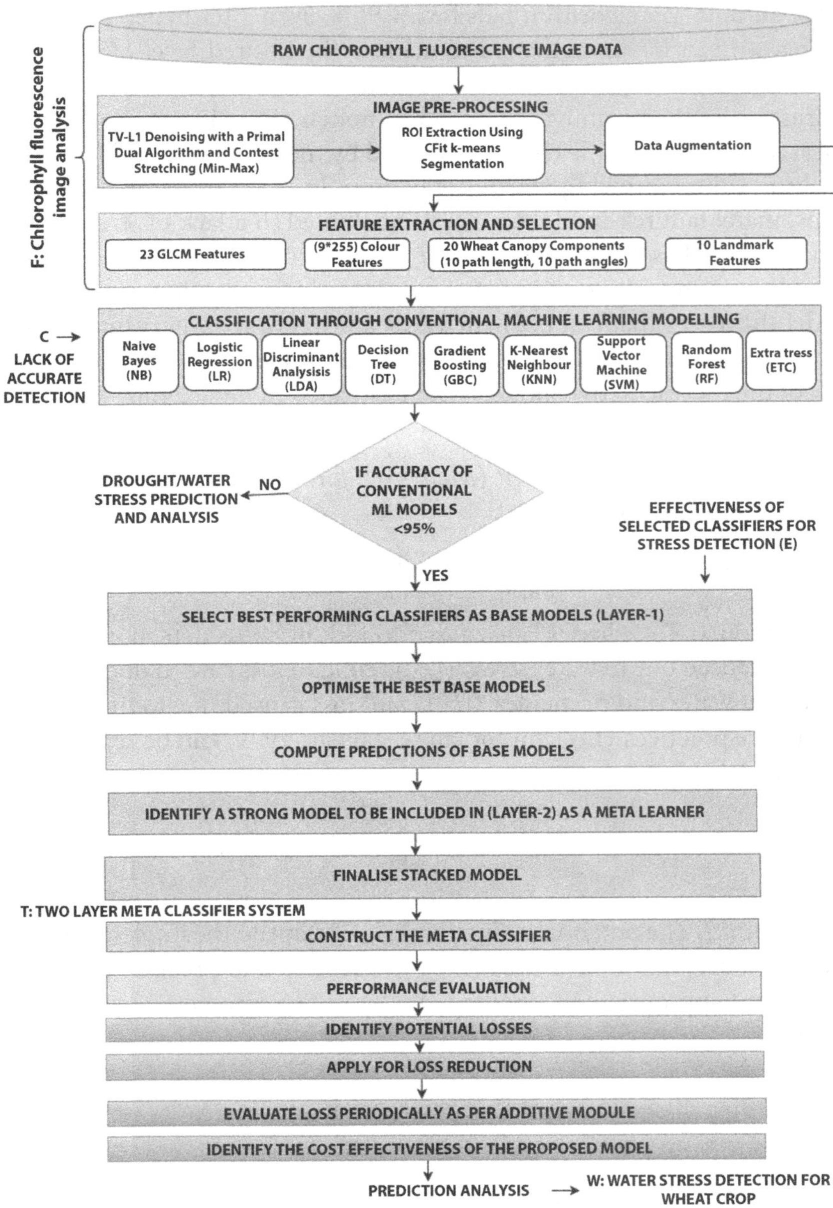

Figure 1.8 Model for water stress detection for the identification of losses due to water stress.

of the automation algorithm falls below 95%, even after hyper-parameter tuning, it must be strengthened to achieve the desired level of effectiveness, by the development of a custom meta-classifier–based model, that addresses the shortcomings of previous models. The ultimate goal of this research is to address the challenges posed by the rising demand for wheat resulting from a growing global population. In the process of cultivating crops, many failures and losses can be attributed to a lack of stress detection 'C' as well as an unscientific and non-systematic approach. This section presents an additive model that may be used to formalize, explain, and model the losses that can be incurred as a result of the interventions that can be carried out in order to have a sustainable future in agriculture. The flow of the representative model is demonstrated in Figure 1.8.

1.6.1.1 Model Development (Meta-ML for Sustainable Agronomics: For Loss Reduction)

The proposed model is based on the following assumptions:

Assumption 1. Total losses caused due to lack of stress detection 'C' contain loss caused due to water stress (L_{ws}), contain loss caused due to excessive use of water and pesticides (L_{eu}), and loss caused due to unscientific agriculture practices (L_{up}), in Indian wheat variety 'V' can be represented by the variable L as

$$L = L_{ws} + L_{eu} + L_{up} \tag{1.1}$$

Assumption 2. The potential reduction in losses due to the implementation of water conservation measures (R_{wcm}), scientific and precise agriculture (R_{spa}), and appropriate technology (R_{at}) can be represented by the variable R as

$$R = R_{wcm} + R_{spa} + R_{at} \tag{1.2}$$

Assumption 3. The implementation of each measure leads to a proportional reduction in losses. The proportional reduction $(p_{wcm}, p_{spa},$ and $p_{at})$ in losses due to the implementation of different measures, i.e., wcm, spa, and at respectively are calculated as

$$p_{wcm} = \frac{R_{wcm}}{L_{ws}} \tag{1.3}$$

$$p_{spa} = \frac{R_{spa}}{L_{eu}} \tag{1.4}$$

$$p_{at} = \frac{R_{at}}{L_{up}} \tag{1.5}$$

Assumption 4. The implementation of each measure; wcm, spa, and at respectively has some cost associated represented by c_{wcm}, c_{spa}, and c_{at}. The total cost per unit area for implementing each of the three measures can be represented as C_{wcm}, C_{spa}, and C_{at}. 'A' is the total agricultural land used for wheat production RAJ 3765 in India.

$$C_{wcm} = c_{wcm} * A \tag{1.6}$$

$$C_{spa} = c_{spa} * A \tag{1.7}$$

$$C_{at} = c_{at} * A \tag{1.8}$$

Assumption 5. The effective cost (EC) of each measure is proportional to the proportional reduction in losses divided by its cost of implementation EC_{wcm}, EC_{spa}, and EC_{at}.

$$EC_{wcm} = \frac{p_{wcm}}{C_{wcm}} \tag{1.9}$$

$$EC_{spa} = \frac{p_{spa}}{C_{spa}} \tag{1.10}$$

$$EC_{at} = \frac{p_{at}}{C_{at}} \tag{1.11}$$

Using these assumptions, we have developed an additive model to estimate the potential reduction in losses due to the implementation of each measure in a cost-effective manner utilizing Equations 1.1 to 1.11.

1.6.1.2 *Effective Cost of Implementation*

To implement the proposed model, one can use computer simulation techniques to identify the areas in which water conservation measures (wcm), scientific and precise agriculture (spa), and appropriate technology (at) are required. This can be achieved by analyzing 'F' to detect 'W' and 'P' using existing machine learning algorithms. After analyzing the effectiveness of different classifiers in identifying water stress 'E', it can be inferred that there is a need to develop a two-layer meta-classifier 'T' to estimate the potential reduction in losses due to each measure in a cost-effective manner.

Assuming that the use of machine learning algorithms can help to reduce losses by 10% for example, we can express the new potential reduction in losses due to the implementation of wcm, sp, and at as

$$\text{New R} = 0.9 * \text{R} = 0.9 * (\text{R}_{\text{wcm}} + \text{R}_{\text{spa}} + \text{R}_{\text{at}}) \tag{1.12}$$

Furthermore, assuming that the use of machine learning algorithms increases the effective cost of each measure by 20%, we can express the effective cost of each measure as

$$New\ EC_{wcm} = 1.2 * EC_{wcm} = 1.2 * (\frac{p_{wcm}}{C_{wcm} * A}) \tag{1.13}$$

$$New\ EC_{wcm} = 1.2 * EC_{wcm} = 1.2 * (\frac{p_{wcm}}{C_{wcm} * A}) \tag{1.14}$$

$$New\ EC_{wcm} = 1.2 * EC_{wcm} = 1.2 * (\frac{p_{wcm}}{C_{wcm} * A}) \tag{1.15}$$

By using these updated values in the model given by Equations 1.12, 1.13, 1.14, and 1.15, we can estimate the potential reduction in losses and effective cost of each measure, accounting for the impact of machine learning algorithms. This can help to inform decisions on which measures to implement and where to maximize the reduction in losses while minimizing costs (all costs are computed per unit area under cultivation in rupees).

The proposed additive model can be utilized for sustainably identifying effective agricultural practices. By integrating machine learning and image processing techniques, the model enables the detection of stress in crops, thereby addressing losses caused by water scarcity. The cost-effectiveness of these practices has been identified, making them feasible for implementation. Leveraging machine learning and image processing technologies, such practices contribute to sustainable agriculture by conserving water resources. This approach not only ensures the efficient use of water but also facilitates the adoption of sustainable practices to mitigate the impact of water stress on crop productivity.

1.7 Conclusion

This study demonstrates that existing machine learning models may be utilized to construct models for water stress detection and classification using CHF images and that deep learning models are not required at this scale. Our results highlight the significance of utilizing stacked models, implementing data augmentation techniques, and performing image pre-processing to enhance the precision of machine model estimation. Among the six different preprocessing pipelines used, pipeline two, which incorporates TV-L1 denoising with a primal-dual algorithm (TV-L1 with a PDA) and contrast stretching(min-max) (CS (min-max)), is the most effective in improving the quality of the images. The results were validated using three performance metrics, *viz*, PSNR, MSE, and SSIM, to demonstrate that the selected pipeline outperforms all other tested pipelines. The application of pre-processing techniques such as TV-L1 denoising with primal-dual algorithm and min-max contrast stretching has been found to significantly enhance the segmentation accuracy of the Cfit K-means algorithm [9].

This improvement in segmentation accuracy further leads to an increase in the estimation accuracy of machine learning models [60]. The best combination of features was selected and was determined by analyzing the set of five different types of features, out of which a total of four feature sets are included following the proper feature selection procedure, which includes 23 GLCM features, (9*255) color features [nine bands (blue, green, red, lightness, green-magenta, blue-yellow, hue, saturation, and value)], 20 wheat canopy components, and 10 landmarks features give the best combination. These features are given as input to nine machine learning techniques (LR, LDA, KNN, DT, NB, RF, SVM, ETC, and GBC) for water stress detection. It was found that the accuracy of existing algorithms did not exceed 93.064 percent on the current dataset based on the findings from AUC-ROC testing and training accuracy. Even after applying hyper-parameter tuning through the grid search method, conventional classification methods proved to be ineffective. After conducting extensive testing, it was discovered that utilizing a stacking or ensemble approach with multiple classifiers yields superior performance in the classification task. Specifically, the two-layer meta-classifier model based on stacking demonstrated exceptional performance. To address the limitations of existing base learners, a custom ensemble was developed, incorporating various classifier strategies to enhance the overall classification performance. The two-layer meta-model has been developed by stacking RF and NB at layer 1, and KNN at layer 2. Upon conducting a comprehensive review and validation process, the custom-built machine learning model exhibited a mean accuracy of 95.892% with a standard deviation of 0.025. These results indicate that the model consistently achieved a high level of accuracy across multiple iterations, with a relatively low variation or inconsistency. The combination of high mean accuracy and low standard deviation suggests that the model is both stable and reliable in its performance, providing confidence in its ability to accurately classify and predict outcomes.

Furthermore, an application of the meta-classifier–based model has been proposed, where an additive model Meta-ML for Sustainable Agronomics: Using Image Processing and Machine Learning for Loss Reduction can be used to estimate the potential reduction in losses due to water stress and inappropriate agricultural practices in India. The model can also be used to identify the most sustainable and effective agricultural practices and estimate the cost-effectiveness of these practices. The use of image processing and machine learning algorithms has emerged as a promising strategy for addressing these issues, as they facilitate the development of water conservation.

References

1. S. Mittal, "Wheat and Barley Production Trends and Research Priorities: A Global Perspective," 2022, pp. 3–18. doi: 10.1007/978-981-16-4449-8_1.
2. P. Giraldo, E. Benavente, F. Manzano-Agugliaro and E. Gimenez, "Worldwide Research Trends on Wheat and Barley: A Bibliometric Comparative Analysis," *Agronomy*, vol. 9, no. 7. Multidisciplinary Digital Institute (MDPI), Jul. 03, 2019. doi: 10.3390/AGRONOMY9070352.

3. Population Reference Bureau, “2019 World Population Data Sheet,” 2019 World Popul. Data Sheet, 2019.
4. S. Kumar *et al.*, “Indian Wheat Genomics Initiative for Harnessing the Potential of Wheat Germplasm Resources for Breeding Disease-Resistant, Nutrient-Dense, and Climate-Resilient Cultivars,” *Front. Genet.*, vol. 13. Frontiers Media S.A., Jan. 01, 2022. doi: 10.3389/FGENE.2022.834366.
5. A. Pandey, R. Khobra, H. M. Mamrutha, Z. Wadhwa, G. Krishnappa and G. Singh, “Elucidating the Drought Responsiveness in Wheat Genotypes,” *Sustainability*, vol. 14, no. 7. Multidisciplinary Digital Institute (MDPI), Mar. 27, 2022. doi: 10.3390/SU14073957.
6. W. Ashfaq, G. Brodie, S. Fuentes and D. Gupta, “Infrared Thermal Imaging and Morpho-Physiological Indices Used for Wheat Genotypes Screening under Drought and Heat Stress.,” *Plants*, vol. 11, no. 23. Multidisciplinary Digital Institute (MDPI), Nov. 28, 2022. doi: 10.3390/PLANTS11233269.
7. R. L. de Paulo, A. P. Garcia, C. K. Umezu, A. P. de Camargo, F. T. Soares and D. Albiero, “Water Stress Index Detection Using a Low-Cost Infrared Sensor and Excess Green Image Processing.,” *Sensors*, vol. 23, no. 3. Multidisciplinary Digital Institute (MDPI), Jan. 24, 2023. doi: 10.3390/S23031318.
8. “Annual Report 2022-2023,” Jan. 2023.
9. A. Gupta, L. Kaur, G. Kaur and V. Gahlaut, “Inferring Agronomical Insights for Wheat Canopy Using Image-Based Curve Fit K -Means Segmentation Algorithm and Statistical Analysis,” *Int. J. Genomics*, vol. 2022. SAGE Publishing, pp. 1–15, Jan. 31, 2022. doi: 10.1155/2022/1875013.
10. H. Cen *et al.*, “Current status and future perspective of the application of deep learning in plant phenotype research,” *Nongye Gongcheng Xuebao/Trans. Chin. Soc. Agric. Eng.*, vol. 36, 2020, doi: 10.11975/j.issn.1002-6819.2020.09.001.
11. Bhawana Sharma and Jay Kant Pratap Singh Yadav, “Predict Growth Stages of Wheat Crop using Digital Image Processing,” *Int. J. Recent Technol. Eng. (IJRTE)*, vol. 8, no. 5. Blue Eyes Intelligence Engineering and Sciences Engineering and Sciences Publication - BEIESP, pp. 3026–3035, Jan. 30, 2020. doi: 10.35940/IJRTE.E6201.018520.
12. V. Mishra, A. Asoka, K. Vatta and U. Lall, “Groundwater Depletion and Associated CO 2 Emissions in India,” *Earth's Future*, 2018, doi: 10.1029/2018EF000939.
13. D. K. Liyanage, I. Chathuranga, B. A. Mori and M. S. Thilakarathna, “A Simple, Semi-Automated, Gravimetric Method to Simulate Drought Stress on Plants,” *Agronomy*, vol. 12, no. 2. Multidisciplinary Digital Institute (MDPI), Jan. 29, 2022. doi: 10.3390/AGRONOMY12020349.
14. B. Kashyap and R. Kumar, “Sensing Methodologies in Agriculture for Soil Moisture and Nutrient Monitoring,” *IEEE Access*, vol. 9, 2021, doi: 10.1109/ACCESS.2021.3052478.
15. J. Su *et al.*, “Machine learning-based crop drought mapping system by UAV remote sensing RGB imagery,” *Unmanned Syst.*, vol. 8, 2020, doi: 10.1142/S2301385020500053.

16. K. Du *et al.*, "Modeling Soil Water Content and Crop-Growth Metrics in a Wheat Field in the North China Plain Using RZWQM2," *Agronomy*, vol. 11, no. 6. Multidisciplinary Digital Institute (MDPI), Jun. 19, 2021. doi: 10.3390/AGRONOMY11061245.
17. J. Jakobi, J. A. Huisman, H. Vereecken, B. Diekkrüger and H. R. Bogena, "Cosmic Ray Neutron Sensing for Simultaneous Soil Water Content and Biomass Quantification in Drought Conditions," *Water Resour. Res.*, vol. 54, no. 10. American Geophysical Union, pp. 7383–7402, Oct. 01, 2018. doi: 10.1029/2018WR022692.
18. P. Stowell, S. Fargher, C. Steer and L. F. Thompson, "Scintillating thermal neutron detectors for cosmic ray soil moisture monitoring," *J. Instrum.*, vol. 16, no. 11. IOP Publishing, Nov. 01, 2021. doi: 10.1088/1748-0221/16/11/P11039.
19. Z. Yang *et al.*, "Coupled soil water stress and environmental effects on changing photosynthetic traits in wheat and maize," *Agric. Water Manage.*, vol. 282. Elsevier, May 01, 2023. doi: 10.1016/J.AGWAT.2023.108246.
20. M. A. Javed, S. R. Ahmad, W. K. Awan and B. A. Munir, "Estimation of Crop Water Deficit in Lower Bari Doab, Pakistan Using Reflection-Based Crop Coefficient," *ISPRS Int. J. Geo-Inf.*, vol. 9, no. 3. Multidisciplinary Digital Institute (MDPI), Mar. 13, 2020. doi: 10.3390/IJGI9030173.
21. T. Kuromori, M. Fujita, F. Takahashi, K. Yamaguchi-Shinozaki and K. Shinozaki, "Inter-tissue and inter-organ signaling in drought stress response and phenotyping of drought tolerance.," *Plant J.*, vol. 109, no. 2. Wiley, pp. 342–358, Dec. 16, 2021. doi: 10.1111/TPJ.15619.
22. T. S. Magney, L. A. Vierling, J. U. H. Eitel, D. R. Huggins and S. R. Garrity, "Response of high frequency Photochemical Reflectance Index (PRI) measurements to environmental conditions in wheat," *Remote Sens. Environ.*, vol. 173. Elsevier, pp. 84–97, Feb. 01, 2016. doi: 10.1016/J.RSE.2015.11.013.
23. W. Jiao, Q. Chang and L. Wang, "The Sensitivity of Satellite Solar-Induced Chlorophyll Fluorescence to Meteorological Drought," *Earth's Future*, vol. 7, 2019, doi: 10.1029/2018EF001087.
24. A. Gupta, L. Kaur and G. Kaur, "Comparitive Analysis of Segmentation Methods for Wheat Canopy Extraction," Jan. 2021. doi: 10.21467/PROCEEDINGS.114.7.
25. B. K. Sinha, G. Chand, R. na and V. Kumar, "Different Degree of Drought Stress Affects Morpho-Reproductive Growth in Wheat (Triticum aestivum L)," *Int. J. Curr. Microbiol. Appl. Sci.*, vol. 8, no. 06. Excellent Publishers, pp. 1047–1053, Jun. 10, 2019. doi: 10.20546/IJCMAS.2019.806.128.
26. N. Fahlgren *et al.*, "A Versatile Phenotyping System and Analytics Platform Reveals Diverse Temporal Responses to Water Availability in Setaria.," *Mol. Plant*, vol. 8, no. 10. Elsevier, pp. 1520–35, Jun. 20, 2015. doi: 10.1016/J.MOLP.2015.06.005.
27. H. Bacher, Y. Sharaby, H. Walia and Z. Peleg, "Modifying root-to-shoot ratio improves root water influxes in wheat under drought stress.," *J. Exp. Bot.*,

vol. 73, no. 5. Oxford University Press, pp. 1643–1654, Mar. 02, 2022. doi: 10.1093/JXB/ERAB500.
28. W. U. HASAN, C. ROY, T. CHATTOPADHYAY, R. D. RANJAN and N. DE, "Effects of heat and drought stress on yield and physiological traits in wheat (Triticum aestivum L.)," *J. Crop Weed*, vol. 17, no. 1. AkiNik Publications, pp. 203–210, Jan. 01, 2021. doi: 10.22271/09746315.2021.V17.I1.1424.
29. M. F. Qaseem, R. Qureshi and H. Shaheen, "Effects of Pre-Anthesis Drought, Heat and Their Combination on the Growth, Yield and Physiology of diverse Wheat (Triticum aestivum L.) Genotypes Varying in Sensitivity to Heat and drought stress.," *Sci. Rep.*, vol. 9, no. 1. Springer Nature, May 06, 2019. doi: 10.1038/S41598-019-43477-Z.
30. O. A. Montesinos-López *et al.*, "Predicting grain yield using canopy hyperspectral reflectance in wheat breeding data," *Plant Methods*, vol. 22, pp. 2–23, 2017, doi: 10.1186/s13007-016-0154-2.
31. N. Lu *et al.*, "Improved estimation of aboveground biomass in wheat from RGB imagery and point cloud data acquired with a low-cost unmanned aerial vehicle system.," *Plant Methods*, vol. 15. Springer Nature, Jan. 01, 2019. doi: 10.1186/S13007-019-0402-3.
32. J. Ma *et al.*, "Estimating growth related traits of winter wheat at seedling stages based on RGB images and convolutional neural network," *Nongye Gongcheng Xuebao/Trans. Chin. Soc. Agric. Eng.*, vol. 35, pp. 183–189, 2019, doi: 10.11975/j.issn.1002-6819.2019.05.022.
33. E. S. Ober *et al.*, "Wheat root systems as a breeding target for climate resilience," *Theor. Appl. Genet.*, vol. 134, 2021, doi: 10.1007/s00122-021-03819-w.
34. M. Hong, D. J. Bremer and D. Merwe, "Thermal Imaging Detects Early Drought Stress in Turfgrass Utilizing Small Unmanned Aircraft Systems," *Agrosyst. Geosci. Environ.*, vol. 2, no. 1. Wiley, pp. 1–9, Jan. 01, 2019. doi: 10.2134/AGE2019.04.0028.
35. D. Gao *et al.*, "In-field chlorophyll estimation based on hyperspectral images segmentation and pixel-wise spectra clustering of wheat canopy," *Biosyst. Eng.*, vol. 217. Elsevier, pp. 41–55, May 01, 2022. doi: 10.1016/J. BIOSYSTEMSENG. 2022.03.003.
36. P. D. Dao, Y. He and C. Proctor, "Plant drought impact detection using ultra-high spatial resolution hyperspectral images and machine learning," *Int. J. Appl. Earth Obs. Geoinf.*, vol. 102, 2021, doi: 10.1016/j.jag.2021.102364.
37. S. El-Hendawy, N. Al-Suhaibani, W. Hassan, M. Tahir and U. Schmidhalter, "Hyperspectral reflectance sensing to assess the growth and photosynthetic properties of wheat cultivars exposed to different irrigation rates in an irrigated arid region," *PLoS One*, vol. 12, pp. 1–22, 2017, doi: 10.1371/journal.pone.0183262.
38. Y. Xie *et al.*, "Hyperspectral monitor on chlorophyll density in winter wheat under water stress," *Agron. J.*, vol. 112, no. 5. Wiley, pp. 3667–3676, Jul. 13, 2020. doi: 10.1002/AGJ2.20306.

39. I. Moya, H. Loayza, M. L. López, R. Quiroz, A. Ounis and Y. Goulas, "Canopy chlorophyll fluorescence applied to stress detection using an easy-to-build micro-lidar," *Photosynth. Res.*, vol. 142, 2019, doi: 10.1007/s11120-019-00642-9.
40. Z. Kandylakis, A. Falagas, C. Karakizi and K. Karantzalos, "Water Stress Estimation in Vineyards from Aerial SWIR and multispectral UAV data," *Remote Sens.*, vol. 12, 2020, doi: 10.3390/RS12152499.
41. K. K. Patro, A. Jaya Prakash, M. Jayamanmadha Rao and P. Rajesh Kumar, "An Efficient Optimized Feature Selection with Machine Learning Approach for ECG Biometric Recognition," *IETE J. Res.*, 2020, doi: 10.1080/03772063.2020.1725663.
42. S. Kumar and T. Kaur. "Efficient solar radiation estimation using cohesive artificial neural network technique with optimal synaptic weights." *Proc. Inst. Mech. Eng., Part A: J. Power Energy,* vol.234 2020, 862-873. doi: 10.1177/0957650919878318.
43. K. Yu, N. Kirchgessner, C. Grieder, A. Walter and A. Hund, "An image analysis pipeline for automated classification of imaging light conditions and for quantification of wheat canopy cover time series in field phenotyping," *Plant Methods*, vol. 13, 2017, doi: 10.1186/s13007-017-0168-4.
44. N. S. Chandel, Y. A. Rajwade, K. Dubey, A. K. Chandel, A. Subeesh and M. K. Tiwari, "Water Stress Identification of Winter Wheat Crop with State-of-the-Art AI Techniques and High-Resolution Thermal-RGB Imagery.," *Plants*, vol. 11, no. 23. Multidisciplinary Digital Institute (MDPI), Dec. 02, 2022. doi: 10.3390/PLANTS11233344.
45. M. M. Hasan, J. P. Chopin, H. Laga and S. J. Miklavcic, "Detection and analysis of wheat spikes using Convolutional Neural Networks," *Plant Methods*, vol. 14, 2018, doi: 10.1186/s13007-018-0366-8.
46. S. Saha, B. Kundu, G. C. Paul and B. Pradhan, "Proposing an ensemble machine learning based drought vulnerability index using M5P, dagging, random sub-space and rotation forest models," *Stochastic Environ. Res. Risk Assess.*, Springer Nature, Mar. 06, 2023. doi: 10.1007/S00477-023-02403-6.
47. A. Ahmad *et al.*, "Screening of Wheat (Triticum aestivum L.) Genotypes for Drought Tolerance through Agronomic and Physiological Response," *Agronomy*, vol. 12, no. 2. Multidisciplinary Digital Institute (MDPI), Jan. 23, 2022. doi: 10.3390/AGRONOMY12020287.
48. M. K. Azameti *et al.*, "Transcriptome profiling reveals the genes and pathways involved in thermo-tolerance in wheat (Triticum aestivum L.) genotype Raj 3765.," *Sci. Rep.*, vol. 12, no. 1. Springer Nature, Sep. 01, 2022. doi: 10.1038/S41598-022-18625-7.
49. Y. Yang, "Rapid and Nondestructive Evaluation of Wheat Chlorophyll under Drought Stress Using Hyperspectral Imaging," *Int. J. Mol. Sci.*, vol. 24, no. 6, p. 5825, Feb. 2023, doi: 10.3390/ijms24065825.

50. J. Yang, L. Du, W. Gong, S. Shi, J. Sun and B. Chen, "Analyzing the performance of the first-derivative fluorescence spectrum for estimating leaf nitrogen concentration.," *Opt. Express*, vol. 27, no. 4. Optica Publishing Group, pp. 3978–3990, Feb. 18, 2019. doi: 10.1364/OE.27.003978.
51. W. Wang, Y. Cheng, Y. Ren, Z. Zhang and H. Geng, "Prediction of Chlorophyll Content in Multi-Temporal Winter Wheat Based on Multispectral and Machine Learning.," *Front. Plant Sci.*, vol. 13. Frontiers Media S.A., Jan. 01, 2022. doi: 10.3389/FPLS.2022.896408.
52. Nyaupane, S., Poudel, M. R., Panthi, B., Dhakal, A., Paudel, H., & Bhandari, R. (2024). Drought stress effect, tolerance, and management in wheat–a review. *Cogent Food & Agriculture,* 10(1), 2296094.
53. I. Spyroglou, K. Rybka, R. Maldonado Rodriguez, P. Stefański and N. M. Valasevich, "Quantitative estimation of water status in field-grown wheat using beta mixed regression modelling based on fast chlorophyll fluorescence transients: A method for drought tolerance estimation," *J. Agron. Crop. Sci.*, vol. 207, 2021, doi: 10.1111/jac.12473.
54. Sukhjit Sandhu, "Plant Stress Analysis Based on Chlorophyll Fluorescence and Image Processing," p. 01, 2019, doi: 10.17632/jnjd835ncg.2.
55. L. K. Ankita gupta, "Automatic Water Stress detection in wheat crop canopy using Chlorophyll fluorescence image dataset," *Dataset.* Eslevier, Aug. 30, 2022. doi: 10.17632/2mpd7d3vry.2.
56. A. Gupta, L. Kaur and G. Kaur, "Impact of Image Pre-processing Operations on Wheat Canopy Segmentation," in *Proceedings of the 2022 Fourteenth International Conference on Contemporary Computing*, 2022, pp. 396–403.
57. Gupta, Ankita, Kaur, Dr. Gurmeet, Kaur, Dr. Lakhwinder (2022), "ROI extraction of chlorophyll Fluorescence wheat canopy images using novel Curve Fit Based K- means segmentation Algorithm for automatic drought stress detection using machine learning ", *Mendeley Data*, V4, doi: 10.17632/crb5tkbvpb.4.
58. A. Gupta, L. Kaur and G. Kaur, "Drought stress detection technique for wheat crop using machine learning," *PeerJ Comput. Sci.*, vol. 9. PeerJ, pp. 2–24, May 19, 2023. doi: 10.7717/PEERJ-CS.1268.
59. M. A. Gehan *et al.*, "PlantCV v2: Image analysis software for high-throughput plant phenotyping.," *PeerJ*, vol. 5. PeerJ, Jan. 01, 2017. doi: 10.7717/PEERJ.4088.
60. B. Sharma *et al.*, "Drought stress and its management in wheat (Triticum aestivum L.): a review," *Agric. Sci. Technol.*, vol. 14, no. 1. Trakia University, pp. 3–14, Mar. 01, 2022. doi: 10.15547/10.15547/AST.2022.01.001.
61. M. M. Daniel Tsegai, "DROUGHT IN NUMBERS 2022 - restoration for readiness and resilience," *UNCCD's 15th Conference of Parties (CoP15)*, United Nations, May 2022.

62. S. Yan *et al.*, "Effects of water and fertilizer management on grain filling characteristics, grain weight and productivity of drip-fertigated winter wheat," *Agric. Water Manage.*, vol. 213. Elsevier, pp. 983–995, Mar. 01, 2019. doi: 10.1016/J.AGWAT.2018.12.019.
63. F. Muhammad *et al.*, "Ameliorating Drought Effects in Wheat Using an Exclusive or Co-Applied Rhizobacteria and ZnO Nanoparticles.," *Biology*, vol. 11, no. 11. Multidisciplinary Digital Institute (MDPI), Oct. 25, 2022. doi: 10.3390/BIOLOGY11111564.
64. "MANUAL FOR December 2016 Department," Department of Agriculture, Cooperation & Farmers Welfare Ministry of Agriculture & Farmers Welfare Government of India New Delhi, India, Dec. 2016.

2

Comparison of Geo-Statistical Techniques: A Study Based on Mapping Soil Properties for a Lower Himalayan Watershed

Sahil Sharma*, Shankar Yadav, Vinay Meena and Deepak Swami

Indian Institute of Technology, Mandi, Himachal Pradesh, India

Abstract

Soil moisture is the main driving factor for agricultural and hydrological processes and is essential for organisms living in the rhizosphere. Soil moisture varies spatially and temporally depending on soil and climate characteristics. Several approaches to measurement, ranging from point to continental scales, are used to capture variations in soil moisture. The moisture distribution varies substantially with land use land cover (LULC) topographic features, and organic content. The geographic information system is a very efficient tool for spatially representing parameter variability over long-time scales. The distribution of soil moisture with soil organic content as a secondary variable is seasonally mapped in the Himalayan watershed Suketi area of 422 km^2 at different LULCs. The spatial interpolation method universal kriging (UK) was used to prepare the soil moisture distribution maps. Statistical analysis, namely, percentage bias, correlation (R), root mean square error (RMSE), and un-bias root mean square error (ubRMSE), were also performed. The present study has shown that forest landforms have greater spatial variability than agricultural and grassland landforms. This chapter will provide a holistic view of the use of geo-statistics and GIS in spatial interpolation techniques for soil moisture distribution. The research findings show that interpolation methods understand the relationship between soil characteristics.

Keywords: Soil moisture, IDW, universal kriging, spatial interpolation, GIS

**Corresponding author*: sharmasahilbuce@gmail.com

Disha Thakur, Sanjay Kumar, Har Amrit Singh Sandhu and Chander Prakash (eds.) Sustainable Development Using Geospatial Techniques, (41–58) © 2024 Scrivener Publishing LLC

2.1 Introduction

Soil moisture is an important driving factor for agricultural and hydrological processes. It impacts various processes in the rhizosphere, such as soil structure, plant growth, nutrient availability, soil erosion, infiltration, runoff, and many others. The parameters affecting the moisture content of the soil can be categorized as physiographic factors (topography and climate), cropping patterns and types, soil characteristics (texture and classification), quantity (spatial and temporal extent), and quantity of measurement [1]. The landform and the presence of organic matter (OM) in the soil are two major factors that greatly affect the field's soil moisture. LULC is one of the most important driving factors that can modify the soil structure and lead to a change in hydraulic attributes that affect spatiotemporal variability in soil moisture [2]. Organic matter-rich soil can boost water-holding capacity by enhancing its ability to retain moisture. It can improve soil structure and prevent moisture loss through evaporation, acting as a protective layer on the soil surface.

Soil moisture measurement studies are carried out by researchers from field and catchment to continental scale, which use different methods of measurement [3, 4]. Geophysical methods and sensor-based measurement can be used for field and catchment levels, but for continental-scale studies, remote sensing-based measurements have been found suitable due to the cost and feasibility of the method [5, 6]. However, remote sensing data needs to be validated with field data to check accuracy. The field-scale variability makes the validation difficult. Thus, upscaling the data at the point and watershed scales is done to minimize the scale difference errors. The techniques for soil moisture measurement can be divided basically into two categories direct and indirect measurements. The gravimetric method is the only direct method for soil sampling where soils are weighed before and after oven drying [7]. The indirect methods are classified based on the mechanism of measurement: contact (invasive) and contact-free (proximal). Contact-based methods kept sensors in contact with soil, while contact-free methods worked remotely without coming into contact [7].

To obtain continuous spatial data, spatial interpolation of real-time observations from locations is required to estimate attribute values at unsampled locations [8]. Many studies suggest the best interpolation methods for various data types, i.e., soil physical properties, but choosing the best method without prior knowledge of datasets can be challenging [8, 9]. The interpolation techniques are simple derivation functions that determine data at the unknown location from known data sources using statistical tools [10]. Geostatistical methods such as ordinary kriging (OK), simple kriging, co-kriging, deterministic, i.e., inverse distance weighting (IDW), and local polynomials are commonly used interpolation techniques [10, 11]. IDW estimates values based on the difference between two known points while kriging uses variance at every predicting point. The performance of these methods depends not only on the characteristics but also on variables such as data variability and sampling design [9]. Most techniques better predict soil properties in uniform terrain but very limited techniques give equally good results for complex terrain [12, 13]. IDW and kriging are two of the most widely used interpolation techniques for complex terrain and hence are selected for our study [14, 15].

2.2 Materials and Methods

2.2.1 Study Area

The Suketi catchment is located in the Mandi district of Himachal Pradesh, which experiences subtropical climate falls in the lower Himalayan region as shown in Figure 2.1. Suketi River is one of the important tributaries of Beas River with a catchment area of 422 km^2 located between the latitude 31°27′ 08′:31°45′ 00′ and longitude 76°48′20′:77°03′09′.

The catchment area is surrounded by high mountains of elevation varying from 739 to 2914 m with centrally located Balh Valley [16]. The basin receives sufficient rainfall from June to September during the monsoon season, but due to the steep slope and impermeable lithology, it is lost as runoff. It also receives snowfall during the peak winter season on mountain peaks.

2.2.2 Field and Remotely Sensed Datasets

The digital elevation data from the Shuttle Radar Topography Mission (SRTM) was used to prepare the elevation map and generate the study area. The land use land cover (LULC) map was downloaded from the Bhuvan portal to extract different landforms. The soil sampling was done using a 10-cm diameter and 15-cm height core cutter in different landforms in gridded locations shown in Figure 2.2. The mean soil moisture values at a depth of 10 cm were used as input in the investigated interpolation approaches for each location.

Extensive fieldwork was carried out from January 2021 to February 2022 to obtain soil moisture data from the field. During the fieldwork,

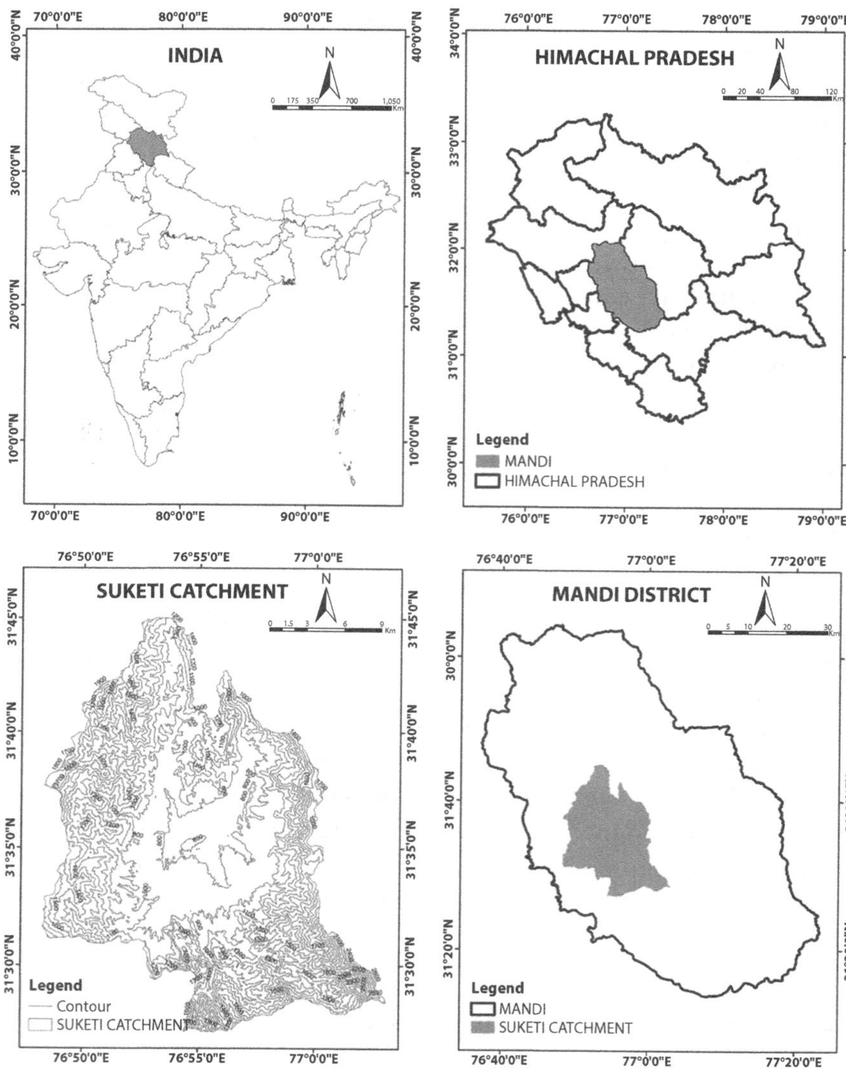

Figure 2.1 Clockwise representation of the location of the Suketi River catchment in Himachal Pradesh India.

high-quality soil moisture data were collected from 104 sampling points over 8 days using Hydrago sensor equipment for the catchment area, covering various landforms. In Figure 2.2a, all 104 points are shown and divided into two categories of observed and predicted points. In Figure 2.2b, the spatial variability in sampling points used for prediction in different landforms has been shown. The data were processed and analyzed

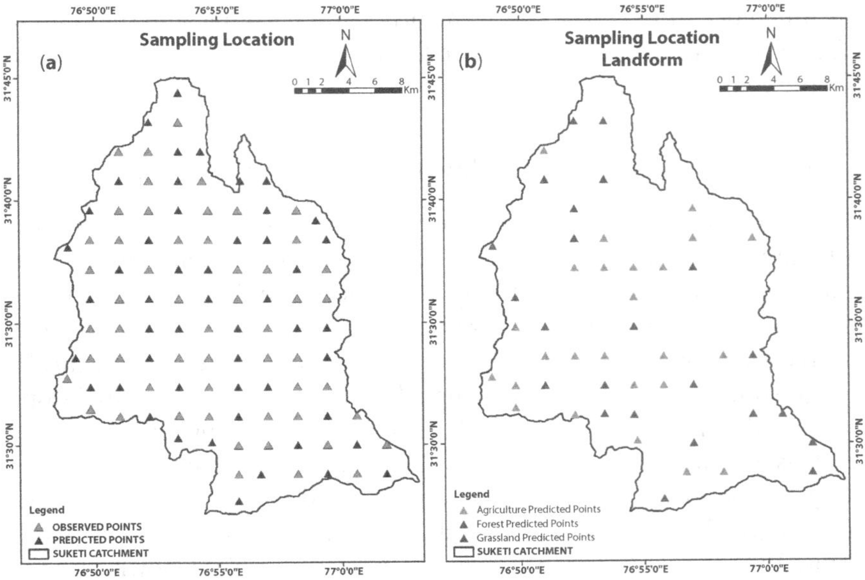

Figure 2.2 (a) The observed and predicted points in gridded location for measurement of soil moisture and soil sampling locations and (b) predicted points based on different landforms.

using ArcGIS v10.8. Statistical analysis was done using ArcGIS v10.1 spatial analyzer and Microsoft Excel.

2.2.3 Soil Physical Property Identification

Soil moisture was determined using mean soil moisture values from the measurements taken at 104 locations (agricultural, 55%; forest, 24%; grassland, 21%) of the study indicated in Figure 2.2. Furthermore, soil samples were also collected from the same location for soil organic content estimation.

This study was conducted for measurements of soil moisture using a Hydrago sensor instrument which provides the average volumetric water content (VMC) and electrical conductivity (EC) along with latitude and longitude. The organic matter content in soil samples was measured using commonly used the muffle furnace method. This method involves the loss on ignition (LOI) of organic compounds by heating the soil sample in a muffle furnace at a high temperature (550°C to 600°C) until all the organic matter in the sample has been burned off. The remaining material, which consists mainly of inorganic components such as minerals, is weighed.

The difference in weight between the original soil sample and the residue left after ignition is used to calculate the LOI, which is an estimate of the organic content of the soil.

2.2.4 Nature of LULC and Elevation of the Catchment

The LULC and elevation map of the Suketi catchment shows are shown in Figure 2.3(a and b, respectively). The catchment is mountainous with undulating landforms dominated by shrubs, forest, and agriculture. In general, the southeast and northeastern parts have a higher elevation which is covered with forest landform while mid-elevation is covered with shrubs and grasses. The valley region having a lower elevation and gradual slope is mostly dominated by agricultural fields. Currently, shrubland classes are taken as grassland only so that generalizations as per the objectives of the study can be made. Soil moisture was collected from three major landforms, agriculture, forest, and grassland, to study variation in the landforms. The major urban area is around Sundernagar, Ner Chowk, and the northern part of the Suketi catchment in Mandi town.

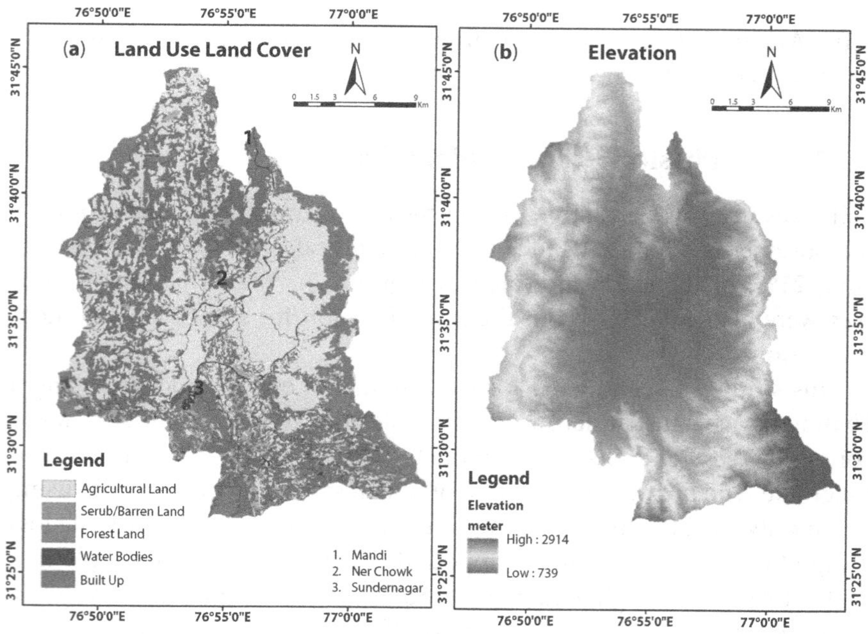

Figure 2.3 (a) LULC map. (b) Elevation map of the Suketi watershed.

2.2.5 Spatial Interpolation Methods

Spatial interpolation is an essential tool for estimating spatial data at unknown locations using known observed data on the Earth's surface. This study focuses on IDW and kriging methods which are commonly used for interpolating techniques for hydrological and soil parameter analysis [17]. The kriging method performs better for low- to high-elevation ranges than IDW [18]. The IDW method estimates the weights by an exponential of the inverse function of the distance to the observed data point while kriging uses a fitted variogram [19]. The deterministic (i.e., IDW) and geo-statistical (i.e., kriging) interpolation methods can be applied to estimate attributes in unsampled locations as per the type of datasets. In this case study, we compared IDW and universal kriging, commonly called KED, in different landforms. The OC and SM collected from different landforms and elevations were interpolated and evaluated to understate the variability of SM and OC in the catchment.

2.2.5.1 Inverse Distance Weighting (IDW)

The IDW method utilizes the weight depending on an exponent of two observed points distance [19]. It estimates the value of unsampled points using Tobler's law which considers that the attribute decreases with increasing distance [20]. The predicted value of unsampled points would be expressed as [21]

$$\hat{\mathbf{Z}}(x_0) = \frac{\sum_{i=1}^{N} w(x_i) \times Z(x_i)}{\sum_{i=1}^{N} w(x_i)} \tag{2.1}$$

where $Z(x_i)$ is the observation at x_i and $w(x_i)$ is the weight associated with observation at x_i location. $w(x_i)$ can be calculated using Equation 2.2

$$w(x_i) = \frac{1}{d(x_i - x_0)^{-p}} \tag{2.2}$$

where d represents the distance between x_i and x_0, and p is a distance-dependent exponent parameter. p is the main factor that affects the accuracy of IDW [22]. Weight decreases with distance, resulting in high values and influencing nearby observation points [23]. In the inverse square distance, the power value is taken as 2, which has a more satisfactory result when compared with a p-value as high as 4 [24].

2.2.5.2 Kriging

Kriging analysis using a semi-variogram is a commonly applied interpolation method for attribute value estimation in unsampled points [25, 26]. The kriging interpolation method estimates the values in the variogram structure based on non-observation points of proximity to the observed points. In this method, the confidence degree of the predicted value is calculated with variance at every point [25, 27].

OK without prior knowledge of the distribution's mean interpolates linearly and calculates a value at a known point on a variogram. This technique employs a random function model, which allows for the estimation of bias and error variance. Weights are then selected for the nearby observed points, ensuring that the model's average error is zero with minimum modeled variance. The sum of these weights must be equal to one for the estimator to be unbiased in applying OK [28, 29].

In place of being a function of the spatial coordinates, the universal kriging includes the neighborhood trend as a linear function of a second variable that is continuously changing [30]. Kriging offers various advantages but also limitations. It offers the most precise linear unbiased prediction. It also includes an assessment of the error/uncertainty at the unsampled location. Although this assumption can be modified with certain kriging methods, it assumes that the data are stationary, which is typically not the case [8].

In this interpolation, the predicted value of unsampled points would be expressed as

$$\hat{Z}(x) - m(x) = \sum_{i=1}^{N} \lambda[Z(x_i) - m(x_i)] \tag{2.3}$$

where $m(x)$ and $m(x_i)$ are predicted values of $Z(x)$ and $Z(x_i)$, respectively, and λ is the weight assigned to the point $Z(x_i)$; $m(x_i)$ is estimated by minimizing the error variance"

$$\sigma_E^2(x) = \mathrm{Var}(\hat{Z}(s) - Z(s)) \tag{2.4}$$

$$\gamma = \frac{1}{1|N(h)|} \Sigma_{N(h)} [Z(x_i) - Z(x_i + h)]^2 \tag{2.5}$$

where γ the semi-variance, n is the number of pairs, and h is the distance between pairs of points.

2.2.6 Statistical Criteria

For the selection of better-performance interpolation techniques, the error estimates are calculated based on bias, root mean square error (RMSE), unbiased RMSE (ubRMSE), and mean absolute error (MAE) were calculated in this study. Percentage bias (PBIAS) measures the predicted and observed value deviation. A positive and negative value indicates over and under-estimation between predicted and observed values [31]. RMSE is one more important statistical parameter used for the assessment of the accuracy of spatial analysis, which signifies the standard deviation (SD) of the sample and the alterations between observed and predicted values [32]. The lower value of RMSE shows better prediction accuracy [33]. ubRMSE can accurately evaluate the performance of predictive models, account for bias, enable model comparison, and provide a basis for model improvement. To determine the accuracy of continuous variables, MAE measures the average magnitude of the errors in a set of estimates without considering their direction [34]. It is a quantity utilized to measure how closely predictions match observed results and a typical gauge of a model's prediction error [31].

2.3 Results and Discussion

The soil physical parameters prepared using the interpolation techniques are compared to the actual field data. The error estimates of the parameters like soil moisture, organic content, and elevation are evaluated using the spatially prepared maps. Furthermore, the results are also compared based on different landforms under study.

2.3.1 Soil Moisture and Organic Content Distribution Dynamics in the Catchment

The spatially generated maps of soil moisture, organic content, and elevation are shown in Figures 2.4 and 2.5. The interpolation techniques applied

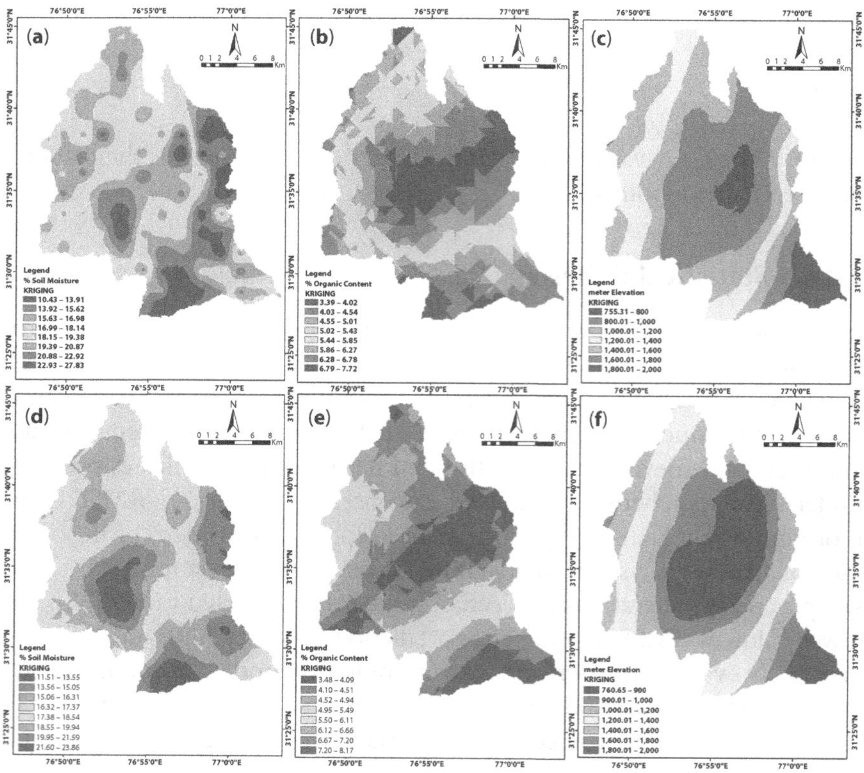

Figure 2.4 Spatial maps of soil moisture, organic content, and elevation prepared using kriging interpolation (a to c) using all 104 points as predicted points and (d to f) using 50 predicted points.

with higher observed points for each parameter show a more detailed output of spatial variation of attributes, as shown in Figures 2.4(a to c) and 2.5(a to c). The mean soil moisture in the grassland landform was found to be higher than in the agriculture and forest landforms. The organic content is found more in forest landforms than in agriculture and grassland landforms.

2.3.2 Spatial Interpolation Methods

The efficacy and reliability of the interpolation methods used for the soil physiographic parameters like soil moisture, elevation, and organic content were assessed in this study. The spatial interpolation findings for all three parameters and landforms were examined using statistical performance

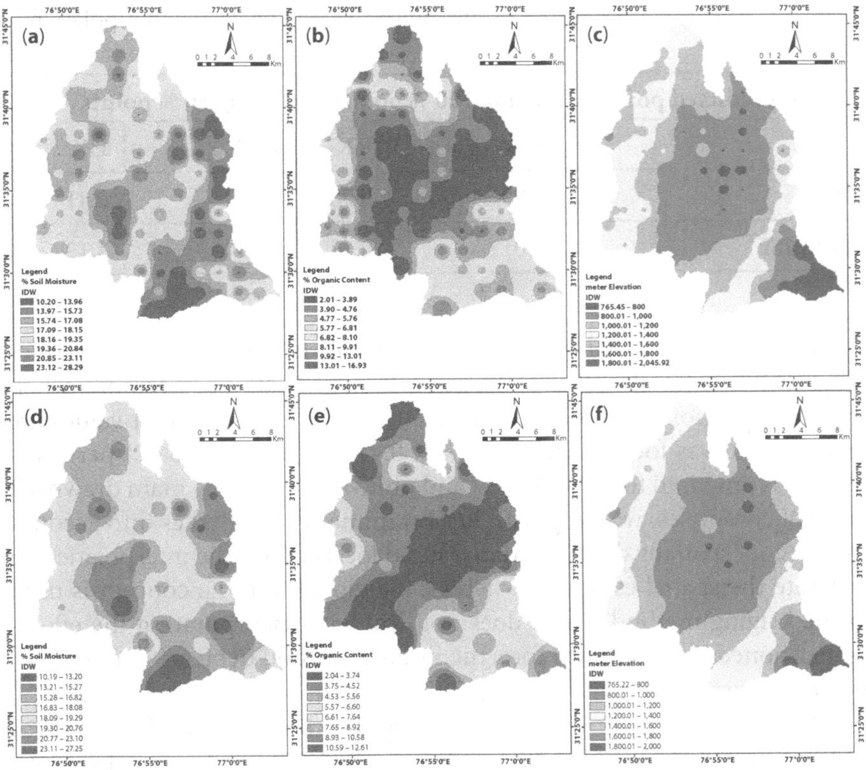

Figure 2.5 Spatial maps of soil moisture, organic content, and elevation prepared using IDW interpolation (a to c) using all 104 points as predicted points and (d to f) using 50 predicted points.

measures such as BIAS, R, RMSE, and ubRMSE. The finding of the work was that universal kriging predicts soil moisture better when compared to IDW. The model variogram for the UK was selected based on the cross-validation results with the lowest RMSE. It can be seen from the visual representation of the interpolated maps prepared using both methods in Figures 2.4(a and d) and 2.5(a and d) that the UK has a better prediction than IDW. While the IDW technique smoothed the map, resulting in a crude estimate and relatively basic soil moisture profile, the UK has both captured greater regional diversity in soil moisture.

The IDW interpolation technique performs better than universal kriging, but it found that the IDW does not interpolate the discrete data spatially when using data points [35]. Table 2.1 shows that the UK performs better with fewer sampling points and predicts more accurate values in predicted points.

Table 2.1 Performance statistics of soil moisture based on spatial interpolation methods.

No. of predicted points	Using 100 % points		Using 50 % points	
	IDW	UK	IDW	UK
R	0.99	0.98	0.37	0.41
ubRMSE	0.012	0.18	3.14	3.12

The interpolated map of soil moisture using all 104 points shows a higher R (ubRMSE) 0.99 (0.012) in IDW, like UK 0.98 (0.18). Furthermore, when the observed points are decreased by 50% and then the spatial values are predicted, it was found that the UK (R 0.41 and ubRMSE 3.12) performs better as compared to IDW (R 0.37 and ubRMSE 3.14). The main drawback of IDW comprises, working with local minima where it takes into account the effect of spatially distributed measured locations on the interpolation to generate the spatial estimates whereas, the global minima concept is used in the kriging method which does not deal with incorporating the effects as mentioned in IDW. Therefore, the kriging method depicts better performance as compared to the former.

IDW performed worse in interpolating soil moisture as it has a higher RMSE value (4.55) in agriculture, whereas the UK has a lower value (2.4), as shown in Table 2.2. At least, according to our investigation, IDW is more sensitive and significantly over-predicted in all three landforms in terms of % BIAS (Table 2.3). The UK techniques accurately estimate soil moisture with minimal bias and RMSE, as shown by an analysis of RMSE and % BIAS. As a result, it can be claimed that the UK method performed better at predicting soil moisture values than the other methods.

The soil moisture, organic content, and elevation spatial variability map prepared by using UK and IDW have been shown in Figures 2.4 and 2. 5, respectively. Tables 2.2, 2.3, and 2.4 show % BIAS, R, RMSE, and ubRMSE for both methods (IDW and UK) for generated spatial variability maps from 104 and 50 points in the catchment. Figures 2.4 and 2.5 clearly show that increasing the number of predicting points has greatly impacted interpolation accuracy in both methods. These generated maps can be used as an inventory for local validation of satellite data with great accuracy.

The agricultural field shows a better correlation in soil moisture when compared with other landforms, as there are more observation points from individual landforms. The Forest landform has less accuracy than the other landforms due to the canopy of trees and fewer observed points in both methods. The interpolated map depicts a wide range of soil moisture between different landforms and elevations. In Figures 2.4(a and d), the valley region of the study area, which is grassland and agricultural dominant, has a high interpolated value compared to higher elevation. The UK method has better-predicted soil moisture in all the landforms using 104 predicted points compared to IDW. When all observed points were used for interpolation, it was found that agricultural landform has a better correlation (0.98) compared to forest (0.92) and grassland (0.95) (Tables 2.2 and 2.3). The R value for soil moisture is low for IDW compared to the UK method in all three landforms when 50 points were used for prediction, which revealed that errors were more likely to occur in the IDW than in the UK. Similarly, different landforms also show that soil moisture is

Table 2.2 Grassland landform error estimates.

	Using 100 % points						**Using 50 % points**					
Attributes	**Elevation**		**Organic content**		**Soil moisture**		**Elevation**		**Organic content**		**Soil moisture**	
Methods	**UK**	**IDW**	**UK**	**IDW**	**UK**	**IDW**	**UK**	**IDW**	**UK**	**IDW**	**UK**	IDW
BIAS	-3.11	-0.07	-0.13	-0.001	0.009	0.0007	-40.74	-56.62	0.47	0.15	-1.75	-1.73
R	0.91	0.95	0.39	0.95	0.95	0.954	0.84	0.83	0.19	0.34	0.11	0.07
RMSE	80.01	0.53	3.16	0.007	0.15	0.009	114.82	138.76	2.19	1.97	4.59	4.55
ub RMSE	79.95	0.53	3.16	0.007	0.15	0.009	107.34	126.68	2.14	1.97	4.24	4.21

Table 2.3 Forest landform error estimates.

	Using 100 % points						**Using 50 % points**					
Attributes	**Elevation**		**Organic content**		**Soil moisture**		**Elevation**		**Organic content**		**Soil moisture**	
Methods	**UK**	**IDW**	**UK**	**IDW**	**UK**	**IDW**	**UK**	**IDW**	**UK**	**IDW**	**UK**	**IDW**
BIAS	4.31	-0.03	-0.03	-0.0002	0.002	0.0004	-77.44	-144.56	0.86	1.05	-0.67	-0.23
R	0.95	0.98	0.49	0.98	0.981	0.98	0.79	0.75	0.56	0.61	0.17	0.03
RMSE	65.88	0.47	2.36	0.011	0.171	0.02	195.21	253.09	1.42	1.56	3.17	3.31
ub RMSE	65.74	0.46	2.36	0.011	0.171	0.02	179.19	207.75	1.14	1.16	3.11	3.31

Table 2.4 Agriculture landform error estimates.

	Using 100 % points						Using 50 % points					
Attributes	**Elevation**		**Organic content**		**Soil moisture**		**Elevation**		**Organic content**		**Soil moisture**	
Methods	**UK**	**IDW**	**UK**	**IDW**	**UK**	**IDW**	**UK**	**IDW**	**UK**	**IDW**	**UK**	**IDW**
BIAS	-4.004	-0.13	-0.08	0.0009	0.0023	-0.0001	-49.71	-54.45	-1.06	-1.15	0.15	-1.73
R	0.94	0.96	0.37	0.96	0.95	0.96	0.82	0.78	0.03	0.05	0.57	0.07
RMSE	79.63	0.48	1.83	0.0056	0.21	0.01	152.91	177.04	3.44	3.54	2.4	4.55
ub RMSE	79.53	0.46	1.83	0.005	0.21	0.01	144.59	169.26	3.27	3.34	2.49	4.20

predicted better using the UK than IDW. In all three landforms, grassland, forest, and agriculture UK performed better than IDW, but the agriculture landform is predicted well with a 0.56 correlation value.

The interpolation Techniques IDW and UK are greatly affected by landform and the number of observed points in different soil attributes. the UK method performs better for all the study landforms based on the R and RMSE values (Table 2.3). The R values are better in the case of continuous data (elevation) compared to the derived soil properties (organic matter and soil moisture) in agricultural fields compared to forest and grassland but there is no specific improvement in the prediction results. Forest landforms are predicted well with R values 0.56 and 0.61 in the UK and IDW, respectively, as they are more uniformly distributed than agricultural landforms. Soil moisture predicted well in all landforms using UK methods with R (ubRMSE) values 0.11 (4.24), 0.17 (3.11), and 0.57 (2.49) than IDW 0.07 (4.21), 0.03 (3.31), and 0.07 (4.20) in grassland, forest, and agriculture (Tables 2.2 to 2.4). When all observed points were interpolated, IDW and UK performed better in continuous data like elevation with a high correlation value of 0.98, while in the case of organic matter and soil moisture, both methods showed a lower correlation.

2.4 Conclusion and Future Scope

This study identified the necessity for understanding the links between soil parameters and the distribution of soil moisture utilizing GIS-based spatial interpolation techniques. This study presents the value of soil properties, i.e., OC and SM, at a small spatial scale based on the extensive ground truth calibrated data used for interpolation. The average soil moisture distribution over the research region demonstrates that soil moisture content is not uniformly distributed over space and decreases with elevation. Even though the distribution of organic content profiles varies, they are generally low in grasslands and high in agricultural zones. Several factors, including the number of samples, design, density, and distribution of data in the region, data quality, approach to choosing data, or implementation methodologies, affect how well the spatial interpolation algorithm functions. The distribution, accuracy, variance, range, and spatial correlation are other factors that affect data quality and performance. Due to limitations, this study did not consider all the parameters separately. When employing cross-validation procedures, the research showed that soil moisture based on data from measurements on the ground showed that universal kriging performed better than IDW in terms of model correctness and performance. Srivastava *et al.* (2019) show that kriging is a better predictor of soil moisture than IDW. Similarly, Xie *et al.* (2020) predicted that kriging predicts soil moisture better near the surface of the IDW.

For accuracy, the spatial mapping of discrete attributes like soil moisture needs to be validated using more ground control points. Furthermore, methods like kriging with external drift (KED) with secondary variables highly correlated with soil moisture can significantly enhance interpolation accuracy.

References

1. L. Brocca, R. Morbidelli, F. Melone and T. Moramarco, "Soil moisture spatial variability in experimental areas of central Italy," *J. Hydrol.*, vol. 333, no. 2–4, pp. 356–373, Feb. 2007, doi: 10.1016/j.jhydrol.2006.09.004.
2. N. N. Kourgialas, G. C. Koubouris and Z. Dokou, "Optimal irrigation planning for addressing current or future water scarcity in Mediterranean tree crops," *Sci. Total Environ.*, vol. 654, pp. 616–632, 2019.
3. W. Hu and B. C. Si, "Soil water prediction based on its scale-specific control using multivariate empirical mode decomposition," *Geoderma*, vol. 193–194, pp. 180–188, Feb. 2013, doi: 10.1016/j.geoderma.2012.10.021.
4. G. Zucco, L. Brocca, T. Moramarco and R. Morbidelli, "Influence of land use on soil moisture spatial–temporal variability and monitoring," *J. Hydrol. (Amst.)*, vol. 516, pp. 193–199, 2014.
5. D. Entekhabi *et al.*, "The soil moisture active passive (SMAP) mission," *Proc. IEEE Inst. Electr. Electron. Eng.*, vol. 98, no. 5, pp. 704–716, 2010.

6. B. P. Mohanty, J. S. Famiglietti and T. H. Skaggs, "Evolution of soil moisture spatial structure in a mixed vegetation pixel during the Southern Great Plains 1997 (SGP97) Hydrology Experiment," *Water Resour. Res.*, vol. 36, no. 12, pp. 3675–3686, 2000.
7. H. Vereecken, J. A. Huisman, H. Bogena, J. Vanderborght, J. A. Vrugt and J. W. Hopmans, "On the value of soil moisture measurements in vadose zone hydrology: A review," *Water Resour. Res.*, vol. 44, no. 4, Apr. 2008, doi: 10.1029/2008wr006829.
8. P. K. Srivastava, P. C. Pandey, G. P. Petropoulos, N. N. Kourgialas, V. Pandey and U. Singh, "GIS and Remote Sensing Aided Information for Soil Moisture Estimation: A Comparative Study of Interpolation Techniques," *Resources*, vol. 8, no. 2, p. 70, Apr. 2019, doi: 10.3390/resources8020070.
9. X. Yao, B. Fu, Y. Lü, F. Sun, S. Wang and M. Liu, "Comparison of Four Spatial Interpolation Methods for Estimating Soil Moisture in a Complex Terrain Catchment," *PLoS One*, vol. 8, no. 1, p. e54660, Jan. 2013, doi: 10.1371/journal.pone.0054660.
10. S. M. Lesch and D. L. Corwin, "Prediction of spatial soil property information from ancillary sensor data using ordinary linear regression: Model derivations, residual assumptions and model validation tests," *Geoderma*, vol. 148, no. 2, pp. 130–140, 2008.
11. A. Govaerts and A. Vervoort, "Geostatistical interpolation of soil properties in boom clay in Flanders. *geoENV VII-Geostatistics for Environmental Applications*," pp. 219–230, 2010.
12. V. Pandey and P. K. Pandey, "Spatial and temporal variability of soil moisture," *Int. J. Geosci.*, vol. 01, no. 02, pp. 87–98, 2010.
13. G. C. Sigua and W. H. Hudnall, "Kriging analysis of soil properties: implication to landscape management and productivity improvement," *J. Soils Sediments*, vol. 8, pp. 193–202, 2008.
14. J. Li and A. D. Heap, "A review of comparative studies of spatial interpolation methods in environmental sciences: Performance and impact factors," *Ecol. Inform.*, vol. 6, no. 3–4, pp. 228–241, 2011.
15. B. Huaxing, L. Xiaoyin, L. Xin, G. U. O. Mengxia and L. I. Jun, "A case study of spatial heterogeneity of soil moisture in the Loess Plateau, western China: A geostatistical approach," *Int. J. Sediment Res.*, vol. 24, no. 1, pp. 63–73, 2009.
16. A. M. Pophare and U. S. Balpande, "Morphometric analysis of Suketi river basin, Himachal Himalaya, India," *J. Earth Syst. Sci.*, vol. 123, no. 7, pp. 1501–1515, 2014.
17. R. A. Cook, S. Mostaghimi and J. B. Campbell, "Assessment of methods for interpolating steady-state infiltration," *Trans. ASAE*, vol. 36, no. 5, 1993.
18. P. V. Arun, "A comparative analysis of different DEM interpolation methods," *Egypt. J. Remote Sens. Space. Sci.*, Dec. 01, 2013. https://doi.org/10.1016/j.ejrs.2013.09.001

19. R. Webster and M. A. Oliver, *Geostatistics for Environmental Scientists*. Melbourne, Australia: Wiley, 2001.
20. W. R. Tobler, "A Computer Movie Simulating Urban Growth in the Detroit Region," *Econ. Geogr.*, vol. 46, p. 234, Jun. 1970, doi: 10.2307/143141.
21. R. S. Bivand, E. J. Pebesma, V. Gomez-Rubio and E. J. Pebesma, *Applied spatial data analysis with R*, vol. 747248717. New York: Springer, 2008.
22. E. H. Isaaks and M. R. Srivastava, *An Introduction to Applied Geostatistics*. New York, NY, USA: Oxford University Press, 1989.
23. N. N. Kourgialas, G. C. Koubouris, G. P. Karatzas and I. Metzidakis, "Assessing water erosion in Mediterranean tree crops using GIS techniques and field measurements: the effect of climate change," *Nat. Hazards (Dordr.)*, vol. 83, no. S1, pp. 65–81, 2016.
24. A. Barbulescu, C. L. Şerban and M.-L. Indrecan, "Computing the Beta Parameter in IDW Interpolation by Using a Genetic Algorithm," *Water*, Mar. 22, 2021. https://doi.org/10.3390/w13060863.
25. J. P. Delhomme, "Kriging in the hydrosciences," *Adv. Water Resour.*, vol. 1, no. 5, pp. 251–266, 1978.
26. S. R. Vieira, J. L. Hatfield, D. R. Nielsen and J. W. Biggar, "Geostatistical Theory and Application to Variability of Some Agronomical Properties," *Hilgardia*, vol. 51, pp. 1–75, 1983.
27. E. Sertel, H. Demirel and Ş. Kaya, "Predictive Mapping Air Pollutants: A Spatial Approach," in *Proceedings CD of the Fifth International Spatial Data Quality Symposium*, vol. 17, Enschede, Netherland, 2012.
28. G. M. Laslett, "Kriging and Splines: An Empirical Comparison of their Predictive Performance in Some Applications," *J. Am. Stat. Assoc.*, vol. 89, no. 426, pp. 391–400, Jun. 1994, doi: 10.1080/01621459.1994.10476759.
29. Z. Dokou, N. N. Kourgialas and G. P. Karatzas, "Assessing groundwater quality in Greece based on spatial and temporal analysis," *Environ. Monit. Assess.*, vol. 187, no. 12, p. 774, 2015.
30. T. C. Coburn and P. Goovaerts, "Geostatistics for natural resources evaluation," *Technometrics*, vol. 42, no. 4, p. 437, 2000.
31. S. Kumar and T. Kaur. "Efficient solar radiation estimation using cohesive artificial neural network technique with optimal synaptic weights." *Proc. Inst. Mech. Eng., Part A: J. Power Energy*, vol.234, no. 6, 2020, 862-873.
32. H. Bourennane and D. King, "Using multiple external drifts to estimate a soil variable," *Geoderma*, vol. 114, no. 1–2, pp. 1–18, May 2003, doi: 10.1016/s0016-7061(02)00338-5.
33. U. Mishra, R. Lal, B. Slater, F. Calhoun, D. Liu and M. Van Meirvenne, "Predicting soil organic carbon stock using profile depth distribution functions and ordinary kriging," *Soil Sci. Soc. Am. J.*, vol. 73, no. 2, pp. 614–621, 2009.

34. H. Adab, R. Morbidelli, C. Saltalippi, M. Moradian and G. A. F. Ghalhari, "Machine Learning to Estimate Surface Soil Moisture from Remote Sensing Data," *Water*, vol. 12, no. 11, p. 3223, Nov. 2020, doi: 10.3390/w12113223.
35. B. Xie, X. Jia, Z. Qin, C. Zhao and M. Shao, "Comparison of interpolation methods for soil moisture prediction on China's Loess Plateau," *Vadose Zone J.*, vol. 19, no. 1, Jan. 2020, doi: 10.1002/vzj2.20025.

3

Monitoring Crop Conditions of Punjab State Using Big Data Analytics

Harpinder Singh[1]*, Ajay Roy[2], Shashikant Patel[1] and Brijendra Pateriya[1]

[1]*Punjab Remote Sensing Centre (PRSC), P.A.U. Campus, Ludhiana, Punjab, India*
[2]*Lovely Professional University (LPU), Phagwara, Punjab, India*

Abstract
Monitoring the health of crops is essential for estimating the effects of climatic variability and biotic and abiotic stressors for the agricultural state of Punjab. Large-scale, continuous information about vegetation can be obtained using satellite remote sensing. Mapping a state's crop stress is a difficult task that requires a lot of resources. This study uses the computational capability of the Google Earth Engine (GEE) platform to map crop stress in the state of Punjab using the enhanced vegetation index (EVI) dataset from the MOD13A1.006 Terra Vegetation collection. The deviation of the EVI of the current (2019–2020) period from the reference period (2014–2019) has been analyzed to classify the crop stress as severe, high, no-change, and improved. The results show that the Gurdaspur district experienced maximum crop stress in the rabi cropping season of 2019–2020. Amritsar, Patiala, Pathankot, and Fatehgarh Sahib followed the Gurdaspur district. The least stress was identified in the Patiala, Nawan Shehar, and SAS Nagar (Mohali) districts.

Keywords: Crop, EVI, Google earth engine, Punjab, MODIS

3.1 Introduction

The primary threat to food security is the biotic and abiotic stress in crops. Biotic stress occurs due to the harm done to the crops by living organisms like insects, fungi, viruses, and weeds. Abiotic stress is usually caused

**Corresponding author*: harpinder13@gmail.com

Disha Thakur, Sanjay Kumar, Har Amrit Singh Sandhu and Chander Prakash (eds.) Sustainable Development Using Geospatial Techniques, (59–72) © 2024 Scrivener Publishing LLC

by human activity. Examples of abiotic stress are floods, drought, nutrient deficiency, salinity, and low and high temperatures. These stress factors are the main reasons for crop yield decline [1].

The traditional way to monitor crop stress is by field surveys and proximal sensing techniques, but this process involves time, costs, and human resources. Satellite remote sensing is an alternative by providing continuous coverage over large areas [2]. Remote sensing can aid in continuous monitoring of the crops, quickly diagnosing the problem, and applying timely measures.

Gogoi *et al.* [3] review many remote sensing techniques for detecting various plant diseases. Techniques like multispectral and hyperspectral imaging, multiband and fluorescence spectroscopy, visible, and infrared are discussed. Various remote sensing tools for crop health assessment like popular vegetation indices and platforms like UAV have been discussed [4]. With the help of machine learning(ML) and remote sensing, [5] provides a thorough analysis of the extensively used and preferred crop water stress monitoring techniques.

A remote sensing researcher requires many resources to download, process, analyze, and store BIG earth observation datasets [6]. The most widely used cloud-based remote sensing platform currently is Google Earth Engine (GEE), which offers ready-to-use high computing capabilities for planetary-scale geospatial big data analytics [7]. This platform is free for research work. GEE includes all the popular remote sensing datasets and the algorithms for processing and analyzing them. A review [8] discusses numerous application areas in which GEE is being used, like, land use, land cover and crop mapping, disaster management, climate change, and agriculture.

A range of spatial, spectral, temporal, and radiometric resolutions are available for the remote sensing data. The datasets required for a crop stress study depend on the nature of the study (parameters to be identified, e.g. nutrients, water, and salinity) and its scale, i.e., if the analysis has to be done for a field, village, district, state, or country. For this research, the whole Punjab state was analyzed for crop stress. As the volume of remote sensing datasets was large, the GEE was exploited for this research. The enhanced vegetation index (EVI) MODIS VI satellite imageries are used for this research. The Terra and Aqua satellites serve as the platform for MODIS (moderate resolution imaging spectroradiometer) [9]. Two MODIS vegetation indices (VI), the normalized difference vegetation index (NDVI) and the EVI, are calculated at 1-km and 500-m resolutions with 16-day compositing periods. These indices offer reliable temporal and spatial data about the vegetation canopy's structure, chlorophyll content, and greenness. The MODIS data have been compared with other remote/proximal datasets by Huete *et al.* [10]. Their results showed a strong correlation between top-of-canopy reflectance values from the MODIS sensor and airborne-measured reflectance at four extensively investigated test locations representing semi-arid grass/shrub, savanna, and tropical forest biomes. The generated field biophysical measurements validated the significance of the MODIS data for scientific research.

The EVI increases sensitivity across densely covered vegetation and reduces canopy-soil variation. The formula of EVI [9] is given in Equation 3.1,

$$EVI = G((NIR - R)/(NIR + C1 * R - C2 * B + L)) \tag{3.1}$$

where the surface reflectance is represented by the near-infrared (NIR), red (R), and blue (B) bands; the aerosol resistance coefficients are C1 and C2; the canopy background adjustment is L; and the scaling factor is G. The coefficient values for the algorithm given above are L=1, G=2.5, C1=6, and C2=7.5.

MODIS EVI is used for various vegetation and agricultural-related applications like crop mapping, yield estimation, agricultural drought monitoring, salinity mapping, water, and other environmental stress. The time series of EVI was used by Li *et al.* [11] to map crop cycles in China. NDVI and EVI products were compared by Son *et al.* [12] for the estimation of large-scale rice yield. They discovered that the EVI-based models had an edge over NDVI-based models in terms of accuracy. Even at low salinity levels, the study [13] demonstrated that phenological parameter extraction from MODIS EVI is a potential method for detecting soil salinity in areas that produce rice at regional scales. In order to map crop phenology, identify cropping systems, identify land-use change, and assess the danger of drought in Ethiopia, Kibret *et al.* [14] employed MODIS EVI data in the GEE. A study by Orimoloye *et al.* [15] found that the 16-day Terra-MODIS composite and EVI products are responsive to drought-related stresses, and these products can help monitor drought events.

3.2 Objectives

This research work aims to map the crop stress in the Punjab state using remotely sensed EVI datasets from the MOD13A1.006 Terra collection. The deviation of the EVI of the current (2019–2020) period from the reference period (2014–2019) has been analyzed to classify the crop stress as severe, high, no-change, and improved.

3.3 Study Area and Data

This section provides an overview of the study area, followed by information on the satellite datasets.

3.3.1 Study Area

The whole Indian state of Punjab is covered in the study area (Figure 3.1). It spans 50,362 km^2 and is located between latitudes 29°33′N & 32°31′N and longitudes 73°53′E & 76°55′E. Three distinct seasons occur in Punjab: winter season (October–March), summer season (April–June), and monsoon season (July–September). The majority of the state's crops are rice, wheat, and cotton. The vegetation in the rabi season (October–March) has been analyzed for this research. Wheat is the major rabi crop grown in Punjab state.

3.3.2 Data

The MODIS VI satellite data MOD13A1.006 Terra Vegetation Indices makes available the global vegetation conditions. Normalized difference vegetation index (NDVI) and EVI are two indices that comprise this

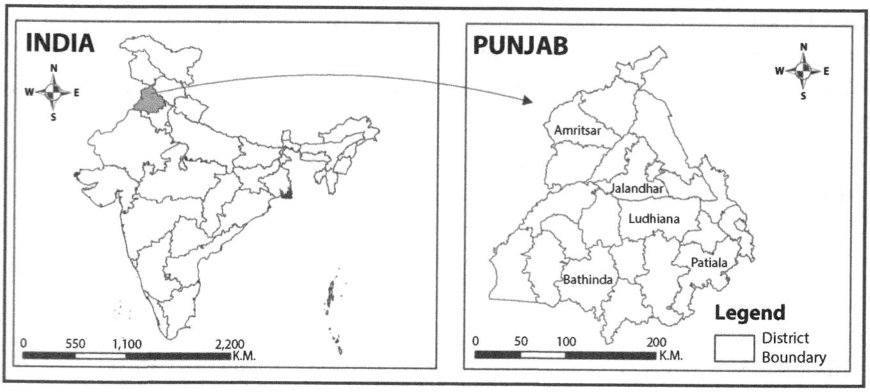

Figure 3.1 Study area.

dataset. Having a 500-m spatial resolution, these datasets become available every 16 days. For this research, the EVI dataset for the rabi season of the period 2014–2020 has been analyzed.

ESA WorldCover 10 m v100 [16] is a product by ESA, the European Space Agency. The 2020 edition of WorldCover, a 10-m resolution map of the world's land cover, is based on Sentinel 1 and Sentinel 2 satellite data. This dataset has been used for masking the nonagricultural areas.

3.4 Methods

For this study, image processing and analysis were done in GEE, a modern cloud-based remote sensing platform. Filtering the MODIS EVI images according to their geographic extent, cropping season, and desired periods is the first step in the process. Other steps include calculating the deviation from the reference and current maximum images, reclassifying the deviation raster, masking off non-cropland areas, and calculating the stressed areas district-wise. Figure 3.2 depicts the research process in detail.

3.4.1 Preparing the MODIS EVI Dataset

The MOD13A1.006 Terra Vegetation Indices images were filtered based on three parameters index (EVI), study area (Punjab state), and acquisition dates. The images acquired between November 2014 and March 2019 were selected to create the reference dataset. The current dataset consisted of images between the period November 2019 and March 2020.

The reference dataset was prepared by computing the maximum (pixel value) of the seasonal (rabi season) images for the past 5 years (2014–2019) and then calculating the mean of all five images. The current dataset was created by taking a maximum of all the pixels of the EVI images of 2019–2020. Figure 3.3 represents the generated datasets.

3.4.2 Deviation

The deviation or anomaly raster was calculated by subtracting the reference from the current dataset. The deviation represents the current crop conditions compared to the conditions from the past 5 years. Figure 3.4 represents the deviation dataset. This raster was reclassified into the following classes which are listed in Table 3.1.

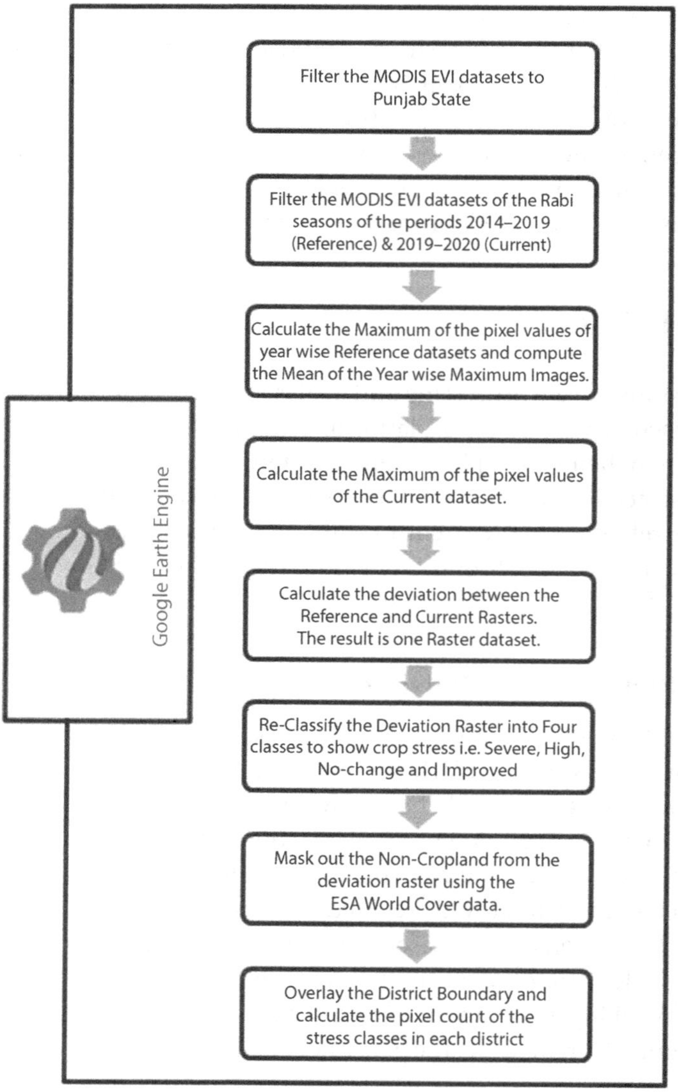

Figure 3.2 Research methodology.

Table 3.1 Deviation classes.

S. no.	Deviation value	Crop conditions
1	-0.25 to -1.0	Severe stress
2	-0.1 to -0.25	High stress
3	-0.1 to 0.1	Normal (no change)
4	0.1 to 0.25	Improvement in crop conditions
5	0.25 to 1.0	Very good conditions

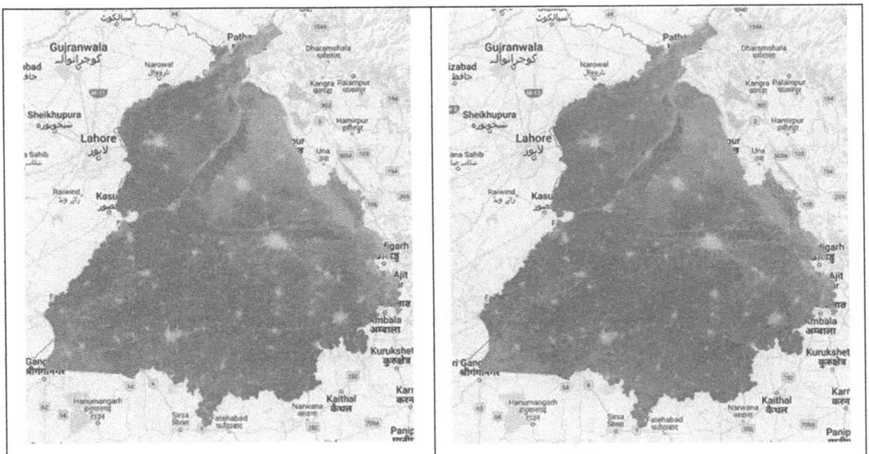

(a) (b)

Figure 3.3 Generated datasets: (a) reference mean of 2014–2019 and (b) current maximum of 2019–2020.

3.4.3 Masking the Non-Cropland Area

With the use of the ESA WorldCover 10 m v100 dataset, the non-cropland regions were masked. This process was incorporated to ensure that the final statistics only included agricultural vegetation and excluded other regions such as forests, built-up areas, and water bodies.

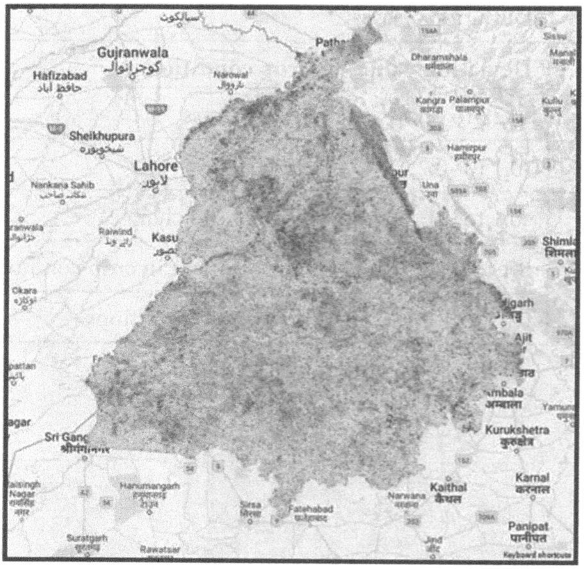

Figure 3.4 Deviation raster (red-green color ramp shows areas of negative to positive deviations).

3.4.4 District-Wise Analysis

The Punjab district layer was overlayed on the masked output to generate crop condition class-wise statistics for all the districts.

3.5 Results and Discussions

The research's findings are presented as maps and tables. According to the results, the Gurdaspur district experienced maximum crop stress in the rabi cropping season of 2019–2020. Amritsar, Patiala, Pathankot, and Fatehgarh Sahib followed the Gurdaspur district. The least stress was identified in the Patiala, Nawan Shehar, and SAS Nagar (Mohali) districts. The district-wise area statistics are provided in Table 3.2. The "very good class" was not found in any district during the analysis.

Although the stress pixels (red color) were scattered all over the state (Figure 3.5), areas around Dera Baba Nanak (Figure 3.6) in Gurdaspur district and Bassi Pathana (Figure 3.7) in Fatehgarh Sahib experienced extensive stress. The stress pixels near these places were found in large clusters. It was also witnessed that the northwestern part of Punjab comprising districts of Pathankot, Gurdaspur, and Amritsar witnessed more stress than

Table 3.2 District-wise area statistics.

Sr. no.	District name	Severe stress		High stress		No change in crop conditions		Improvement in crop conditions	
		Total pixels	Area (ha)	Total pixels	Area (ha)	Total pixels	Area (ha)	Total pixels	Area (ha)
1	Gurdaspur	2	50	106	2650	2271	56,775	47	1175
2	Amritsar			20	500	2142	53,550	111	2775
3	Patiala			17	425	2836	70,900	150	3750
4	Pathankot			15	375	536	13,400	9	225
5	Fatehgarh Sahib			12	300	1138	28,450	26	650
6	Bathinda			10	250	2953	73,825	128	3200
7	Kapurthala			8	200	1364	34,100	148	3700
8	Ludhiana			8	200	2977	74,425	164	4100
9	Muktsar			6	150	2323	58,075	142	3550
10	Sangrur			6	150	3300	82,500	133	3325
11	Hoshiarpur			6	150	2172	54,300	86	2150
12	Rupnagar			5	125	865	21,625	32	800
13	Mansa			5	125	2062	51,550	33	825
14	Jalandhar			5	125	2169	54,225	221	5525
15	Moga			4	100	2135	53,375	46	1150
16	Faridkot			4	100	1297	32,425	53	1325
17	Firozpur			3	75	2278	56,950	84	2100
18	Tarn Taran			3	75	2288	57,200	74	1850
19	Barnala			2	50	1362	34,050	44	1100
20	Fazilka			2	50	2423	60,575	127	3175
21	SAS Nagar (Mohali)			1	25	742	18,550	40	1000
22	Nawan Shehar			1	25	987	24,675	70	1750
23	Patiala			0	0	11	275	0	0

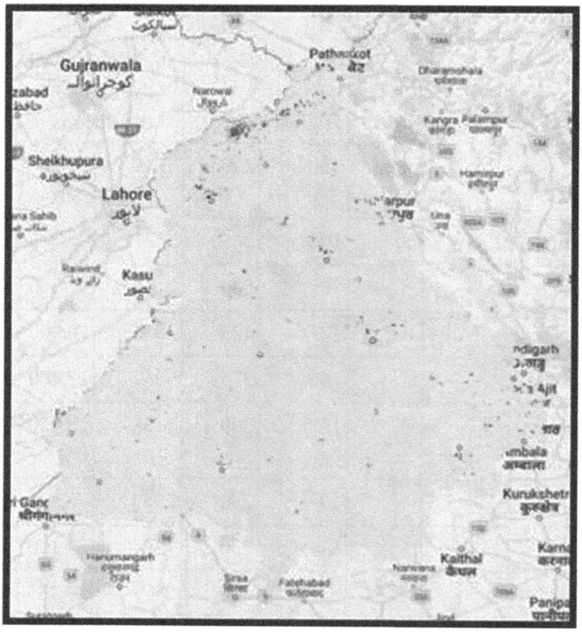

Figure 3.5 Stressed areas (red color) in Punjab state.

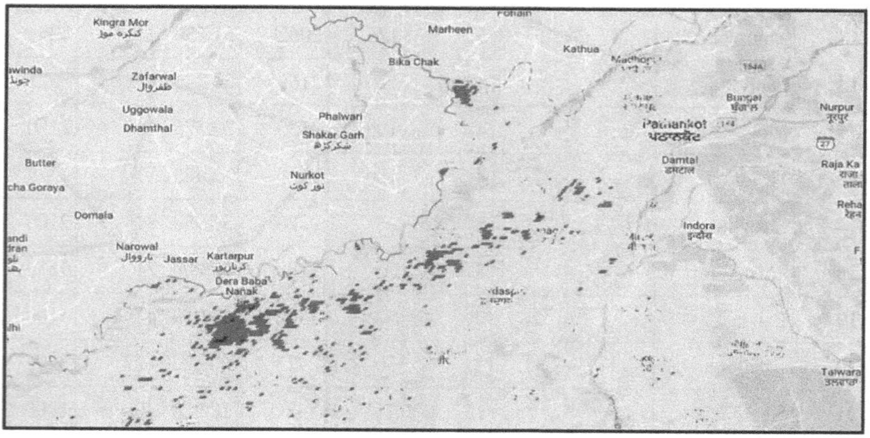

Figure 3.6 Stressed areas (red color) around Dera Baba Nanak in Gurdaspur.

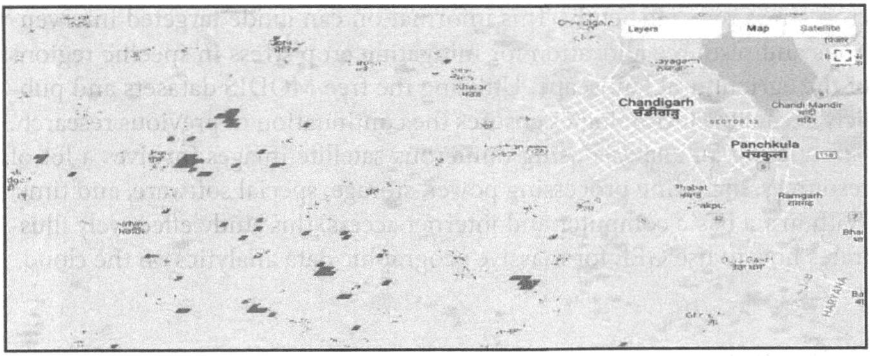

Figure 3.7 Stressed areas (red color) around Bassi Pathanan in Fatehgarh Sahib.

Table 3.3 District-wise wheat yield (per hectare in kilogram) statistics.

Rabi period	Gurdaspur	Amritsar	Patiala	Pathankot	Fatehgarh Sahib
2017–2018	4733	4866	5272	4018	5341
2018–2019	4717	4813	5472	3984	5180
2019–2020	4175	4533	4771	3944	4394

the rest of Punjab. Some parts of the Jalandhar, Ludhiana, and Patiala districts also witnessed improvements in crop conditions.

According to the Statistical Abstract of Punjab, 2020 [17], 2019 [18], and 2018 [19], the wheat crop yield (per hectare in kilogram) for the past few years of Gurdaspur, Amritsar, Patiala, Pathankot, and Fatehgarh Sahib districts is given in Table 3.3. According to the table, it can be seen that the crop yield has decreased in the Gurdaspur, Amritsar, Patiala, Pathankot, and Fatehgarh Sahib districts. This proves the relation between the decrease in the EVI and the crop yield.

3.6 Conclusions

Field surveying has been used to diagnose crop stress and calculate its area statistics. Therefore, creating an effective approach for mapping crop stress using remote sensing datasets is crucial. In conclusion, this research utilizes GEE, a cloud-based earth observation tool to produce a Punjab crop stress map efficiently. This information can guide targeted interventions and resource allocation for mitigating crop stress in specific regions of the agricultural landscape. Utilizing the free MODIS datasets and publicly accessible technologies ensures the continuation of previous research. Conducting an analysis using numerous satellite images involves a lot of resources, including processing power, storage, special software, and time. With just a basic computer and internet access, this study effectively illustrates how to use GEE for massive geographic data analytics on the cloud.

References

1. R. D. Jackson, "Remote Sensing of Biotic and Abiotic Plant Stress," *Annu. Rev. Phytopathol.*, vol. 24, no. 1, pp. 265–287, 1986.
2. M. A. Wulder, N. C. Coops, D. P. Roy, J. C. White and T. Hermosilla, "Land cover 2.0," *Int. J. Remote Sens.*, vol. 39, no. 12, pp. 4254–4284, 2018.
3. N. K. Gogoi, B. Deka and L. C. Bora, "Remote sensing and its use in detection and monitoring plant diseases: A review.," *Agric. Rev.*, vol. 39, no. 4, 2018.
4. R. Ferguson, D. Rundquist, D. K. Shannon, D. E. Clay and N. R. Kitchen, "Remote sensing for site-specific crop management," *Precis. Agric. Basics*, 2018.
5. S. S. Virnodkar, V. K. Pachghare, V. C. Patil and S. K. Jha, "Remote sensing and machine learning for crop water stress determination in various crops: a critical review," *Precis. Agric.*, vol. 21, no. 5, pp. 1121–1155, 2020.
6. M. Chi, A. Plaza, J. A. Benediktsson, Z. Sun, J. Shen and Y. Zhu, "Big data for remote sensing: Challenges and opportunities," *Proc. IEEE*, vol. 104, no. 11, pp. 2207–2219, 2016.
7. N. Gorelick, M. Hancher, M. Dixon, S. Ilyushchenko, D. Thau and R. Moore, "Google Earth Engine: Planetary-scale geospatial analysis for everyone," *Remote Sens. Environ.*, vol. 202, pp. 18–27, 2017.
8. O. Mutanga and L. Kumar, "*Google Earth engine applications*." Multidisciplinary Digital Publishing Institute, 2019.
9. K. Didan, A. B. Munoz, R. Solano and A. Huete, "MODIS vegetation index user's guide (MOD13 series)," *Univ. Arizona Veg. Index Phenol. Lab*, 2015.
10. A. Huete, K. Didan, T. Miura, E. P. Rodriguez, X. Gao and L. G. Ferreira, "Overview of the radiometric and biophysical performance of the MODIS vegetation indices," *Remote Sens. Environ.*, vol. 83, no. 1–2, pp. 195–213, 2002.
11. L. Li, M. A. Friedl, Q. Xin, J. Gray, Y. Pan and S. Frolking, "Mapping crop cycles in China using MODIS-EVI time series," *Remote Sens.*, vol. 6, no. 3, pp. 2473–2493, 2014.

12. N. T. Son, C. F. Chen, C. R. Chen, V. Q. Minh and N. H. Trung, "A comparative analysis of multitemporal MODIS EVI and NDVI data for large-scale rice yield estimation," *Agric. For. Meteorol.*, vol. 197, pp. 52–64, 2014.
13. A. Paliwal, A. Laborte, A. Nelson and R. K. Singh, "Salinity stress detection in rice crops using time series MODIS VI data," *Int. J. Remote Sens.*, vol. 40, no. 21, pp. 8186–8202, 2019.
14. K. S. Kibret, C. Marohn and G. Cadisch, "Use of MODIS EVI to map crop phenology, identify cropping systems, detect land use change and drought risk in Ethiopia–an application of Google Earth Engine," *Eur. J. Remote Sens.*, vol. 53, no. 1, pp. 176–191, 2020.
15. I. R. Orimoloye, J. A. Belle and O. O. Ololade, "Drought disaster monitoring using MODIS derived index for drought years: A space-based information for ecosystems and environmental conservation," *J. Environ. Manage.*, vol. 284, p. 112028, 2021.
16. D. Zanaga *et al.*, "ESA WorldCover 10 m 2020 v100," 2021.
17. Economic Advisor, "Statistical Abstract of Punjab" Department of Planning, Government of Punjab, 2020.
18. Economic Advisor, "Statistical Abstract of Punjab" Department of Planning, Government of Punjab, 2019
19. Economic Advisor, "Statistical Abstract of Punjab" Department of Planning, Government of Punjab, 2018.

4

An Investigation into Use of Ethereum Blockchain Technology to Validate the Reliability and Quality of Stored Satellite Images

Lince Mathew[1], Hamid Jahankhan[1] and Sumesh Dadwal[2]*

[1]Northumbria University London, London, UK
[2]London South Bank University, London, UK

Abstract
The number of uses for satellite imagery has increased dramatically in recent years. Therefore, for precise analysis and decision-making, assuring the dependability and quality of archived satellite photos is essential. The application of Ethereum blockchain technology to verify the dependability and quality of saved satellite photos is explored in this research. The use of the Ethereum blockchain for satellite image validation is examined in several papers and projects. These attempts establish an unchangeable record of satellite image metadata, including acquisition time, location, and processing stages, using smart contracts and blockchain-based timestamping algorithms. One of the main benefits of blockchain-based systems for building trust among stakeholders is their transparency and auditability. Overall, this research offers insightful information about using Ethereum blockchain technology to verify the accuracy and caliber of saved satellite photos. It demonstrates how blockchain technology could improve data integrity, trust, and accountability while also recognizing the difficulties that still need to be overcome before it can be widely adopted. The findings lay a platform for further investigation in this developing sector and add to the body of currently known knowledge.

Keywords: GPS, satellite imagery, blockchain, ethereum, ganache, truffle

**Corresponding author*: dadwals@lsbu.ac.uk

Disha Thakur, Sanjay Kumar, Har Amrit Singh Sandhu and Chander Prakash (eds.) Sustainable Development Using Geospatial Techniques, (73–106) © 2024 Scrivener Publishing LLC

4.1 Introduction

In recent years, the combination of blockchain technology and GPS (Global Positioning System) has been a topic of intense discussion. The satellite-based navigation system GPS is essential for several sectors, including logistics, transportation, mapping, and location-based services [1]. Although GPS has transformed location-based services and navigation but is not without flaws and risks. The vulnerability to signal fraud in standard GPS systems is one of the main issues [2]. Malicious individuals, producing incorrect positional data, can influence GPS signals. In crucial industries like aircraft and marine navigation, where accurate and dependable location data are crucial, this poses considerable threats. The need for strong security measures is highlighted by the potential repercussions of GPS signal fraud, which vary from serious mishaps to navigational blunders.

The concept of sustainability can be seen from many dimensions, for instance, it will be seen from people's bottom line as people, planet, and profit or it can be seen as a city that is equitable, environment-friendly and economic, equitable people oriented means the city is serving the social SoC society, it is people equitably [3]. Equity is not just equal treatment but providing facilities and services in such a way that there is inclusiveness in practice. For example, poor people in many cities may not be able to afford electric cars or may not be able to install solar panels, etc., so how does a smart city enable such people to participate in sustainable goals?

The United Nations established 17 Sustainability Development Goals (SDGs) in 2015, for peace and prosperity for people and the planet, present, and into the future [4]. With this lens, a technological solution to GPS shall efficiently and effectively address its problems such as capacity, poverty, health and education, inequality, decent work and economic growth, affordable and clean energy, responsible consumption and production, peace and justice, partnerships, etc. while tackling climate change and preserving our forests and oceans.

Blockchain technology has gained popularity due to cryptocurrencies like Bitcoin because of how autonomous and open it is [5]. A distributed database called a blockchain keeps a safe and unchangeable record of transactions among numerous nodes [6]. Despite relying on a centralized authority, confidence and integrity in data can be established because of this dispersed design and agreement method. A solid source of truth is provided by the integrity of blockchain records, which ensure that when a transaction gets recorded, it cannot be changed or manipulated.

Blockchain integration with GPS devices has enormous promise for fixing the flaws in traditional GPS. The reliability of GPS information can be improved by utilizing blockchain technology's dispersed and open nature, reducing the possibility of signal faking and data tampering. Several nodes can independently check the validity of GPS data using the consensus method used by blockchains, eliminating dependence on a single reliable source. In addition, by establishing an auditable atmosphere for GPS data, blockchain technology can provide increased dependability and credibility. The blockchain can save each GPS transaction as a block, creating an unchangeable chain of activities [7]. As a result, responsibility and trust among parties are encouraged, while open tracking and inspection of the location, date, and integrity of the GPS information are made possible. Applications where trust is essential, such as supply chain management or self-driving car navigation, can be significantly impacted by such openness and accountability.

Furthermore, a blockchain-based system can aid in effectively managing the utilization of resources such as land, water, energy, transport and housing networks, and public services, as well as safeguarding scarce resources, sharing resources, securing data, and allowing democracy, decentralization, and delegations [8].

Blockchain systems can be used for GPS image correction in addition to their effects on data security and automation. To create georeferenced images, GPS image rectification includes aligning images from satellites with correct GPS coordinates. The georeferenced photos that arise from this method can be more trustworthy and accurate by including blockchain. Potential mistakes or inconsistencies can be found and corrected by checking GPS coordinates utilizing blockchain technology and assuring the validity and usability of geospatial data.

Overall, combining GPS and blockchain systems provides a promising way to overcome the shortcomings and risks of conventional GPS systems while at the same time going for a sustainable way of development. The integrity, dependability, and trustworthiness of GPS information can be improved by utilizing the dispersed and open characteristics of blockchains. Additionally, blockchain integration makes it possible for GPS transactions to be transparently tracked, auditable, and automated, promoting efficiency and accountability. Enhancing the accuracy and validity of georeferenced satellite images is possible with the use of blockchain networks in GPS image correction [9]. With an emphasis on the integrity of data, transparency in automation, and GPS picture correction, this study attempts to investigate how blockchain technology may affect GPS systems.

4.2 Literature Review

Recently, there has been a lot of interest in using Ethereum blockchain technology to verify the validity and caliber of archived satellite photos. The quality and validity of these photos must be guaranteed due to the growing reliance on satellite imagery across numerous businesses. The decentralized and transparent nature of the Ethereum blockchain offers a solid answer to these problems. Each satellite image can be given a distinct identification or hash by utilizing cryptographic hashing algorithms, making verification and validation simple. According to Ali *et al.*, 2021 [9], the consensus mechanism provides general agreement on image validity, while smart contracts on the Ethereum blockchain specify the guidelines and requirements for image validation. The blockchain's immutable audit trail offers accountability, transparency, and tamper-proof records of validation actions. Improved data security, fraud, and compatibility with other decentralized apps are advantages. There are difficulties like scalability and gas costs, but continuing improvements in the Ethereum ecosystem aim to solve these problems. All things considered, the Ethereum blockchain technology provides a decentralized, open, and trustworthy method for assessing the accuracy and dependability of archived satellite photos, promoting confidence and integrity in satellite data applications.

4.2.1 Satellite Imaging Technology

Several automated image-processing techniques have been created in the past. For instance, STORM is a successful illustration that was initially designed for 6.5-m Rapid Eye interpreting, but issues cropped up when the workflow was applied to 2-m World View—2 (WV—2) imagery [11]. These difficulties were related to the improved spatial resolution and were specifically experienced during geometric rectification. Even though the workflow was deemed successful, it was dependent on proprietary applications (i.e., IDL and ATCOR), whose licenses may not always be accessible by image processing labs, especially when applied in low-income nations where smallholder farming is the norm. GeoMultiSens is a comparable autonomous analysis and processing system being developed for multi-sensor satellite data [12]. The system places a strong emphasis on doing global analyses of heterogeneous and multi-sensor pictures. A method created by Morris *et al.*, 2013 [12], can extract vegetative phenological indicators from traffic webcams with minimal to no human involvement. Another piece of work, by Inglada *et al.*, 2015 [13], focuses on verifying and training satellite data, but it also calls for the user to choose reference data to set up a functioning system for types of crop maps. Clewley *et al.*, 2014 [14], developed an open-source, modular system called GEOBIA that can be accessible using Python and has features that are comparable to those of other GEOBIA programs.

More general-purpose technologies that provide access and processing capacities have recently been launched in addition to specific automated workflows. A technology called Google Earth Engine provides parallel computing, which greatly accelerates processes [15]. The GBDX platform, developed by the Digital Globe (DG) corporation (now a division of MDA Corporation), allows users to get post-processed outcomes rather than data, which allays concerns about the cost-effectiveness of obtaining derived picture products [16]. Another approach for automatically extracting near–real-time image information using Big Earth data is ImageQuerying [17]. Through an intuitive graphical user interface, this system assists the user in performing semantic content-based image retrieval and in creating decision rules using both temporal and spatial operators. There are also more general cloud computing solutions like Microsoft Azure [18] and Amazon EC2 [19] accessible. Generally, most of the imaging frameworks in satellite imaging technology use high-end processing and sophisticated algorithms to process the data acquired. However, the final storage of all the metadata collected still uses cloud platforms for their safekeeping.

In several industries, including agriculture, environmental monitoring, disaster management, and urban planning, satellite imaging technology is essential. However, assuring the accuracy and excellence of the satellite photos is crucial for their efficient use. Averin and Averina, 2019 [5], have opined that the integrity and authenticity of saved satellite photos can be verified using Ethereum blockchain technology. The satellite imaging process is made more transparent, accountable, and trustworthy with the help of smart contracts and cryptographic hashing. By utilizing the blockchain's immutability, satellite photos may be securely stored, validated, and accessed, offering a trustworthy source of information for analysis and decision-making across a range of businesses.

4.2.2 Current Technology Used and Its Drawbacks and Impact of Using Blockchain Systems

Satellites orbiting the Earth and cutting-edge sensors are capturing high-resolution images of our world. These technologies comprise optical sensors, radar systems, and other specialized tools that offer useful information for a range of applications. However, there are several shortcomings and restrictions with the available technology.

4.2.2.1 *Drawbacks of Current Technology*

1. Reliability: Cao *et al.*, 2021 [19], have discussed that satellites are susceptible to technical failures or malfunctions, resulting in data gaps and reduced reliability of image capture.
2. Limited Coverage and Resolution: The number of satellites and their imaging capabilities limit the coverage and resolution of satellite images. Some areas may have lower image availability or lower resolution due to limited satellite coverage.
3. Cost: Access to high-quality satellite imagery can be costly, making it challenging for smaller organizations or individuals with limited budgets to obtain the necessary data.
4. Data Privacy and Security: With the increasing volume of satellite imagery data, ensuring data privacy and security is crucial. There are concerns regarding unauthorized access, data breaches, and the protection of sensitive information contained within satellite images [21].

4.2.3 Impact of Using Blockchain Systems

By incorporating blockchain technology, satellite imaging systems can address some of the aforementioned issues and experience several benefits:

1. Data Integrity and Trust: Blockchain's immutability and transparency enhance the trust and integrity of stored satellite images. The decentralized nature of blockchain ensures that once an image is validated and recorded, it cannot be tampered with or altered [22].
2. Secure Data Sharing: Blockchain allows for secure and controlled data sharing among authorized parties, reducing the risks associated with centralized data repositories.
3. Traceability and Accountability: Blockchain's transparent and auditable nature enables traceability of image sources, validation processes, and access history, ensuring accountability throughout the data lifecycle.
4. Incentivization and Crowdsourcing: Blockchain-based systems can incentivize individuals or organizations to

contribute their satellite imagery data, thereby expanding coverage and resolution through crowdsourcing efforts [2].

5. Smart Contract Automation: Smart contracts can streamline and automate various processes, such as image validation, licensing, and payment systems, reducing the complexity and costs associated with traditional methods.

The satellite-imaging sector may be able to overcome obstacles, improve data security and integrity, and promote stakeholder cooperation by utilizing blockchain technology. This might result in more effective and reliable satellite image acquisition, validation, and use [23].

4.2.4 Blockchain Technology

Blockchain technology is a decentralized and distributed ledger system that permits safe and open data sharing between several parties as well as secure transactions. Immutability, transparency, decentralization, and cryptographic security are some of its salient features.

Blockchain networks come in a variety of forms, including consortium, private, and public blockchains. Permissionless participation is possible on public blockchains like those used by Bitcoin and Ethereum, which are accessible to everyone. Private blockchains provide a group or organization more control over access and governance since they are only accessible to that group or organization [10]. A hybrid type of blockchain that has several organizations cooperating cooperatively is known as a consortium blockchain.

A blockchain system's fundamental building blocks include blocks, transactions, and consensus processes. A group of validated transactions is included in blocks. The units of data that are logged on the blockchain are represented by transactions. Consensus methods are protocols that network users employ to agree on the legitimacy and chronological sequence of transactions.

The use of blockchain technology has several advantages across numerous industries. Offering a shared and unchangeable record of transactions improves transparency by lowering the possibility of fraud and manipulation. Decentralization is made possible, removing the need for middlemen and promoting peer-to-peer connections [24]. Using cryptographic methods, blockchain also improves security and guarantees data integrity. Through smart contracts and programmable functions, it may also simplify procedures, cut costs, and allow new business models.

4.2.5 Ethereum Blockchain

The creation and execution of smart contracts are made possible by the decentralized Ethereum blockchain technology. While bringing new features and capabilities, it expands on the core tenets of blockchain technology, including decentralization, immutability, and transparency.

The idea of smart contracts is one of Ethereum's main characteristics. Smart contracts are self-executing contracts having in-code-written predetermined rules and circumstances. When the prerequisites are satisfied, these contracts immediately go into effect, allowing for tamper-free and trustless exchanges between the parties [25]. Decentralized and secure asset trading, the development of decentralized apps (daps), and the application of sophisticated business logic are all made possible by smart contracts on Ethereum.

On the Ethereum network, smart contracts are carried out *via* the Ethereum virtual machine (EVM), a runtime environment. A Turing-complete virtual computer supports arbitrary code execution. By ensuring that smart contracts are performed consistently across all Ethereum nodes, the EVM achieves consensus on each contract's conclusion. The EVM also imposes a "gas" pricing mechanism to deter fraud and equitably distribute computing resources.

Ethereum offers a variety of frameworks and tools to make blockchain development and simulation easier. Solidity, a high-level programming language created expressly for creating smart contracts on Ethereum, is the most popular development framework. Developers may specify the logic and behavior of their contracts using Solidity, as well, as how those contracts communicate with one another and with other data sources [26].

Ethereum features a full development tool called Truffle in addition to Solidity. A set of tools called Truffle is available for creating, evaluating, and implementing smart contracts. Additionally, it has network management and automatic contract transfer tools.

Ethereum provides tools like Ganache (formerly known as TestRPC), which enables developers to build up a local development network for quick contract writing and testing without the requirement for real Ether, enabling blockchain simulation and testing.

4.2.6 Satellite Image Integrity and Accuracy

Several sectors, including agriculture, urban planning, and environmental monitoring, depend heavily on the quality and integrity of satellite images. For making educated judgments, comprehending land use patterns, tracking environmental changes, and assuring the success of diverse applications, accurate and trustworthy satellite imagery is crucial.

Satellite pictures are used in the agricultural industry to measure vegetation growth, track crop health, and improve irrigation and fertilization techniques. Accurate information on crop yields, soil moisture, and plant health enables farmers to implement timely interventions and increase overall production [27].

Satellite photography is a useful tool in urban planning for planning infrastructure, zoning, and disaster management. Urban planners may make educated decisions and promote sustainable urban growth by using accurate maps and images to determine population density, land cover, and infrastructure requirements.

Monitoring of environmental changes in forests, water bodies, and biodiversity mainly relies on satellite photography. Deforestation is made easier to spot, wildlife habitats are monitored, and the effects of climate change are evaluated [28]. Researchers and politicians can implement effective conservation measures and save the environment thanks to accurate and current satellite data.

The integrity and accuracy of satellite photos must be guaranteed, yet there are risks involved. The presence of noise and artifacts in the photographs brought on by atmospheric conditions, sensor limits, and data transmission faults is a significant obstacle [29]. These elements may cause errors and distortions, affecting the accuracy and dependability of the data.

The possibility of deliberate modification or tampering with satellite photos is another difficulty. With image editing tools becoming more widely available, unscrupulous actors can change photos to trick or confuse people. The legitimacy and integrity of satellite images are seriously threatened by this.

As stated by Iyer *et al.*, 2022 [29], different methods for confirming and certifying satellite image data have been developed to solve these issues. Algorithms for image processing can be used to fix distortions and improve image quality. To account for sensor features and atmospheric impacts, calibration processes, including radiometric and geometric adjustments, are used.

A digital watermarking technique can be used to implant undetectable marks into the photographs to assure authenticity and tamper resistance [31]. These stamps serve as proof of provenance and reveal any unauthorized changes. Also, these stamps verify the integrity and validity of satellite images using cryptographic techniques like hash functions and digital signatures.

Additionally, ground truth data and outside observations are routinely used for validation. By comparing satellite photography with measurements and observations done on the ground, the accuracy of the images was assessed and validated. Ground truth data are acquired by field surveys, sensors on the ground, or other remote sensing techniques.

Automated image analysis techniques in machine learning and artificial intelligence (AI) have also benefited from developments. By automatically finding anomalies, classifying land cover, and extracting usable information from satellite photographs, these techniques improve the accuracy and efficacy of image analysis [32].

For a variety of industries, including agricultural, urban planning, and environmental monitoring, the quality and integrity of satellite images are essential. It is essential to overcome obstacles like noise, manipulation, and distortion to provide accurate satellite pictures. The integrity and accuracy of satellite images are being improved through methods like image processing, calibration, watermarking, validation using ground truth data, and the use of advanced machine learning algorithms. This enables their efficient use in decision-making and applications across a variety of domains.

4.2.7 Blockchain-Based Approaches for Satellite Image Safekeeping

The safeguarding of satellite pictures is one application where blockchain technology has shown promise in improving image and data integrity. Numerous research papers and initiatives have investigated blockchain-based systems for satellite picture storage by utilizing the inherent properties of blockchain, such as immutability, transparency, and decentralized consensus.

Supply chain management, intellectual property protection, and digital rights management are just a few of the areas where blockchain technology is being used to ensure the integrity of images and data. In the context of storing satellite photos, blockchain offers a tamper-resistant and auditable ledger that maintains the integrity and authenticity of images throughout their existence [30].

Blockchain technology has the potential to be used for satellite picture storage, as shown by research studies and initiatives. The blockchain may be used to store picture metadata and cryptographic hashes, for example. The cryptographic hashes function as the photos' identities, and the metadata can include details about the satellite, acquisition time, and location. As any alteration to the picture will produce a different hash value, this permits the verification of the image's integrity and validity.

Another line of inquiry is the application of smart contracts on blockchain systems like Ethereum for the preservation of satellite images.

Automating the archiving and retrieval of satellite photos can provide safe, auditable access. Authorized entities can retrieve and check the accuracy of the photographs thanks to access restrictions and permissions that can be established in the smart contract code.

Through decentralized consensus processes, blockchain-based technologies can also make validating and verifying satellite pictures easier. Several network parties, including satellite operators, data suppliers, and picture analyzers may verify picture integrity and correctness. The danger of data tampering is decreased by this distributed verification procedure, which also increases confidence [33].

The development of decentralized marketplaces for satellite imagery is also made possible by blockchain-based methods, allowing image providers to sell their data to interested parties directly and securely. Blockchain enables efficient and transparent transactions, doing away with the need for intermediaries and cutting expenses.

Satellite picture verification and security have shown promise when using an Ethereum blockchain simulation. As a decentralized blockchain platform, Ethereum offers a solid framework for putting verification processes in place and guaranteeing the accuracy of satellite images. By putting information and cryptographic hashes on the blockchain in the Ethereum blockchain simulation, it is possible to verify satellite photos due to the immutability and transparency of the blockchain. Each image's related metadata, including collection information, location, and satellite data, may be stored as transactions on the blockchain. The image's cryptographic hash is a distinctive identifier confirming its authenticity [34]. Any unauthorized change or tampering may be identified by comparing an image's hash with the hash that has been stored on the blockchain.

Additionally, the Ethereum blockchain simulation may make use of smart contracts to enable the safe storage of satellite photos. Smart contracts are self-executing contracts with in-code preset rules. Smart contracts may be used to automate the procedures involved in uploading, verifying, and securely storing photos in the context of satellite image authentication and storage. The smart contracts can specify requirements and circumstances that must be satisfied for an image to be accepted as legitimate. A smart contract can, for instance, demand the availability of certain information, timestamps, or digital signatures from reputable sources. The smart contract may automatically authenticate and save the picture on the blockchain when these requirements are met [24]. Smart contracts can also impose access control methods, limiting who can access and see satellite pictures to only authorized parties. The smart contract can specify access rights, and picture retrieval can be restricted to users who satisfy requirements or who interact with the contract using cryptographic keys or digital identities.

A decentralized and secure system for storing satellite images may be established using smart contracts and Ethereum blockchain simulation. A tamper-resistant storage solution is provided by the immutability of the blockchain, which makes sure that once an image is saved, it cannot be changed or tampered [23]. Additionally, the programmability of Ethereum permits the creation of sophisticated functionality in smart contracts for the archiving of satellite images. For instance, picture-licensing systems in smart contracts may specify and maintain ownership rights and usage permits automatically. Due to this, decentralized markets for satellite photos may be made possible, enabling image producers to sell and license their data to interested parties safely.

4.3 Methodologies and Tools Used for Simulating the Ethereum Blockchain

Various approaches and technologies are frequently used in research projects to simulate the Ethereum blockchain. The performance, scalability, and behavior of the Ethereum network in various circumstances are tested and evaluated using these simulations [20]. The following are some methods and technologies that are often used to simulate the Ethereum blockchain:

- **TestRPC/Ganache:** A well-liked tool for locally emulating Ethereum blockchain networks is TestRPC, currently known as Ganache. To set up private Ethereum networks for testing and experimentation, it offers researchers a lightweight and programmable development environment. Ganache includes functions including pre-defined accounts, transaction scrutiny, and quick mining and enables rapid development.
- **Geth:** Geth, a complete node in the Ethereum network, is the Ethereum protocol's Go implementation. To mimic the behavior of a genuine Ethereum network, researchers can set up a private Ethereum network using Geth [21]. To personalize the simulation environment, Geth offers a command-line interface and a wide range of configuration options.
- **Remix:** Remix is a platform for creating Ethereum smart contracts online. For researchers to create, build, and deploy

smart contracts on a simulated Ethereum network, it offers a complete simulation environment. Researchers may interactively replicate and study the behavior of smart contracts using Remix's different testing and debugging features.

- **Ethereum Virtual Machine (EVM):** Frameworks for modeling EVM, the environment used to perform smart contracts on the Ethereum network: Several simulation frameworks have been created expressly for simulating the EVM. Researchers may examine the performance and security facets of smart contracts using frameworks, such as Sereum and Echidna, which imitate EVM behavior, including gas consumption, opcode execution, and contract interactions [7].

4.4 Evaluation Metrics and Criteria for Assessing the Effectiveness of Ethereum Blockchain in Satellite Image Safekeeping

Researchers and practitioners use a variety of assessment measures and criteria to evaluate the Ethereum blockchain's performance and appropriateness for storing satellite images. The following are some typical evaluation measures and criteria:

- **Data Integrity:** A critical requirement is the blockchain's capacity to guarantee the accuracy of satellite photos. Data integrity may be assessed using metrics including the number of successful verifications, the precision of image integrity checks, and the detection of any tampering or unauthorized adjustments [35].
- **Security:** Security is essential for storing satellite images. The resilience of the blockchain system to assaults, the efficiency of access control measures, the security of private keys, and the avoidance of unauthorized access or data breaches are possible evaluation criteria.
- **Scalability:** Analyzing the scalability of the Ethereum blockchain is crucial since the number of satellite photos might be significant. To determine the blockchain's capacity for handling massive picture storage and retrieval, metrics such as transaction throughput, block confirmation time, and network latency can be evaluated.

- **Performance:** Another crucial element is how well the Ethereum blockchain performs in terms of how quickly transactions are processed, how effectively resources are used, and so on. Performance may be assessed using metrics like transaction confirmation time, gas usage, and block propagation time [36].
- **Cost Efficiency**: Given the possible costs involved with storing and retrieving satellite photos, it is important to assess the Ethereum blockchain's cost-effectiveness. To determine if utilizing the Ethereum blockchain for the storing of satellite images is economically feasible, metrics like transaction fees, storage costs, and computing expenses are considered.
- **Usability:** Important considerations include the blockchain system's usability and user experience. Insights into the viability and usability of the blockchain solution may be gained by assessing characteristics, including the simplicity of picture upload and retrieval, user interface design, and accessibility [29].
- **Decentralization:** Evaluations of the Ethereum blockchain's level of decentralization are common. The number of running nodes, the distribution of processing power, and the degree of consensus reached among network users are some examples of metrics [37].

4.5 Case Studies and Experiments Demonstrating the Integrity and Accuracy of Satellite Images Using Ethereum Blockchain Simulation

By simulating the Ethereum blockchain, several case studies and tests have shown the reliability and accuracy of satellite photos. These investigations show how Ethereum's blockchain technology has the potential to guarantee the reliability of satellite image data. Here are a few noteworthy instances:

- **SpaceMesh Project:** The SpaceMesh project tested utilizing blockchain technology to protect the integrity of satellite pictures. On the Ethereum blockchain, they created a proof-of-space-time consensus mechanism to make sure that satellite photos are not altered or tampered with. The tests demonstrated the Ethereum blockchain simulation's capacity to preserve the integrity of satellite picture data [38].

- **Satella:** A blockchain-based network called Satella attempts to protect the accuracy and provenance of satellite images. To validate and safely store satellite photos, they use Ethereum smart contracts. By contrasting cryptographic hashes that are recorded on the blockchain with real photos, the platform enables users to validate the legitimacy and integrity of satellite photographs.
- **Cervical Cancer Screening:** For the safe and correct storage of cervical cancer screening photos, a research study investigated the usage of an Ethereum blockchain simulation. Smart contracts were used in the study to verify and store the photos, guaranteeing their integrity and facilitating quick retrieval for analysis and diagnosis. The outcomes showed the promise of blockchain technology in preserving the precision and privacy of medical imaging data.

4.6 Research Method

The purpose of this research was to examine how the Ethereum blockchain can be used to verify the accuracy and value of satellite photos that have been saved. How the legitimacy and integrity of satellite images must be ensured! As it has an impact on several applications, including environmental monitoring and catastrophe management. This study explored how satellite picture storage can be verified for accuracy and quality using Ethereum blockchain technology. Various applications, such as environmental monitoring, urban planning, and disaster management, depend on the legitimacy and integrity of satellite photos, which are currently difficult to check. To accomplish the research goals, this study uses a mixed-approaches strategy that combines quantitative and qualitative methods [1].

An extensive and methodical approach is used in the study process methodology to examine the usage of Ethereum blockchain technology to verify the dependability and quality of stored satellite photos. Quantitative techniques, such as surveys, will collect information from a variety of specialists and professionals who work with satellite images. On the other side, qualitative methods, including interviews, will explore the perspectives of important stakeholders, enabling a greater understanding of their ideas, worries, and recommendations.

A combination of primary and secondary sources has been used to collect data for the inquiry of the usage of Ethereum blockchain technology to verify the dependability and quality of stored satellite photos.

4.7 Data Analysis and Critical Discussions

While conducting interviews with participants, some participants stated that blockchain technology is used for judging the authenticity of images. At the same time, other participants have initiated that the image processing approach is used to assess the technical aspects of the satellite images. Furthermore, image processing algorithms are used to eradicate artifacts and make corrections for errors. Another participant has depicted that machine learning algorithms are trained to get the distinct features of the satellite images in high quality. Significantly, another participant responded that the cryptographic approach protects the images from unauthorized access. Additionally, some participants have depicted that digital watermarking is used to verify the authenticity of images as in incurred an embedding piece of information.

Participants were asked to give their opinions about judging the effectiveness of the Ethereum technology and some have argued that the effectiveness is judged with the pathway of decentralization. On the other hand, the expense and cost revolving around the Ethereum blockchain are judged and assessed while keeping the satellite image safe. The participants have depicted that the scalability and the reliable use of resources are assessed to judge the effectiveness of the Ethereum blockchain.

Attempting to explain the tools for simulating the Ethereum blockchain, participants highlighted that Ganache is used to test all the solidity contracts and hinted that Geth is an interface used for effective interaction with the Ethereum network and is used to handle the execution of smart contracts. Significantly, participants have explained that Remix is an efficient tool for deploying and managing smart contracts on the network of Ethereum. It can package smart contracts within a single file, then upload them into the Ethereum network, and handle the deployment across various modes. Others described that the EVM is one of the virtual machines that efficiently deploy the smart contract within the Ethereum network. Finally, some participants responded that Truffle is used for testing and building smart contracts on the Ethereum network.

As per the research work of Lee *et al.*, 2022 [35], it was stated that GPS positioning systems were developed with the primary toll of solidity and Python and implemented using a smart contract of Ethereum to improve the probity of GPS data. It was further mentioned that a prototype system was initiated to trace the GPS data successfully. At the same time, the correction was generated with the use of Ethereum within the survey processor.

4.8 Testing Phase

During the testing phase, we ensured that the prototype functions effectively within the network, including both the main network and the test network. For experimentation purposes, we utilized the Ropsten test network.

During the deployment of a smart contract, Algorithm 1 is intended to input the surveyor's address and store the user's wallet address (Figure 4.1). Algorithm 1 stores NMEA and RTCM data and Algorithm 2 reads operator and GPS data, making up its two main duties (Figure 4.2). When Algorithm 1 is executed, this procedure happens concurrently with the distribution of smart contracts and automatically saves the user's wallet address. Algorithm 4 stores data obtained from Algorithm 1 as appropriate data types. Later access to GPS data and worker details contained within the contract is made possible by Algorithms 1 and 2. Despite its ease of use, Algorithm 3 protects the accuracy of GPS data and operator information *via* a Solidity smart contract on the Ethereum network.

Algorithm 2 initializes and creates a serial communication link with the GPS sensor to enable GPS location (Figure 4.2). As the device runs, it continuously gets a variety of data from satellites. It uses the input caster information to build an NTRIP client to communicate with the NTRIP server. The mount point VRS_V32 on the NGIS server was used in this study. The initial connection is then effortlessly established by extracting

```
Data: Wallet Address
Surveyor = Wallet Address
Function 1 : GPS Positioning
  | Data: NMEA from Receiver, RTCM from Caster
  | NMEA = NMEA from Receiver
  | RTCM = RTCM from Caster
end
Function 2 : Call the data
  | Result: Surveyor, NMEA, RTCM
end
```

Figure 4.1 Smart contract: Ensuring and returning RTCM, NMEA, and surveyor: Algorithm 1.

GNGGA data from the device's location measurements and transmitting it to the server *via* the NTRIP client. For precise GPS positioning, this mechanism ensures real-time data exchange with the server.

Algorithm 3 stores NMEA and RTCM data on the Ethereum blockchain in this proof of concept (Figure 4.3). It must install a smart contract and be connected to the Ethereum network. Entering the test net URL, wallet address, and private key are required steps in the procedure. The contract is deployed using ABI, bytecode, and compiled Solidity code taken from the Web3 instance after establishing a network connection *via* an HTTP provider and creating a Web3 instance.

Algorithm 4 continuously performs blockchain-based GPS positioning for the period provided by the surveyor (Figure 4.4). Algorithm 1 accepts NMEA and RTCM data from GPS surveys and logs the information on the blockchain while storing the transaction hash. Using the linked hash values, anyone can use this method to validate and trace the original data, protecting it from data manipulation. Cross-platform verification may involve developing consensus methods and uniform validation protocols

Data: *Serial, NTRIP Caster*
Result: *Receiver, NTRIP Client*
while *Receiver is not connected* **do**
| *Receiver* = Connect through the *Serial*
end
while *Client is not connected* **do**
| *NMEA* = Parsing from *Receiver*
| *NTRIP Client* = Connect to *NTRIP Caster* using *NMEA*
end

Figure 4.2 Connecting to NTRIP caster and GPS receiver: Algorithm 2.

Data: *Network URL, Wallet Address, Private Key, Contract Code*
Result: *Web3, Contract*
if *Web3 is not connected* **then**
| *Web3* = Connect through HTTP provider with *Network URL*
end
Compiled Code = compile the *Contract Code*
ABI = Parsing from *Compiled Code*
Bytecode = Parsing from *Compiled Code*
TxHash = Transact the contract using *ABI, Bytecode*
while *Tx is not mined* **do**
| *Tx* = Wait for transaction receipt using *TxHash*
end
Contract = Defined using *ABI* and Address of *Tx*

Figure 4.3 Connecting to this Ethereum blockchain and deploying this smart contract: Algorithm 3.

```
Data: Web3, Contract, Receiver, NTRIP Caster, NTRIP Client, Deadline
while RunningTime ≠ Deadline do
    NMEA = Parsing from Receiver
    if NMEA is True then
        RTCM = send NMEA to NTRIP Caster through NTRIP Client
        Tx Hash = Function 1 of Algorithm 1 with NMEA, RTCM using Web3
        NMEA = Send RTCM to Receiver
    end
end
```

Figure 4.4 GPS positioning: Algorithm 4.

that enable the quick validation and comparison of pictures from various sources.

The use of Solidity and Python facilitated the development of smart contracts, ensuring the system's functionality. By conducting experiments on the Ropsten test network, we were able to validate the system's performance and collect data for analysis at specific contract addresses. By fusing blockchain technology with data processing platforms to provide end-to-end certification, it is possible to guarantee the dependability and quality of satellite images from acquisition to processing.

As blockchain technology develops for satellite image validation systems, it is essential to create solid governance structures and dispute resolution processes.

The deployed smart contract technology was noted in block 1, and Tables 4.1 and 4.2 show the first observed coordinate. Then, 56 further transactions were carried out by executing Algorithm 4, 56 times. In Ethereum, the blockchain is composed of a series of blocks, each containing a set of transactions. Blocks serve as containers that store and record transactional data, including the deployment of smart contracts. Each block includes a reference to the previous block, forming a chronological chain of blocks that provides a transparent and immutable ledger. When deploying the smart contract for validating satellite images, a new block is created on the Ethereum blockchain. This block contains information about the contract deployment transaction, ensuring the transparency and traceability of the process.

Table 4.1 The deployed smart contract technology in block 1.

Field	Value
NMEA Data	$GNGGA,025610.00,3735.2330935,N,12700.9212481,E,5,12,0.77,54.482,M,18.504,M,1.0,4095*5B
RTCM Data	\xd3\x00\t? _ \xff\xe8\xde\x15\x18\x80H\xfa\xe4R\xd3\x00\x15 >\xef\xff\x03\xf8\xc5#6\xc5\tR\x1e\x13\x9f\x88\xfd\xabl\xfa\x00\x00h\xe8S\xd3\x00\x14@\x8f\xff \x00\x008\xba\xf0\x8b\xe5%JZ[\xb8\x8fR\xf0B\xc0\x94\x90\xc3\xd3\x00(@\x9f\xff \x0eGPPNULLANTENNA\x00\x00\rTrimbleNetR9\x045.37\x00f\x9c\x84\xd3\x00\x14?\x0f\xff \x0eGPPNULLANTENNA\x00\x00\xc0e\x00\xd3\x00\x0cL\xef\xff\x8f\x03\x8a\x03?\x03P\x03MY\xac \x9d\xd3\x002@f\xc1\xa9\xff\xf0\xa7\x0c4 : \xe0\x80\x90\x00pF\x00@H\x00\x12 \x00\x16\x0c\xac\x08\x08\x00\t\x05\x19\x06\x07\x00\x04\x82m\x02\x82\xc0\x02\x81gA\x81\xc0\x04\x02;

Table 4.2 Automated validation of smart contracts.

Field	Value
Tx Hash	0x205393a698c8fd6a966f88a68b9b8a16ea9e499027174a53c0c8f8cc0c02b85c
Worker ID (Wallet Address)	0x49B99615B39430993Ce8D2fC4577b2585d24a68B
Point ID (Contract Address)	0x4cDD2600bd83A308dF598F5FdA8D3Ef2BdFbf348
Gas Used	348,221
Tx Data	0xee19829500000000000000000000000000000000000000000000000000000000000000000040000000000 000000000000000000000000000000000000000000000000000000c00000000000000000000000000000000000000 0000000000000000000000000005724474e4747412c3032353631302e30302c333733352e323333303933352c4e 2c31323730302e393231323438312c452c352c31322c302e37372c35342e3438322c4d2c31382e3530342c4d2c 312e302c343039352a35420000000000000000000000000000000000000000000000000000000000000000000000 00000000000108d300093f5fffe8de15188048fae452d300153eefff03f8c52336c509521e139f88fdab49fa000068 e853d30014408fff000038baf08be5254a5a5bb88f52f042c09490c3d30028409fff0e4750504e554c4c414e54454 e4e4100000d5472696d626c65204e6574523904352e3337004a9c84d300143f0fff0e4750504e554c4c414e544 54e4e410000c06500d3000c4cefff8f038a033f0350034d59ac9dd300324066c1a9fff0a70c343a00160cac0808000 9051906070004826d0282c00281674181c004023be0809000704600404800020123c74e0d3002c40775e3fff853 060a0b000804840404800482828586800 1c1034181e001c101c080a0006038f0a0b800884c03a69af0000000000 000000000000000000000000000000000000000

Through smart contracts, which are pieces of self-executing code, the Ethereum blockchain can be used to store and maintain the metadata for satellite image data. Information about image acquisition, processing techniques, and quality assessments are all included in the metadata, which may be safely stored and retrieved by authorized parties.

4.8.1 Automated Validation

Automatic information validation is made feasible by smart contracts, which reduces the need for manual verification processes and improves the overall level of quality assurance for archived satellite photographs. The decentralized architecture of the Ethereum blockchain allows for distributed picture verification by several network nodes.

4.8.2 Transaction Information of this GPS Positioning

The verification step includes comparing the saved image data to external sources like real-world data, meteorological data, or other satellite photography. To ensure agreement among network participants and increase dependability, consensus processes might be utilized. Proof of stake (PoS) and proof of authority are two examples.

To confirm the recorded value in the smart contract, we ran Algorithms 1 to 3. Through this procedure, we were able to verify that the measured coordinate matched the value sent into the Ethereum network precisely. The consistency and integrity of the data stored on the blockchain were confirmed by comparing the measured coordinate to the recorded value. A transaction is a fundamental unit of interaction within the Ethereum blockchain. It represents an action or operation, such as the deployment of a smart contract or the transfer of Ether (cryptocurrency). In the context of the implementation, the transaction information includes details related to the deployment of the contract responsible for validating satellite images. In Ethereum, transactions require a certain number of computational resources to be executed, which is measured in gas. Gas represents the fee paid to the network for performing operations on the blockchain. When deploying the contract, the transaction includes the specified gas limit, which determines the maximum amount of gas that can be consumed during contract deployment. The transaction information includes the address of the sender, who initiates the contract deployment, and the recipient address, which typically represents the contract itself.

The use of Ethereum blockchain technology to confirm the reliability and caliber of recorded satellite images is quite promising.

4.8.3 GPS Positioning Data

Smart contracts, decentralized verification processes, and satellite imagery's immutable and transparent nature allow for safe storage of the data. By leveraging tokenization and incentives, the Ethereum blockchain encourages active stakeholder involvement and a community-driven approach to image validation. Additionally, the Ethereum blockchain's collaboration and interoperability enhance the dependability and quality assurance of satellite images in general. It is projected that as blockchain technology advances, it will play an increasingly significant role in assuring the authenticity and reliability of satellite photos, enabling a range of businesses to make decisions based on accurate and high-quality data.

4.8.4 GPS Positioning Data During Transaction Information

Satellite imagery is crucial in many industries, including agricultural, environmental monitoring, urban planning, and disaster management. Maintaining the dependability and quality of the stored satellite images has proven to be a significant challenge. With the development of blockchain technology, a possible substitute, termed Ethereum, has emerged.

This essay looks at how satellite images could be validated for accuracy and value using the Ethereum blockchain. The Ethereum blockchain is the ideal storage solution for satellite images due to its immutable and transparent ledger. By storing each image in a way that is timestamped, salted, and distributed, data integrity may be ensured and unauthorized manipulation can be avoided. By being transparent, stakeholders may verify the validity of the saved photos, promoting reliability and trust.

The input data field contains the encoded information required to deploy the contract. It includes the compiled code of the smart contract, as well as any parameters or initialization data necessary for its execution. Every transaction is assigned a unique transaction hash, which serves as its identifier. The hash is generated through cryptographic hashing algorithms and is used to track and reference the specific transaction on the blockchain. The nonce is a value associated with each transaction and helps maintain the order of transactions from a specific account. It ensures that transactions are processed in the correct sequence, preventing any potential conflicts or double-spending. The transaction information includes the block number in which the transaction is included. This provides the context and position of the transaction within the blockchain, allowing for easy retrieval and referencing. The timestamp represents the date and time when the transaction was added to the block. It provides information about the chronological order of transactions and can be used for auditing or analysis purposes.

4.8.5 Security Analysis of Ethereum Blockchain

Increased data integrity: Each satellite image is individually identifiable and linked to its pertinent data using cryptographic hashing and metadata. This improves the images' data integrity and makes easy work of authenticating and verifying them.

Immutable audit: A thorough and unchangeable audit trail of all picture submissions, validation activities, and outcomes is kept on the Ethereum blockchain. This makes auditing and traceability clear, ensures responsibility, and makes it easier to comply with regulations when necessary.

Trustless verification: By employing the image's unique identifier (hash), users can independently check the validation status and quality of saved satellite photos on the Ethereum blockchain. This encourages a trust-less verification procedure and reduces the need to rely entirely on centralized authorities.

Efficient validation: The workflow for image submission and certification is automated and streamlined *via* smart contracts on Ethereum. This

improves efficiency, decreases manual involvement, and could even hasten the certification process.
Increase participation and collaboration: Due to Ethereum's decentralized structure, a dispersed network of validators can take part in the verification process. This encourages cooperation and guarantees a wide variety of viewpoints for evaluating the dependability and quality of satellite photos.

4.8.6 Evaluation of Ethereum Blockchain

Transaction data are essential for tracking and verifying the data that are recorded when GPS positions are registered on the Ethereum blockchain. Users can obtain transaction information and explore the Ethereum network using the well-known blockchain explorer Etherscan.

- A transaction hash serves as a unique identifier for each transaction on the Ethereum blockchain. This hash acts as the transaction's unique identification and can be used to retrieve and confirm certain transaction information. The transaction hash assigned to each transaction in the event of recording GPS coordinates serves as a point of reference for retrieving and validating the data recording.
- The addresses of the sender and recipient involved in the transaction are included in the transaction information. The sender address in the context of GPS position recording denotes the account or entity that initiated the transaction to record the GPS position. On the other side, the recipient address may be a smart contract or a predetermined location where the GPS position data are stored. These addresses make it easier to identify the people who took part in the recording process.
- The payload or additional data that comes with a transaction is represented by the input data field. The GPS coordinates (latitude and longitude) or other pertinent metadata related to the GPS position may be included in the input data when recording GPS positions. The immutability and transparency of the GPS positions that have been captured are guaranteed by the encoding and storage of this data within the transaction.
- The computational resources needed to process an Ethereum transaction are measured in gas. The maximum amount of gas permitted for the transaction is specified in the gas limit

field in the transaction details. The real amount of gas utilized during the transaction's execution is shown in the gas used field. These fields shed light on the computational expense incurred when GPS positions are stored on a blockchain.

- The block number and associated timestamp for the transaction are contained in the transaction information. The transaction's chronological order and the moment it was added to the blockchain are both provided by this information. Stakeholders can retrieve and confirm the GPS positions that were recorded in the order that they were uploaded to the blockchain by referring to the block number and timestamp.
- Stakeholders can access and examine the transaction data related to logging GPS positions by using Etherscan or another Ethereum blockchain explorer. For computational cost analysis, they may confirm the gas usage, retrieve the input data containing the GPS locations, and validate the sender and recipient addresses. The block number and timestamp also allow for the retrieval and referencing of recorded GPS coordinates in chronological order.

4.8.7 Discussions on Ethereum Blockchain's Smart Contracts

The Ethereum blockchain's smart contracts are essential for streamlining the validation process. The guidelines and requirements for verifying satellite photos are laid out in these self-executing agreements. The picture hashes, metadata, and validation outcomes are securely and openly stored in the smart contract. Validators who are taking part in the consensus procedure can access the smart contract to examine the accuracy and value of the photos. The PoS consensus process, for example, makes sure that validators collectively decide the validity of each image based on preset criteria. This distributed strategy decreases the possibility of fake or altered photographs being verified, does away with the necessity for centralized authorities, and improves transparency. The status of each image as well as the validation results are safely recorded on the Ethereum blockchain. This promotes trust in the image repository by enabling users to independently check and query the validation status of saved satellite photos. There could be issues with network congestion, petrol prices, and scalability. Ethereum 2.0 and layer 2 solutions, among other continuing developments in the platform, aim to address these issues and enhance the blockchain's general performance and scalability.

Using Ethereum blockchain technology to verify the dependability and caliber of saved satellite photos makes several important advancements in the industry. The Ethereum blockchain offers a transparent and immutable data storage option. The integrity of the data is guaranteed and unauthorized manipulation is avoided by timestamping, hashing, and storing satellite photos on the blockchain. By tackling a significant difficulty in the sector, this transparency increases trust and dependability in the stored photos. Second, the Ethereum blockchain's implementation of smart contracts enables the safe administration and storage of image metadata. Authorized parties can keep and access metadata related to satellite images, including acquisition information, processing methods, and quality evaluations. Smart contracts make it possible for metadata to be validated automatically, eliminating the need for manual inspection and improving the overall quality assurance of satellite images that are being kept. The decentralized image verification procedure has made a substantial additional contribution. Multiple network nodes can independently validate saved photos because of the decentralized structure of the Ethereum blockchain. Consensus can be reached among network participants by comparing the picture data with information from external sources, such as ground truth data or other satellite photography. The trustworthiness and credibility of the archived satellite photos are further increased by this decentralized verification method.

The Ethereum blockchain's tokenization and incentive mechanisms aid in the validation process. It is possible to generate tokens that reflect ownership or access rights to particular satellite photos, facilitating the secure and open transfer of image rights between parties. Smart contracts can also be used to give incentives to promote community engagement in the picture verification process. This gamification strategy encourages a lively and involved community that is dedicated to preserving the accuracy and excellence of satellite photos. Collaboration and interoperability are two more benefits made possible by the Ethereum blockchain. Due to its open architecture, all players in the satellite imaging ecosystem may easily communicate information, verify the quality of the images, and build confidence. Collaboration between satellite imagery producers, data analysts, and end users *via* standardized protocols and smart contracts will increase the overall reliability and quality assurance of satellite images.

Decision-making procedures in a variety of industries become more precise and informed because of this collaborative approach. The usage of Ethereum blockchain technology, in conclusion, makes several improvements to the verification of the validity and caliber of archived satellite photos. Its transparent and immutable data storage, smart contracts for managing metadata, decentralized image verification, tokenization and incentive systems, and interoperability for stakeholder collaboration all play significant roles. Blockchain technology is anticipated to have a big impact on maintaining the reliability and integrity of satellite imagery as it develops. These developments open the door for greater precision and well-informed decision-making across businesses that rely on satellite data. With applications ranging from agriculture and environmental monitoring to urban planning and crisis management, the use of Ethereum blockchain technology to verify the dependability and quality of stored satellite photos has enormous potential. Ethereum-based solutions can address the issues related to image manipulation, data integrity, and confidence in the satellite imagery industry by utilizing the decentralized and transparent characteristics of blockchain.

The Ethereum blockchain's transparency encourages stakeholders to trust one another. Users have access to the blockchain and may check the origin, changes, and any associated metadata for each image. The legitimacy of the photos is increased by the transparency, which also assures accountability for the entire image lifecycle. This factor is especially important in fields like agriculture, where precise and trustworthy satellite photos are essential for making decisions. Although using the Ethereum blockchain to validate satellite images has several advantages, it is important to take into account the technology's drawbacks. To achieve widespread acceptance, it is imperative to address two key issues: scalability and cost. The slow transaction speeds and high gas costs of Ethereum could make it difficult to process a huge number of satellite photos. However, continuing initiatives like Ethereum 2.0 are designed to address these scalability problems and enhance the network's overall performance. As a result, several businesses that largely rely on precise and reliable imagery have the potential to be revolutionized by the deployment of Ethereum blockchain technology for confirming the trustworthiness and quality of stored satellite photos. Stakeholders can guarantee the integrity and authenticity of satellite images by utilizing the decentralized nature of blockchain, the transparency and immutability of Ethereum, and the automation capabilities of smart contracts. This will ultimately improve decision-making, increase efficiency, and increase trust in the use of satellite imagery. Even though there are obstacles to be addressed, the advantages provided by the Ethereum blockchain make it a potential option for resolving the crucial problems related to picture tampering and data integrity in the satellite imagery industry.

It is critical to address scalability and performance challenges as the Ethereum blockchain continues to grow in acceptance and popularity.

Future research should concentrate on investigating layer two protocols (such as state channels, side chains, or alternative blockchain platforms) or systems that can effectively handle the growing volume of satellite imaging data. To meet the rising demand for image validation, it will be essential to optimize transaction throughput and lower latency. Satellite image dependability and quality assessment can be improved with the help of AI and machine learning techniques. Future research should examine the automation of the Ethereum blockchain ecosystem using AI algorithms. To increase the precision and effectiveness of the validation process, tools like computer vision, anomaly detection, and deep learning models may be used. While smart contracts offer a safe way to store picture metadata, more investigation is required to create reliable standards and methods for managing metadata. This entails creating ways to assess and verify the authenticity of metadata, defining standardized metadata formats, and guaranteeing compatibility between various satellite imagery platforms. Future research should concentrate on creating thorough frameworks that handle the difficulties of metadata management in a decentralized blockchain setting.

It is essential to maintain the confidentiality and privacy of satellite imagery data, especially when delicate or secret information is involved. Future research should investigate privacy-preserving methods to safeguard the secrecy of satellite images and associated metadata, such as zero-knowledge proofs, secure multi-party computation, or differential privacy. To further protect the integrity and confidentiality of stored data, strong security measures including encryption, access control, and auditing methods should be put in place. Future work should concentrate on creating industry-wide standards for Ethereum blockchain satellite image validation to promote wide adoption and collaboration. The areas that need standardization include data formats, image quality measures, validation procedures, and interfaces for smart contracts. The removal of a single point of failure is one of the main benefits of using the Ethereum blockchain for picture validation.

Traditional centralized systems pose vulnerabilities and the possibility of data tampering because they depend on a central authority or a reliable third party to validate and store images. With Ethereum, the validation process is divided across several nodes, making it extremely resistant to hacking and guaranteeing that the images' integrity is upheld. Additionally, the validation process can be automated using smart contracts built on Ethereum, which would require less manual work and be more effective. Smart contracts can be programmed to autonomously validate incoming satellite photos based on established criteria, such as image resolution, metadata consistency, and source legitimacy. Smart contracts are self-executing contracts with predefined conditions. The validation procedure is streamlined, expenditures are decreased, and the overall dependability and quality assurance of the archived satellite photos are improved thanks to this automation.

The suggestions made above give prospective paths for ongoing research and development into using Ethereum blockchain technology to verify the accuracy and caliber of satellite photos that are being stored. The development and applicability of blockchain-based satellite image validation systems will be aided by addressing scalability, incorporating AI and machine learning, improving metadata management, assuring privacy and security, fostering standardization and interoperability, and conducting real-world deployment studies. Following these suggestions will help the field improve the reliability, validity, and usability of satellite imaging data, opening up a wide range of applications and stimulating innovation across several sectors. Although the theoretical advantages of utilizing the Ethereum blockchain for satellite image validation are encouraging, further real-world implementation and case studies are required to confirm its viability and efficiency.

To evaluate the viability, scalability, and impact of Ethereum blockchain-based validation systems in real-world scenarios, future work should concentrate on building pilot projects in partnership with satellite image suppliers, governmental organizations, and industry players. Future work should concentrate on creating industry-wide standards for Ethereum blockchain satellite image validation to promote wide adoption and collaboration. The areas that need standardization include data formats, image quality measures, validation procedures, and interfaces for smart contracts. To facilitate seamless data interchange and validation throughout the ecosystem, interoperability is also essential between various blockchain platforms and satellite image suppliers.

4.9 Challenges and Future Directions

To improve its efficacy and scalability, using Ethereum blockchain simulation for satellite picture storage has several drawbacks and difficulties that must be resolved. The following list of existing issues and potential future paths for improvement include:

Due to its limited capacity for processing transactions, Ethereum has scalability problems. The blockchain could have trouble keeping up with the growing demand as the number of satellite photos increases. Research can concentrate on investigating scalability options such as layer two protocols (e.g., side chains and state channels) or developing sharing strategies to divide the blockchain network and increase throughput [28].

Large satellite photos might be difficult to store directly on the Ethereum blockchain owing to space restrictions. Future research can look into methods that combine blockchain-based metadata storage with decentralized storage systems (like IPFS and Swarm). In this approach, picture information and cryptographic hashes are saved on the blockchain, but the actual image data is kept off-chain.

Conclusion

It is quite promising to employ Ethereum blockchain technology to verify the dependability and caliber of saved satellite photos. Satellite imagery may be safely stored thanks to its immutable and transparent nature, smart contracts, and decentralized verification procedures. The Ethereum blockchain promotes active stakeholder participation and a community-driven approach to image validation by utilizing tokenization and incentivization. Additionally, the collaboration and interoperability enabled by the Ethereum blockchain improve the general dependability and quality assurance of satellite imagery. As blockchain technology develops, it is anticipated to become more and more important in ensuring the reliability and validity of satellite images, empowering a variety of industries to make decisions based on precise and high-quality data. In several sectors, including agriculture, environmental monitoring, urban planning, and disaster management, satellite imaging is essential. A crucial difficulty has been maintaining the dependability and quality of the archived satellite photos. This research has aimed to highlight how the Ethereum blockchain could be used to verify the accuracy and value of satellite photos that have been recorded. The Ethereum blockchain can be used to store and maintain satellite picture metadata through smart contracts, and self-executing code. Authorized parties can securely store and retrieve metadata, which includes details on image acquisition, processing methods, and quality evaluations. Smart contracts make it possible to automatically validate information, which lessens the need for manual verification procedures and raises the overall level of quality assurance for saved satellite photos.

Future work should concentrate on incorporating decentralized Oracles into the Ethereum blockchain ecosystem to improve the dependability and accuracy of satellite image validation. To verify the legitimacy and caliber of archived satellite photos, decentralized oracles can offer real-time data feeds from reliable sources like ground-based sensors or satellite image libraries.

References

1. Gaspar, J., Ferreira, R., Sebastião, P. and Souto, N., 2020. Capture of UAVs through gps spoofing using low-cost SDR platforms. *Wirel. Pers. Commun.*, *115*, pp.2729-2754.
2. UNECE and ITU (2021) Sustainable Smart Cities | UNECE. Available at: https://unece.org/housing/sustainable-smart-cities (Accessed: 26 May 2022).
3. United Nation (2015) THE 17 GOALS | Sustainable Development. Available at: https://sdgs.un.org/goals (Accessed: 7 December 2022).
4. Averin, A. and Averina, O., 2019, October. Review of blockchain technology vulnerabilities and blockchain-system attacks. In *2019 International Multi-Conference on Industrial Engineering and Modern Technologies (FarEastCon)* (pp. 1-6). IEEE.
5. Ahmad, I., Alqarni, M.A., Almazroi, A.A. and Alam, L., (2021), Real estate management via a decentralized blockchain platform. *Comput. Mater. Contin.*, *66*(2), pp.1813-1822.
6. Alharby, M. and van Moorsel, A., 2020. Blocksim: An extensible simulation tool for blockchain systems. *Front. Blockchain*, *3*, p.28.
7. Xia, W., Chen, X. and Song, C. (2022) 'A Framework of Blockchain Technology in Intelligent Water Management', Frontiers in Environmental Science. *Front. Media S.A.*, 10, p. 753. doi: 10.3389/FENVS.2022.909606/BIBTEX.
8. Sharma, V., Tripathi, A.K. and Mittal, H., 2022. Technological revolutions in smart farming: Current trends, challenges & future directions. *Comput. Electron. Agric.*, p.107217.
9. Ali, O., Jaradat, A., Kulakli, A. and Abuhalimeh, A., 2021. A comparative study: Blockchain technology utilization benefits, challenges and functionalities. *IEEE Access*, *9*, pp.12730-12749.
10. Oštir K.; Čotar K.; Marsetič A.; Pehani P.; Perše M.; Zakšek K.; Zaletelj J.; Rodič T. Automatic Near—Real—Time Image Processing Chain for Very High Resolution Optical Satellite Data. *Int. Arch. Photogramm. Remote Sens. Spat. Inf. Sci.* 2015, XL— 7/W3, 669–676.
11. Scheffler D.; Sips M.; Behling R.; Dransch D.; Eggert D.; Fajerski J.; Freytag J.C.; Griffiths P.; Hollstein A.; Hostert P.; *et al.* Geomultisens—A common automatic processing and analysis system for multi—sensor satellite data. In *Proceedings of the Second joint Workshop of the EARSeL Special Interest Group on Land Use & Land Cover and the NASA LCLUC Program: "Advancing Horizons for Land Cover Services Entering the Big Data Era"*, Prague, Czech Republic, 6–7 May 2016.

12. Morris D.E.; Boyd D.S.; Crowe J.A.; Johnson C.S.; Smith K.L. Exploring the potential for automatic extraction of vegetation phenological metrics from traffic webcams. *Remote Sens.* 2013, 5, 2200–2218. [Google Scholar].
13. Inglada J.; Arias M.; Tardy B.; Hagolle O.; Valero S.; Morin D.; Dedieu G.; Sepulcre G.; Bontemps S.; Defourny P.; Koetz B. Assessment of an Operational System for Crop Type Map Production Using High Temporal and Spatial Resolution Satellite Optical Imagery. *Remote Sens.* 2015, 7, 12356–12379. [Google Scholar].
14. Clewley D.; Bunting P.; Shepherd J.; Gillingham S.; Flood N.; Dymond J.; Lucas R.; Armston J.; Moghaddam M.. A Python—Based Open Source System for Geographic Object—Based Image Analysis (GEOBIA) Utilizing Raster Attribute Tables. *Remote Sens.* 2014, 6, 6111–6135. [Google Scholar].
15. DigitalGlobe Platform—Actionable Insights. Global Scale Available online: https://platform.digitalglobe.com/gbdx (accessed on 14 June 2023).
16. Tiede D.; Baraldi A.; Sudmanns M.; Belgiu M.; Lang S.. ImageQuerying—Automatic real—time information extraction and content—based image retrieval in big EO image databases. In *Proceedings of the Second joint Workshop of the EARSeL Special Interest Group on Land Use & Land Cover and the NASA LCLUC Program: "Advancing Horizons for Land Cover Services Entering the Big Data Era"*, Prague, Czech Republic, 6–7 May 2016.
17. Microsoft Azure (2017) Global Trusted. Hybrid. Available online: https://azure.microsoft.com/en—us (accessed on 14 July 2017).
18. Amazon EC2(2017)—Secure and Resizable Compute Capacity in the Cloud Launch Applications When Needed without Upfront Commitments. Available online: https://aws.amazon.com/ec2 (accessed on 14 July 2017).
19. Cao, S., Dang, S., Zhang, Y., Wang, W. and Cheng, N., 2021. A blockchain-based access control and intrusion detection framework for satellite communication systems. *Comput. Commun.*, *172*, pp.216-225.
20. Faria, C. and Correia, M., 2019, July. BlockSim: blockchain simulator. In *2019 IEEE International Conference on Blockchain (Blockchain)* (pp. 439-446). IEEE.
21. Fu, Z., Dong, P. and Ju, Y., 2020. An intelligent electric vehicle charging system for new energy companies based on consortium blockchain. *J. Cleaner Prod.*, *261*, p.121219.
22. Hassanien, A.E., Torky, M., Goda, E., Snasel, V. and Gaber, T., 2021. Proof of Space Transactions: A Novel Blockchain Protocol for Secure Authentication of Satellite Transactions.
23. Javaid, M., Haleem, A., Singh, R.P., Khan, S. and Suman, R., 2021. Blockchain technology applications for Industry 4.0: A literature-based review. *Blockchain: Res. Appl.*, 2(4), p.100027.
24. Mohammed, A.H., Abdulateef, A.A. and Abdulateef, I.A., 2021, June. Hyperledger, Ethereum and blockchain technology: a short overview. In *2021*

third International Congress on Human-Computer Interaction, Optimization and Robotic Applications (HORA) (pp. 1-6). IEEE.
25. Oliva, G.A., Hassan, A.E. and Jiang, Z.M., 2020. An exploratory study of smart contracts in the Ethereum blockchain platform. *Empirical Software Eng., 25*, pp.1864-1904.
26. Zhang, X., Yan, H., Zhang, L. and Wang, H., 2020. High-resolution remote sensing image integrity authentication method considering both global and local features. *ISPRS Int. J. Geo-Inf., 9*(4), p.254.
27. Ibrahim, H., Shouman, M.A., El-Fishawy, N.A. and Ahmed, A., 2021, July. Literature Review of Blockchain Technology in Space Industry: Challenges and Applications. In *2021 International Conference on Electronic Engineering (ICEEM)* (pp. 1-8). IEEE.
28. Liu, Y., Yang, X., Wang, Z., Lu, C., Li, Z. and Yang, F., 2019. Aquaculture area extraction and vulnerability assessment in Sanduao based on richer convolutional features network model. *J. Oceanol. Limnol., 37*(6), pp.1941-1954.
29. Iyer, V., Mehmood, A., Babu, B. and Reddy, Y.B., 2022, October. Spatial Blockchain: Smart Contract Using Multiple Camera Censuses. In *Proceedings of the Future Technologies Conference (FTC)* 2022, Volume 2 (pp. 55-66). Cham: Springer International Publishing.
30. Jain, S., Ahuja, N.J., Srikanth, P., Bhadane, K.V., Nagaiah, B., Kumar, A. and Konstantinou, C., 2021. Blockchain and autonomous vehicles: Recent advances and future directions. *IEEE Access, 9*, pp.130264-130328.
31. Nalepa, J., Myller, M. and Kawulok, M., 2019. Validating hyperspectral image segmentation. *IEEE Geosci. Remote Sens. Lett., 16*(8), pp.1264-1268.
32. Xie, G., Liu, Y., Xin, G. and Yang, Q., 2021. Blockchain-based cloud data integrity verification scheme with high efficiency. *Secur. Commun. Netw., 2021*, pp.1-15.
33. Zubaydi, H.D., Varga, P. and Molnár, S., 2023. Leveraging Blockchain Technology for Ensuring Security and Privacy Aspects in Internet of Things: A Systematic Literature Review. *Sensors, 23*(2), p.788.
34. Razzaq, A., Mohsan, S.A.H., Ghayyur, S.A.K., Alsharif, M.H., Alkahtani, H.K., Karim, F.K. and Mostafa, S.M., 2022. Blockchain-Enabled Decentralized Secure Big Data of Remote Sensing. *Electronics, 11*(19), p.3164.
35. Lee, S., Seok, H.W., Lee, K.R. and In, H.P., 2022. B-GPS: Blockchain-Based Global Positioning System for Improved Data Integrity and Reliability. *ISPRS Int. J. Geo-Inf., 11*(3), p.186. Availableat: https://doi.org/10.3390/ijgi11030186 (Accessed 25 August 2023).
36. Wang, Y., Su, Z., Ni, J., Zhang, N. and Shen, X., 2021. Blockchain-empowered space air-ground integrated networks: Opportunities, challenges, and solutions. *IEEE Commun. Surv. Tutorials, 24*(1), pp.160-209.

37. Tang, Y., Li, T. and Ma, X., 2016. Creep and recovery behavior analysis of space mesh structures. *Acta Astronaut.*, *128*, pp.455-463.
38. Saijare, M., Medeni, T.D., Medeni, I.T. and Soylu, D., 2020. Digital Maps and Blockchain, Simplification of Information Sharing. In *IVUS* (pp. 6-15).

5

Urban Expansion and Traffic Congestion: A Geographical Study of Shimla

Akshat Sharma and Amardeep Boora*

Department of Civil Engineering, Jaypee University of Information Technology, Waknaghat, Himachal Pradesh, India

Abstract
The study examines the phenomenon of urban sprawl and its consequences on traffic congestion in the Shimla subdistrict. Through the analysis of land use/cover maps, travel distance, travel time, and congestion levels in various areas of the city, the study highlights the challenges associated with rapid urbanization. The results indicate a noteworthy rise in the built-up area over the years, accompanied by a decline in vegetation cover and agricultural land. This expansion has contributed to traffic congestion in key areas such as Mall Road, Sanjauli Chowk, and Circular Road. The implications of urban sprawl and traffic congestion are discussed, emphasizing the significance of effective urban planning and regulatory interventions. The conclusion highlights the importance of sustainable land use practices, improved transportation systems, and the preservation of open spaces to address these challenges. The paper concludes by emphasizing the significance of holistic approaches to urban development, considering the unique context of Shimla, to ensure a liveable and sustainable future for the city's residents.

***Keywords*:** Urban sprawl, traffic congestion, urban planning, transportation, sustainability

5.1 Introduction

Urban sprawl and traffic congestion are two interrelated issues that have significant impacts on the liveability and sustainability of cities. As cities

**Corresponding author*: amardeepboora9@gmail.com

Disha Thakur, Sanjay Kumar, Har Amrit Singh Sandhu and Chander Prakash (eds.) Sustainable Development Using Geospatial Techniques, (107–128) © 2024 Scrivener Publishing LLC

grow and expand, they often experience urban sprawl, which is distinguished by the unplanned and excessive expansion of urban areas beyond their core boundaries [1]. Urban areas around the world are expanding rapidly, both in terms of their geographical size and population density. This growth is particularly evident in developing countries, where cities experiencing rapid change [2]. As per the United Nations' 2014 report on global urbanization [3], by 2050, it is anticipated that 66% of the world's population will reside in cities, with 90% of this expansion taking place in emerging nations. Asia and Africa will collectively concentrate around 73% of the world's urban population [4]. The most recent information available on the United Nations world population dashboard indicates that India's population has surpassed China's population [5]. According to World Bank United States Census Bureau estimates, India now has a population of 1.4076 billion (140.76 crore) [6]. India's proportion of people living in cities has consistently increased since the country's independence, making it the world's second-largest urbanized country, with over 30% of its population living in urban areas [2]. It is estimated that by 2025, half of India's population will be urbanized. Large cities in India are experiencing rapid growth, leading to a significant concentration of the urban population compared to smaller urban centers. The majority of India's cities have their populations concentrated in large CBDs (Central Business Districts) [7].

Shimla, a picturesque hill station in northern India, is rapidly becoming more urbanized, which is causing urban expansion and making traffic worse. This frequently results in more reliance on private automobiles and lengthier travel distances, which contributes to traffic congestion. The major environmental impacts of urban sprawl are increasing pollution, including air pollution and water pollution, due to increased vehicle use and runoff from impervious surfaces [8, 9]. Problems associated with urban sprawl include traffic congestion, air and noise pollution, water scarcity, and loss of green spaces. The unplanned development in Shimla has caused natural ecosystems to become fragmented and the loss of biodiversity [10]. Furthermore, the area's fast urbanization and growth have raised property values and reduced the availability of cheap housing for the local populace, making it challenging for many to find suitable accommodation. The increasing demand for land and resources is also resulting in the encroachment of forest areas and agricultural land, leading to environmental degradation and soil erosion [11]. In addition to these challenges, Shimla's hilly terrain poses significant challenges for urban planning and infrastructure development. The construction of buildings on sloping terrain can lead to the erosion of soil and landslides during the monsoon season, resulting in property damage and loss of life [12]. In recent times, there has been an uncontrolled spread of urban development in some areas around Shimla, while traffic congestion is the result of increased demand for transportation services that exceeds the available supply [13]. The relationship between urban sprawl and traffic congestion is complicated and varies depending on several factors such as land-use patterns, transportation infrastructure, and population density. Studying the correlation between urban sprawl and traffic congestion in Shimla is important because it can help policymakers understand the extent of the problem and formulate policies and strategies to manage urban growth and traffic congestion sustainably and equitably. The present study aims to utilize GIS applications to pinpoint locations where urban sprawl has occurred in Shimla and analyze traffic flow and density to determine the extent of traffic congestion. Consequently, different satellite imagery, maps, land-use data, and other relevant information will be analyzed to identify areas that have experienced urban sprawl. By comparing the data on urban sprawl and traffic congestion, the present study aims to determine the influence of urban expansion on traffic congestion and vice versa. Traffic congestion can have a significant economic impact on Shimla, with increased travel times and reduced productivity. The study analyzes traffic congestion in Shimla using traffic flow and density data to identify areas that are prone to traffic congestion and formulate different guidelines to analyze future growth patterns based on different scenarios such as population growth projections or zoning laws. A comparison between urban sprawl and traffic congestion data was conducted to identify any correlation between them if any. Interpretation of the results of the analysis and modeling draw implications for urban planning in Shimla. Based on the study findings, several recommendations have been formulated to effectively manage urban sprawl and alleviate traffic congestion in Shimla. The results of this study permitted policymakers to create policies and strategies for controlling urban expansion and traffic congestion sustainably and equitably about the effects of urban sprawl on traffic congestion.

Land usage and land cover have changed significantly because of urbanization, resulting in changes in urban structures and patterns. In order to understand the spatial characteristics of metropolitan regions, landscape metrics were used to quantify the arrangement of the categories of land cover and the overall landscape [14, 15]. GIS tools enabled the calculation of urban landscape metrics, such as patchiness and density, to analyze spatial distribution and changes in urban areas. The analysis of sprawl patterns and landscape metrics cannot be viewed independently, as the identification of the pattern provides an initial understanding of the landscape configuration [15–17]. Additionally, the correlation between the two variables was done which will help in understanding the changes in urban structures over time. Studies addressed the significance of landscape metrics and GIS tools in analyzing urbanization patterns and changes in land use, which could aid in effective urban planning and management. Urban sprawl was formerly measured by [18] and evaluated using an indicator-based methodology, which was predicated on the idea that sprawl was a multidimensional phenomenon requiring a multiple-indicator approach. The GIS-based tool provided a quantitative derivation of certain sprawl types, enabling researchers and urban planners to identify specific patterns and trends in urban sprawl. In a recent study [4], the multidimensional index method was utilized to quantify the urban sprawl in the US (United States). Earlier [8], a range of indicators were proposed to measure sprawl in the city of Istanbul. Consequently, the correlation between urban sprawl and health, examining considerations, including pollution in the air, traffic accidents, pedestrian accidents, and fatalities, the amount and purity of water, mental wellness, and communal wealth were explored. Through their analysis, the authors found that there were both health benefits and costs associated with sprawl. Researchers noted that the negative effects of sprawl did not affect all populations equally and emphasized the need for special attention to those most affected. The authors emphasized the importance of incorporating health considerations into policy-making at all levels, from personal transportation decisions to federal regulations, arguing that it was crucial due to the uneven distribution of health effects of sprawl [19]. A recent study [20] presented an analysis of forest cover change (FCC) and land surface temperature (LST) in a specific study area from 2000 to 2021. The research revealed a noteworthy decline of 5.59% in heavily forested regions and a total decrease of approximately 2.99 km^2 in pasture and non-forest regions. However, during the period from 2000 to 2021, there was a substantial increase in these areas, with a rise of 7.25% for dense forest and 0.22% for pasture and non-forest areas [21]. Earlier, two different algorithms, namely, the single channel (SC) and split window (SW) algorithms, were adopted to retrieve LST. Based on the findings, an average difference of ± 1 K between the years 2000, 2013, and 2021 was observed due to rapid urbanization [22]. Furthermore, the studies revealed a strong negative correlation coefficient between the LST retrieved using the SC algorithm and the NDVI (normalized difference vegetation index) throughout the years 2000 and 2013 ($R^2 = 0.791$ and 0.750, respectively), When compared to the NDVI in 2021, the LST generated by utilizing the SW technique has an optimal negative correlation coefficient ($R^2 = 0.646$) [23]. Earlier [1], a study was conducted to provide a comprehensive understanding of urban sprawl in Israel through a land-use survey of 78 urban settlements over 15 years. The study employed factor analysis to calculate an integrated sprawl index that accounted for various measures of urban sprawl, including density, fragmentation, and land-use mix. The findings demonstrated that urban sprawl is a multidimensional phenomenon requiring a multifaceted approach for effective quantification. The results also indicated a significant increase in urban sprawl in Israel over the 15 years studied, attributed to population growth, economic development, and inadequate urban planning and management strategies [1]. The use of satellite data for analyzing the outcomes of land use/land cover (LULC) variations on the temperature of the land retrieval has been extensively explored in recent studies. Frumkin [9], Thakur *et al.* [24], and Negassa *et al.* [25] utilized Landsat series datasets with TIRS sensors to examine the relationship between LULC and LST. This technology plays a crucial role in understanding land-surface interactions, temperature changes, global warming, LULC, surface energy flow, and soil moisture. Previous studies [23, 26] also supported the use of satellite data for LST retrieval.

5.2 Study Area

Shimla is a picturesque hill station that serves as the capital of Himachal Pradesh, a northern Indian state. In accordance with the 2011 Census of India, Shimla has 1,69,578 residents. It is a bustling metropolis that has grown significantly throughout the years. Shimla is located in the Himalayan foothills, between 31°3'N and 31°8'N latitudes and 77°7'E and 77°13'E longitudes. The city of Shimla is spread over an area of 30 km^2, the length of which follows a ridge with seven spurs. Shimla is one of the highest cities in India, having a mean elevation above mean sea level of 2195 m. The city's elevation gives it a unique climate, with cool temperatures prevailing throughout the year. It is true that Shimla's pleasant weather during the summer months, with average temperatures ranging from 19°C to 28°C, attracts many people to the city. Shimla's popularity as a tourist destination and a hill station has led to a significant rise in the number of individuals desiring to reside in urban areas permanently, which has caused some problems, including traffic congestion. Urban sprawl and traffic congestion are critical issues that affect the quality of life in many cities around the world, and Shimla is no exception. The city's unique geographical location, coupled with its growing population and popularity as a tourist destination, has resulted in significant urbanization and expansion in recent years, leading to an increase in traffic congestion and urban

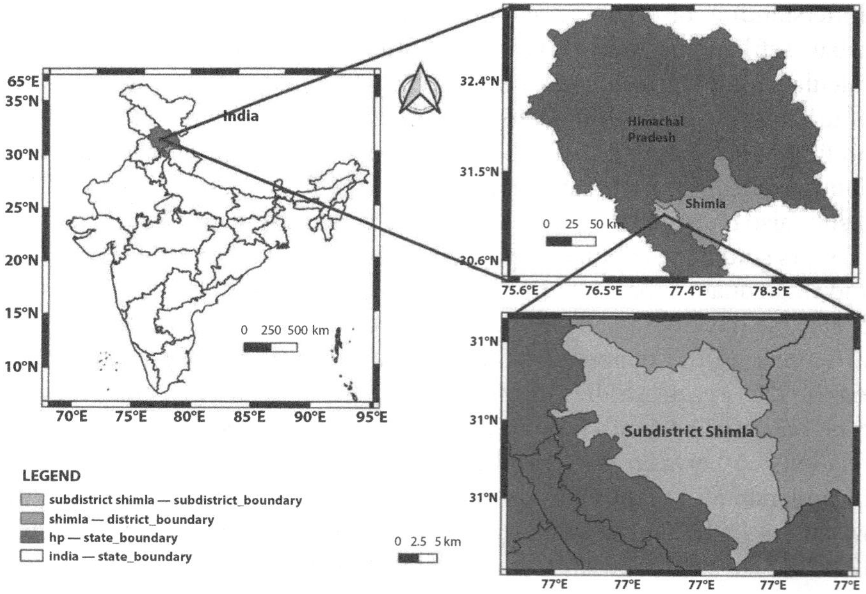

Figure 5.1 Geographic overview: Study area location map.

Table 5.1 Satellite imagery and data sources.

S. no.	Satellite imagery data	Date of acquisition	Scale and spatial resolution	Source
1	Landsat 05 TM C2 L2	10-08-1993	30 m	https://earthexplorer.usgs.gov/
2	Landsat 07 ETM+ C2 L2	24-09-2003	29.8 m	https://earthexplorer.usgs.gov/
3	Landsat 08TIRS C2 L2	24-09-2013	28 m	https://earthexplorer.usgs.gov/
4	Landsat 09 TIRS C2 L2	28-10-2022	30 m	https://earthexplorer.usgs.gov/
5.	Shapefile of the entire country up to subdistrict level	15-04-2022	1:50,000	Survey of India

sprawl. As a result, Shimla has become an ideal study area for urban sprawl and traffic congestion. The location map depicted in Figure 5.1 provided an overview of the specific area used for the study. It illustrated the geographical context and spatial boundaries of India, Shimla, and Shimla Subdistrict, enabling a clear understanding of its location about surrounding regions or landmarks. Table 5.1 provided a comprehensive reference for the satellite imagery data and other relevant information that was utilized in the study. It included essential details regarding the source of the satellite imagery, such as the satellite series and sensors used, which could include the Landsat series with TIRS sensors. This table also specified the specific years or timeframes for which the imagery data was obtained, such as 1993, 2003, 2013, and 2022.

5.3 Methodology

The primary focus of this research work is to investigate the urban sprawl in Shimla subdistricts, spanning a timeframe of 30 years, specifically from 1993 to 2022. The study employed satellite imagery and geospatial tools to examine the patterns and extent of urban expansion in the region. The methodology for using GIS to analyze urban sprawl involved several steps, and collecting relevant data was the first step. Data included satellite imagery, maps, land-use data, population data, transportation data, building footprints, and other relevant information. The collected data were then integrated into GIS software. Integration of the data was necessary to facilitate the analysis. The integrated data were then analyzed using QGIS spatially to identify areas that had experienced urban sprawl. Spatial analysis also helped in identifying areas that were at risk of urban sprawl. Spatial analysis was done using tools like buffer analysis, density analysis, and hotspot analysis. GIS was used to build models to simulate future growth patterns based on different scenarios. The study utilized cloud-free Landsat satellite information regarding multi-temporal pictures collected from the Bhuvan (Indian Geo Platform of ISRO) and the US Geological Survey's Earth Explorer portal. To aid in analysis, the information collected was preprocessed and then reflected onto the Universal Transverse Mercator (UTM) projecting framework. QGIS software was used to calculate the urban sprawl for four different periods: 1993, 2003, 2013, and 2022. The Semi-Automatic Classification Plugin (SCP) tools were utilized for image classification, enabling the identification of different land cover categories according to the spectral signatures of the pixels. The HCMGIS (Ho Chi Minh City Geospatial Information System) served as the base map for

Table 5.2 Overview of the classifications for land cover and usage.

Class ID	Class name	Description
1	Urban or built-up areas	Includes dwelling, industrial, and commercial areas.
2	Forest area (vegetation)	Areas covered by trees, shrubs, and other plants.
3	Water bodies	Rivers, lakes, ponds, and other bodies of water.
4	Agricultural land	Areas used for growing crops, including apples and pears orchards.
5	Open space	Areas that are not developed include parks, glaciers, snow-capped mountains, and wilderness areas.

reference. Utilizing this as a base map offered access to satellite imagery procured from various sources, including Google and other platforms. The inclusion of this base map proved particularly advantageous when grappling with challenges related to training inputs as this is georeferenced. To perform the classification, the pixel colors were assigned training inputs based on the RGB (Red-Green-Blue) index. In this study, a 5-4-3 index pattern was utilized to identify and highlight vegetation and built-up areas. On the other hand, the 6-5-2 index was employed to differentiate agriculture, open space, and water bodies. The classified images were then imported into QGIS, where they were analyzed and processed to derive the urban sprawl. The analysis involved comparing the land cover classes of the different periods to identify changes in the urban areas. The urban sprawl was then calculated by measuring the expansion of the built-up area over time. Table 5.2 describes the LULC classes employed in this research. The LULC classification was determined by interpreting the satellite imagery using various spectral bands and indices.

5.4 Result and Discussion

The analysis of the LULC maps of different years (1993, 2003, 2013, and 2022) reveals notable changes in the urban landscape of Shimla.

Figure 5.2(a) represents the map of Shimla's land usage and cover in 1993, showcasing the distribution of various land use categories surrounding the geographical region. The built-up area, which represents the urbanized and developed land, has experienced a substantial increase over the years. In 1993, the built-up area covered 100.40 km^2, corresponding to about 20.08% of the entire area of the Shimla subdistrict, which is approximately 500.12 km^2 as given in Table 5.3. Figure 5.2(b) displays the corresponding map for the year 2003, reflecting the changes in land use over a decade. The trend remained similar in 2003, with a built-up area of 105.12

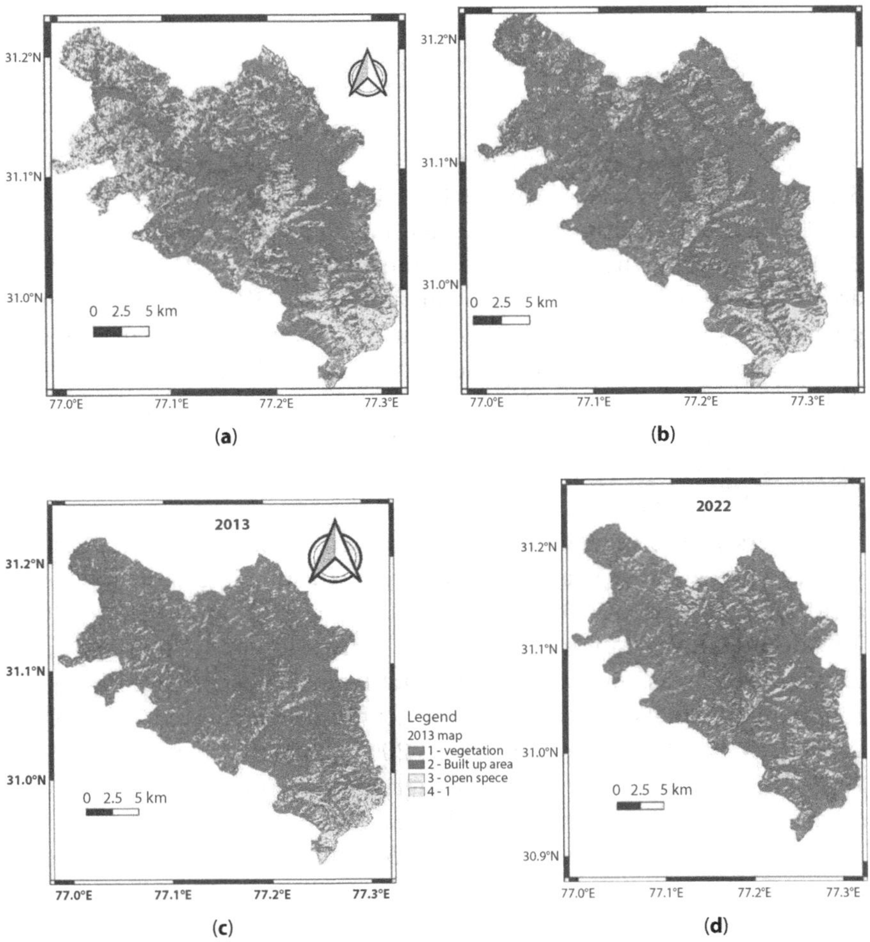

Figure 5.2 (a) Land use and land cover map of Shimla subdistrict in 1993. (b) Land use and land cover map of Shimla subdistrict in 2003. (c) Land use and land cover map of Shimla subdistrict in 2013. (d) Land use and land cover map of Shimla subdistrict in 2022.

Table 5.3 Population and land use distribution in different years.

Year	Population	Urban area (km^2)	Agricultural land (km^2)	Forest land (km^2)	Open space (km^2)
1993	1,02,186 [27]	100.40	35.60	288.67	75.45
2003	1,42,555 [28]	105.12	31.87	281.84	81.29
2013	1,69,578 [29]	185.32	14.25	281.21	19.34
2022	8,89,813	199.14	6.34	251.46	43.18

km^2, also representing 21.02% of the entire region as presented in Figure 5.2(b). However, as time progressed, the built-up region increased significantly. Figure 5.2(c) provides insights into the LULC map of Shimla in 2013, revealing further alterations in the landscape. In 2013, the built-up area had increased to 185.32 km^2, constituting 37.05% of the entire region given in Figure 5.2(c). Besides this, Figure 5.2(d) presents the most recent LULC map of Shimla, reflecting the land use dynamics up to the year 2022. This trend continued in 2022, with the built-up area reaching 199.02 km^2, accounting for 39.82% of the entire region as depicted in Figure 5.2(d). This indicates a substantial growth in urbanization and development within the Shimla subdistricts.

Conversely, the vegetation cover witnessed a notable decline over the years. In 1993, the vegetation cover accounted for 57.72% of the overall area, but it reduced to 56.35% in 2003, 56.23% in 2013, and 50.28% in 2022. This decline in vegetation highlights the conversion of natural land cover to urbanized areas. Similarly, agricultural land also experienced a significant reduction in its extent. In 1993, agricultural land covered 7.12% of the entire area, but it declined to 6.37% in 2003, 2.85% in 2013, and 1.27% in 2022. This indicates the transition of agricultural land into urban areas or other land uses. These findings indicated Shimla's fast urban sprawl and increase of built-up regions over time. Agricultural land and plant cover have suffered as a result of the expansion of built-up regions.

This urban sprawl has led to various environmental and social issues, including traffic jams, environmental impact, and the decline of green cover [30]. Figure 5.2 shows the different land use/cover classes in Shimla in 1993, which serves as a baseline for the subsequent analysis. This considerable growth in the populated region is indicative of rapid urbanization and urban sprawl in the research area.

As the population of the city has grown in recent years, there has been a corresponding increment in the necessity for housing and infrastructure, leading to the transformation of agricultural land, open spaces, and water bodies into built-up areas. This tendency is anticipated to continue in the future since the city's population is expected to grow further. This emphasizes the importance of good urban expansion planning and management to encourage sustainable growth and minimize the negative impact that urbanization has on the environment and the standard of living for its citizens. The land use map of 1993 has been included in the research paper to provide a baseline for the study of changing land use and urban sprawl in Shimla over the past few decades. It permits an understanding of the variations in land cover and usage that have taken place in the research area over time and provides insights into the drivers of urbanization and sprawl in the region. The land use map of 1993 provides a snapshot of the study area before the significant changes that have occurred in recent years and serves as a reference point for the analysis of subsequent years.

Table 5.4 provides a comparison of land use statistics for different years and presents the measurements of variations over time. Regarding the percentage of urban area, there was an increase from 20.08 km^2 in 1993 to 39.82 km^2 in 2022, indicating a growth of 0.94 km^2 (4.68%) from 1993 to 2003 and 18.04 km^2 (85.47%) from 2003 to 2013. The percentage change from 2013 to 2022 was 2.76%, this expansion can be correlated to factors including population growth, urbanization, growth of the economy, and the growth of the tourism industry. The agricultural land witnessed a decline over the years due to the transformation of agricultural land for urban development, changing land use priorities, and a shift away from traditional agricultural practices with 7.12 km^2 in 1993 decreasing to 1.27 km^2 in 2022. This represents a decrease of 0.75 km^2 (-10.54%) from 1993 to 2003, 3.52 km^2 (-55.23%) from 2003 to 2013, and 1.58 km^2(-55.44%) from 2013 to 2022. Factors contributing to this decline include deforestation, encroachment, and resource extraction activities. These factors have impacted the forested areas, leading to the reduction in forest land. Similarly, the forest land decreased from 57.72 km^2 in 1993 to 50.28 km^2 in 2022. The percentage change was -1.37% from 1993 to 2003, -0.13% from 2003 to 2013, and -5.95% from 2013 to 2022. Open space, which includes public parks and green areas, experienced fluctuations. It expanded from 15.09 km^2 in 1993 to 16.25 km^2 in 2003, representing a growth of 1.17 km^2 (7.76%). However, there was a significant decrease to 3.87 km^2 in 2013

Table 5.4 Comparison of land use statistics for different years and change analysis.

Area type	1993	2003	2013	2022	Percentage change (1993–2003)	Percentage change (2003–2013)	Percentage change (2013–2022
Urban area in (km^2)	20.08	21.02	37.06	39.82	0.94	16.04	2.76
Agricultural land (km^2)	7.12	6.37	2.85	1.27	-0.75	-3.52	-1.58
Forest land (km^2)	57.72	56.35	56.23	50.28	-1.37	-0.13	-5.95
Open space (km^2)	15.09	16.25	3.87	8.63	1.17	-12.39	4.77

(-16.16%) before a slight recovery to 8.63 km^2 in 2022. The percentage change was -12.39% from 2003 to 2013 and 4.77% from 2013 to 2022. The changes in open space can be attributed to urban development, land scarcity, and changes in development policies.

5.5 Accuracy Assessment

Projection accuracy was examined using the error matrix and Kappa coefficient technique to ensure the correctness of the land cover categorization. An error matrix is a table that compares classified data to reference data, which is a set of independently collected data with known ground truth. For each land cover class, the matrix computes the entire quantity of properly categorized pixels (i.e., diagonal components) and the percentage of misclassified pixels (i.e., off-diagonal elements). The Kappa coefficient is a statistical measure that provides a standardized degree of conformity between reference and categorized data. It considers the probability of the agreement occurring by chance and provides a measure of the overall accuracy of the classification. A coefficient of 0 indicates agreement that is no better than chance, whereas a Kappa value of 1 indicates total agreement. The reliability of the categorization findings may be assessed and confirmed by employing the error matrix and Kappa coefficient approach. This information can be used to improve the classification methodology, if necessary, and to provide confidence in the accuracy of the final land cover categorization. The classification precision was assessed utilizing ground truth data and statistical methods, as in the error matrix and Kappa coefficient. The land use and land cover information were used to examine urban sprawl, land use changes, and their impact on traffic congestion in the study area.

Table 5.5 presents the point accuracy, user/producer accuracy, and Kappa coefficient values for the classification of different land use classes. A measure of producer accuracy is the degree to which actual land cover classes may be characterized using errors of omission. The likelihood that a recognized pixel would match the kind of land cover at its corresponding real-world location is indicated by the user's accuracy. The classification accuracy for the urban/built-up region in the current study was observed to be 88.7% which indicates a relatively high accuracy in identifying and mapping urban areas. The forest area exhibited a higher accuracy of 94.5%, indicating a more precise classification of forested regions. Water bodies were classified with an accuracy of 97.8%, demonstrating a high level of accuracy in identifying water features. The agricultural land class achieved an accuracy of 89.3%, indicating a reasonable level of precision in mapping agricultural areas. Open space areas were classified with an accuracy of 92.1%, signifying good conformity between the categorized map and reference data for open space regions.

Table 5.5 Results of land use/cover mapping accuracy.

Category	**Point/ overall accuracy (%)**	**Kappa coefficient**	**User accuracy (%)**	**Producer accuracy (%)**
Urban/ built-up area	88.7	0.83	90.5	88.9
Forest area	94.5	0.91	95.7	95.2
Water bodies	97.8	0.96	98.0	98.6
Agricultural land	89.3	0.84	91.2	88.7
Open space	92.1	0.88	93.5	92.5

Point/overall accuracy = the number of ground truth points divided by the total number of points

To analyze and simulate the urban growth trend in the region of interest, several landscape measures were calculated using statistical data on demographics and built-up areas. One of the primary metrics used was Shannon entropy (H_n), which is commonly utilized to assess urban growth phenomena. The formula for calculating Shannon entropy involves the summation of all types of land use and cover in the study region multiplied by its natural logarithm. As part of the analysis, calculations were performed for map density, and buffering along major roads. These landscape metrics help to provide a thorough knowledge of the urbanization pattern in the research area, which can aid in the improvement of effective urban planning and management techniques. The value of Shannon entropy is directly related to the uncertainty or diversity of land use/cover classes in a specified region. A higher value of entropy indicates a greater degree of uncertainty in the information conveyed, which can be interpreted as a higher level of diversity or fragmentation in the land use/cover layouts. In the context of urban sprawl, a high value of entropy may suggest that the city is expanding in an uncontrolled or unplanned manner, with a greater degree of fragmentation in the land use/cover patterns.

The equation for calculating Shannon entropy (H_n) is as follows:

$$H_n = \sum_i^n P_i log_e \frac{1}{P_i} \tag{5.1}$$

H_n is the Shannon entropy value

P_i is the percentage of each land use/cover class in the research area, and it is determined by dividing the class's total area by the study area as a whole.

$$P_i = \frac{x_i}{\sum_i^n x_i} \tag{5.2}$$

To analyze urban growth patterns in different subdistricts of the research area, Shannon entropy was calculated for each ward within a subdistrict. In this case, each ward was considered as an independent spatial unit. The formula for calculating Shannon entropy for each ward is given in Equation 5.3.

Table 5.6 Confusion matrix for land use/cover classification.

	Built-up area	**Forest area**	**Water bodies**	**Agricultural land**	**Open space**
Built-up area	150	5	3	10	2
Forest area	6	180	1	8	3
Water bodies	1	0	15	2	0
Agricultural land	9	7	2	30	5
Open space	2	4	0	7	25

$$H_i^n = \Sigma_i^n P_i log_e \frac{1}{P_i} \quad (5.3)$$

Table 5.6 depicts a potential confusion matrix (error matrix) utilized in the research for LULC categorization. The columns are the ground truth, showing which true classes the pixels in the validation set. In contrast, the rows show the classes into which the picture pixels were classified throughout the classification process. The matrix's diagonal represents the correctly identified pixels. Outside the diagonal, the off-diagonal items reflect misclassifications that occurred during the assignment process between distinct land-cover classes. These components indicate the presence of pixels that were incorrectly allocated to their respective classes, emphasizing the misunderstanding between different land-cover classes.

5.6 Traffic Congestion Analysis

The locations of Mall Road, Sanjauli Chowk, Victory Tunnel, Circular Road, Tutikandi, Chotta Shimla, and Kasumpti in Shimla were included in the study for the assessment of traffic congestion. These locations were selected based on their significance in terms of traffic flow and urban development in Shimla. By analyzing the traffic congestion and urban sprawl patterns in these areas, the study aimed to gain insights into the dynamics of urbanization's effects on transportation infrastructure in Shimla. Travel time and distance between different parts of the city and their correlation with traffic congestion are presented in Table 5.7.

Table 5.7 provides valuable insights into the travel distances, travel times, and congestion levels of different areas in Shimla, highlighting the varying degrees of traffic congestion experienced in these locations. These data were then used to compare travel time and distance across different parts of the city and to explore how traffic congestion levels may be related to travel time and distance. Mall Road, situated 2.5 km away, experiences a travel time of approximately 15–20 min and has a congestion level of 5, indicating high traffic congestion. Sanjauli Chowk, situated at a distance of 5 km, had a travel time of 30–40 min and a congestion level of 4. Similarly, Victory Tunnel, with a travel distance of 3 km, requires approximately 20–25 min of travel time and has a congestion level of 4. Circular Road, spanning 6 km, has a longer travel time of 40–50 min and a congestion level of 5, suggesting significant traffic congestion in the area. Tutikandi, located at a distance of 8 km, requires a travel time of 60–70 min and has a congestion level of 4. Chotta Shimla, covering a travel distance of 6 km, has a relatively shorter travel time of 30–40 min and a congestion level of 2, indicating lower traffic congestion compared to other areas. Kasumpti, spanning 7 km, has a travel time of 50–60 min and a congestion level of 3. As depicted in Table 5.8, Mall Road and Tutikandi are identified as areas with high congestion levels, indicating heavy traffic and potential traffic jams. These areas also exhibit high levels of urban sprawl, indicating significant urban expansion and development. Sanjauli Chowk and Chotta Shimla have a moderate congestion level, suggesting relatively smoother traffic

Table 5.7 Areas in Shimla with their travel distance and travel time during peak traffic hours.

Area	Travel distance (km)	Travel time (minutes)	Congestion level (1–5)
Mall Road	2.5	15–20	5
Sanjauli Chowk	5	30–40	4
Victory Tunnel	3	20–25	4
Circular Road	6	40–50	5
Tutikandi	8	60–70	4
Chotta Shimla	6	30–40	2
Kasumpti	7	50–60	3

Table 5.8 Spatial distribution of traffic congestion and urban sprawl in different parts around Shimla.

City	Congestion level	Urban sprawl level
Mall Road	High	High
Sanjauli Chowk	Moderate	High
Victory Tunnel	Low	Low
Circular Road	High	Moderate
Tutikandi	High	High
Chotta Shimla	Moderate	Moderate
Kasumpti	Moderate	Low

Table 5.9 Characteristics of land use categories and their contribution to urban sprawl.

Land use type	Characteristics	Contribution to urban sprawl
Residential	Low density	High
	Large lot sizes	
	Single-use zoning	
Commercial	Low density	Moderate
	Large parking lots	
	Single-use zoning	
Industrial	Low density	Moderate
	Large industrial parks	
	Single-use zoning	
Mixed-use	High density	Low
	Small lot sizes	
	Mixed-use zoning	
Open space	Low density	Low
	Public parks, green spaces	

flow compared to the high-congestion areas. However, they still exhibit a high level of urban sprawl, indicating considerable urban development.

Table 5.9 summarizes the key characteristics of different land use types and their contribution to urban sprawl. As the population has grown, the urban area has expanded, leading to a decrease in agricultural land, forest land, and open space. Residential land use is typically associated with low density and large lot sizes, which can lead to urban sprawl. Commercial and industrial land use also tends to be low-density, with large parking lots as well as single-use zoning, which may have an impact on urban expansion. Mixed-use zoning and high-density development can help to reduce urban sprawl, while open space typically has a low contribution to urban sprawl.

5.7 Conclusion

The study of traffic congestion and urban sprawl in Shimla has highlighted the challenges that the city is facing due to its quick urbanization. The analysis of land use/cover maps from different years has revealed a notable expansion of the built-up area, indicating the expansion of urban infrastructure and the encroachment of natural and agricultural land.

Urban expansion has led to various consequences, including a reduction in the quality of the natural environment, loss of agricultural land, and reduction in open spaces have adverse effects on the city's sustainability and liveability. The increased traffic congestion not only hampers the efficiency of transportation but also leads to increased air pollution, accidents, as well as a drop in the inhabitants' general standard of living.

The findings indicate that urban sprawl in Shimla is driven by several factors. The growing population, influx of tourists, infrastructure development, availability of quality education and employment opportunities, favorable environment, standard of living, and economic growth have resulted in the expansion of residential, commercial, and industrial areas.

To address these challenges, effective urban planning policy interventions are crucial. The integration of sustainable land-use practices, mixed land-use zoning, and the promotion of compact, walkable neighborhoods can help manage urban sprawl and reduce dependence on private vehicles. Furthermore, conservation and restoration efforts should be undertaken to protect and enhance open spaces, green areas, and the natural landscape. Therefore, it is crucial for policymakers, urban planners, and stakeholders, to consider the long-term implications of urban sprawl and traffic congestion in Shimla. By adopting a holistic and integrated approach, focused on sustainable development, environmental preservation, and efficient transportation systems, the city can strive toward a more liveable, resilient, and environmentally friendly future.

References

1. Frenkel and M. Ashkenazi, "Measuring urban sprawl: How can we deal with it?" *Environ. Plan. B Plan. Des.*, vol. 35, no. 1, pp. 56–79, 2008, doi: 10.1068/b32155.
2. M. K. Jat, P. K. Garg and D. Khare, "Monitoring and modelling of urban sprawl using remote sensing and GIS techniques," *Int. J. Appl. Earth Obs. Geoinf.*, vol. 10, no. 1, pp. 26–43, 2008, doi: 10.1016/j.jag.2007.04.002.
3. S. Farooq and S. Ahmad, "Urban sprawl development around Aligarh city: A study aided by satellite remote sensing and GIS," *J. Indian Soc. Remote Sens.*, vol. 36, no. 1, pp. 77–88, 2008, doi: 10.1007/s12524-008-0008-0.
4. M. Jia, H. Zhang and Z. Yang, "Compactness or sprawl: Multi-dimensional approach to understanding the urban growth patterns in Beijing-Tianjin-Hebei region, China," *Ecol. Indic.*, vol. 138, no. January, p. 108816, 2022, doi: 10.1016/j.ecolind.2022.108816.
5. Department of Social and Economic Affairs, "India overtakes China as the world's most populous country," UN DESA Policy Br. No. 153.
6. J. Chamie, "Population Growth Diversity Continuing in the Twenty-first Century," *Popul. Levels, Trends, Differ.* Springer, Cham., 2022, doi: https://doi.org/10.1007/978-3-031-22479-9_38.
7. Rahman, S. P. Aggarwal, M. Netzband and S. Fazal, "Monitoring Urban Sprawl Using Remote Sensing and GIS Techniques of a Fast Growing Urban Centre, India," *IEEE J. Sel. Top. Appl. Earth Obs. Remote Sens.*, vol. 4, no. 1, pp. 56–64, 2011, doi: 10.1109/JSTARS.2010.2084072.
8. H. Frumkin, "Urban sprawl and public health," *Public Health Rep.*, vol. 117, no. 3, pp. 201–217, 2002, doi: 10.1016/S0033-3549(04)50155-3.
9. S. He and J. Wu, "Relationships of groundwater quality and associated health risks with land use/land cover patterns: A case study in a loess area, Northwest China," *Hum. Ecol. Risk Assess.*, vol. 25, no. 1–2, pp. 354–373, 2019, doi: 10.1080/10807039.2019.1570463.
10. M. S. Boori, M. Netzband, K. Choudhary and V. Voženílek, "Monitoring and modeling of urban sprawl through remote sensing and GIS in Kuala Lumpur, Malaysia," *Ecol. Process.*, vol. 4, no. 1, pp. 1–10, 2015, doi: 10.1186/s13717-015-0040-2.
11. K. S. Sunil Sankhala, "Evaluation of Urban Sprawl and Land Use Land cover Change using Remote Sensing and GIS Techniques: A Case Study of Jaipur City, India," *Int. J. Emerg. Technol. Adv. Eng.*, vol. 3, no. 1, pp. 1–5, 2014, [Online]. Available: http://www.krishisanskriti.org/Publication.html.

12. R. Gupta, M. Sharma, G. Singh and R. K. Joshi, "Characterizing urban growth and land surface temperature in the western Himalayan cities of India using remote sensing and spatial metrics," *Front. Environ. Sci.*, vol. 11, no. January, pp. 1–15, 2023, doi: 10.3389/fenvs.2023.1122935.
13. M. A. Mahboob, I. Atif and J. Iqbal, "Remote Sensing and GIS Applications for Assessment of Urban Sprawl in Karachi, Pakistan," *Sci. Technol. Dev.*, vol. 34, no. 3, pp. 179–188, 2015, doi: 10.3923/std.2015.179.188.
14. P. Cabral, H. Geroyannis, J. P. Gilg and M. Painho, "Analysis and modeling of land-use and land-cover change in Sintra-Cascais area," *Proc. 2005 - 8th Agil. Int. Conf. Geogr. Inf. Sci. Agil.* 2005, 2005.
15. M. Herold, G. Menz and K. C. Clarke, "Remote Sensing and Urban Growth Models – Demands and Perspectives," *Regensburg. Geogr. Schriften*, vol. 35, pp. 78–88, 2001, [Online]. Available: http://www.eo.uni-jena.de/~c5hema/pub/herold_menz_clarke.pdf.
16. H. S. Sudhira, T. V Ramachandra and K. S. Jagadish, "Map India 2003 Municipal GIS Urban Sprawl pattern recognition and modeling using GIS Map India 2003," map India, pp. 28–31, 2003.
17. H. S. Sudhira, T. V. Ramachandra and K. S. Jagadish, "Urban sprawl: Metrics, dynamics and modelling using GIS," *Int. J. Appl. Earth Obs. Geoinf.*, vol. 5, no. 1, pp. 29–39, 2004, doi: 10.1016/j.jag.2003.08.002.
18. S. Siedentop and S. Fina, "Monitoring urban sprawl in Germany: Towards a gis-based measurement and assessment approach," *J. Land Use Sci.*, vol. 5, no. 2, pp. 73–104, 2010, doi: 10.1080/1747423X.2010.481075.
19. L. Feng, "Applying remote sensing and GIS on monitoring and measuring urban sprawl. A case study of China," *Int. J. Appl. Earth Obs. Geoinf.*, vol. 10, no. 1, pp. 47–56, 2008.
20. P. K. Thakur, "Spatiotemporal changes of forest cover and land surface temperature using geo-spatial techniques in Talra Wildlife Sanctuary, Shimla, North-Western Himalaya," 2023.
21. P. Jain, I. Singh and M. Jain, "Built-up Analysis of the Hill Town of Nainital," vol. 10, pp. 3872–3878, 2023.
22. M. P. Johnson, "Environmental impacts of urban sprawl: A survey of the literature and proposed research agenda," *Environ. Plan. A* vol. 33, no. 4, pp. 717–735, 2001, doi: 10.1068/a3327.
23. P. K. Thakur, M. Kumar and V. E. Gosavi, Monitoring and Modelling of Urban Sprawl Using Geospatial Techniques—A Case Study of Shimla City, India. 2020. doi: 10.1007/978-981-15-2097-6_8.
24. M. D. Negassa, D. T. Mallie and D. O. Gemeda, "Forest cover change detection using Geographic Information Systems and remote sensing techniques: a spatio-temporal study on Komto Protected forest priority area, East Wollega Zone, Ethiopia," *Environ. Syst. Res.*, vol. 9, no. 1, pp. 1–14, 2020, doi: 10.1186/s40068-020-0163-z.
25. M. Chowdhury, M. E. Hasan and M. M. Abdullah-Al-Mamun, "Land use/land cover change assessment of Halda watershed using remote sensing

and GIS," *Egypt. J. Remote Sens. Sp. Sci.*, vol. 23, no. 1, pp. 63–75, 2020, doi: 10.1016/j.ejrs.2018.11.003.

26. R. Nikam, F. Ibragimov, A. Chouksey, V. Garg and S. P. Aggarwal, "Retrieval of land surface temperature from Landsat 8 TIRS for the command area of Mula irrigation project," *Environ. Earth Sci.*, vol. 75, no. 16, 2016, doi: 10.1007/s12665-016-5952-3.
27. Census of India, "Decadal variation in population since 1901," 1993.
28. Census Commission of India, "Census of India 2001," 2001.
29. Census of India, "Shimla city census 2011 data," 2011.
30. S. Kumar, T. Kaur, M. K. Arora and S. Upadhyay. "Resource estimation and sizing optimization of PV/micro hydro-based hybrid energy system in rural area of Western Himalayan Himachal Pradesh in India." *Energy Sources Part A* vol. 41, no. 22, 2019 pp. 2795-2807, doi; https://doi.org/10.1080/15567036.2019.1576075

6

Landslide Susceptibility Analysis for Sustainable Development in the Indian Himalayas

Ankur Sharma* and Har Amrit Singh Sandhu

Department of Civil Engineering, Punjab Engineering College, Chandigarh, India

Abstract

The Indian Himalayas experience a high number of landslides due to the terrain's fragility and intense seismicity. In recent years, the increasing population and growing demand for urban land in the Himalayas have further increased the risk of landslides. The present study employs the frequency ratio method to generate a landslide susceptibility zonation (LSZ) map of the study area and demarcate areas that are prone to landslides. Accurately delineating such areas can aid in circumventing crucial developmental projects in the prone regions, thereby mitigating landslide exposure. Conversely, these areas can conveniently serve a host of other functions like biodiversity preservation, soil conservation, afforestation, eco-tourism, enhancement of ecosystem services, etc. The efficacy of the generated LSZ map is evaluated utilizing the area under the receiver operating characteristic (ROC) curve and the LSZ map is found suitable for its intended use. The generated LSZ map classifies the study area into five susceptibility classes: "very high" (0.23%), "high" (8.41%), "moderate" (35.13%), "low" (34.35%), and "very low" (21.88%), respectively. The study highlights the importance of sustainable development in the Himalayan regions to decrease the susceptibility of mountain communities to landslides. It also underscores the value of being prepared to handle future events.

***Keywords*:** Sustainable development, Indian Himalayas, landslide susceptibility mapping, geographical information systems, frequency ratio method

**Corresponding author*: ankursharma.phdciv20@pec.edu.in

Disha Thakur, Sanjay Kumar, Har Amrit Singh Sandhu and Chander Prakash (eds.) Sustainable Development Using Geospatial Techniques, (129–154)

6.1 Introduction

The geological setting and the topography of the Indian Himalayas make them a global hotspot for landslides [12, 24]. Climate change and human interference with the environment have further exacerbated the risk of landslides in these mountainous regions with severe consequences for the exposed population, infrastructure, and the mountain economy [6, 26, 27]. Changing hydrological regimes and land use changes may have grave repercussions for water supply, agriculture, industry, hydroelectric power, carbon sequestration, biodiversity, etc. The negative impact of climate change in these mountainous regions is strikingly eminent because of low infrastructure redundancy leading to key infrastructure becoming increasingly vulnerable, for example, the disruption of the regional transportation network by a single slope failure can have a cascading effect on other critical infrastructure [34]. Hence, rapid unplanned urbanization in mountainous regions, particularly in areas prone to slope failures, can compound the risk of natural disasters [2, 20, 40]. The use of geospatial technologies presents a viable solution for the sustainable development of high-mountain regions to minimize the adverse impacts of such disasters while simultaneously enhancing the overall resilience of these regions.

An increasing awareness of the detrimental effects of landslides, coupled with the pressing need for urban development, has led to a significant increase in research on landslide-related topics, particularly landslide susceptibility zonation (LSZ) mapping which is a technique that categorizes the land surface into distinct stability zones based on the likelihood of the occurrence of landslides. LSZ maps are valuable instruments for planners and field engineers, as they provide crucial probabilistic information about the tendency of slope failure in the form of areal maps of landslide susceptibility (LS) [3, 4]. They can contribute to planning key infrastructure development; formulating hazard mitigation strategies and geo-environmental conservations; and adopting sustainable development strategies in landslide-prone terrains. The use of accurate LSZ maps can aid in circumventing the occurrence of landslides in regions where crucial developmental projects are at stake, consequently mitigating the level of exposure to such risks [13]. On the other hand, landslide-prone areas can be utilized for a variety of other purposes, such as preserving biodiversity, encouraging afforestation, promoting ecotourism, conserving soil, and enhancing ecosystem services.

Broadly, LSZ mapping approaches are classified into qualitative, quantitative, deterministic, and process-based approaches [9, 15, 18, 19].

Qualitative methods rely on expert judgment and knowledge of the area to evaluate various factors that contribute to LS. These factors may include slope steepness, geological characteristics, land cover, and historical landslide records. Qualitative methods provide a general understanding of the susceptibility level but lack quantitative measurements. The quantitative methods, on the other hand, involve the use of numerical data and statistical analyses to quantify LS [10, 23, 38]. These methods utilize geospatial data such as digital elevation models (DEMs), soil properties, rainfall patterns, and vegetation indices. By employing statistical techniques, such as logistic regression or artificial neural networks, quantitative methods provide a more objective and precise assessment of LS. Deterministic methods focus on identifying the causal relationships between landslide occurrences and the contributing factors. These methods use physical models and mathematical equations to simulate the behavior of landslides based on parameters such as slope stability, groundwater conditions, and material properties. Deterministic methods require detailed input data and are often used for localized analysis rather than large-scale assessments. Process-based methods integrate the understanding of physical processes and mechanisms that control landslides. These methods consider factors such as slope hydrology, soil mechanics, and erosion processes. Process-based models simulate the dynamic behavior of slopes, considering the interactions between various environmental factors. This approach provides insights into the complex processes driving landslides and makes predictions more accurate.

Each method has its strengths and limitations, and the choice of approach depends on the data availability, the scale of analysis, and the specific objectives of the investigation. Presently, we demonstrate the use of the frequency ratio (FR) method, a popular bivariate statistical technique that is employed in LSZ mapping studies, to create an accurate LSZ map of the selected region in the Indian Himalayas. This approach analyzes the impact of individual causal factors on landslide occurrences by dividing the digital maps of the causal factors into spatial grids and adjusting the weights assigned to each of the grids based on the observed landslide frequency in each grid. The individual weights of all the causal factors are assigned using their relative importance in landslide occurrence using the prediction rates. The LS of the study area is finally generated by multiplying the FR value of each bin of each causal factor by its corresponding weight and summing up the weighted FR values to obtain a susceptibility index for all the grid cells within the study region. Then, this map is classified into different LS zones to generate an LSZ map of the study area. The accuracy evaluation of the generated LSZ map is conducted with the help of the area under the receiver operating characteristic (ROC) value and the map is found suitable for its intended usage. In the following sections, we discuss the data and methodology employed in generating the LSZ map. Additionally, the accuracy assessment procedure and the application of the LSZ map for promoting sustainable development in the challenging topography of the Himalayan mountains are also discussed.

6.2 Study Area

The study area for the present investigation is situated in the lower Shivalik Hills of the Himalayas. Administratively, the study area is spread across two north Indian states of Haryana and Himachal Pradesh. It partially covers the Kalka tehsil of the Panchkula district in the state of Haryana and the Kasauli, Solan, Kandaghat, and Shimla tehsils of the Solan and Shimla districts in the state of Himachal Pradesh. Geographically, the study area is located between the latitudes 30°47′30″N to 31°09′00″N and the longitude 76°53′0″E to 77°13′30″E and covers an area of approximately 610 km^2 (Figure 6.1).

The elevation of the research region varies between a minimum value of 897 m above the mean sea level (MSL) to a maximum of 2204 m above the MSL. The average temperature of the region varies between a minimum of 9.8°C and a maximum of 25.9°C annually, while the average annual rainfall received is approximately 1251 mm. Geomorphologically, the area is composed of highly dissected hills with intermontane valleys. The proximity to the major Himalayan fault systems results in high seismic activity and frequent earthquakes in the region [12]. According to seismic zonation maps published by the Government of India, the study area lies in zone IV with an attributable peak ground acceleration of 0.24g. The major rock groups present in the study area include phyllite and quartzite of the Jutogh group, diamictite and shale of the Baliana group, shale of the Sirmaur/Dharamshala group, and sandstone and siltstone of the Dagshai and Kasauli group [35].

Some of the major urban centers that lie within the study area include Kalka, Parwanoo, Solan, and Shimla. The study area is of significant socio-economic value due to the presence of pharmaceutical and manufacturing industries, as well as eco-tourism point opportunities. The study area is currently witnessing numerous ongoing highway construction projects, with many more in the planning stages [33]. Unfortunately, this development is causing a continuous and detrimental impact on the natural bed

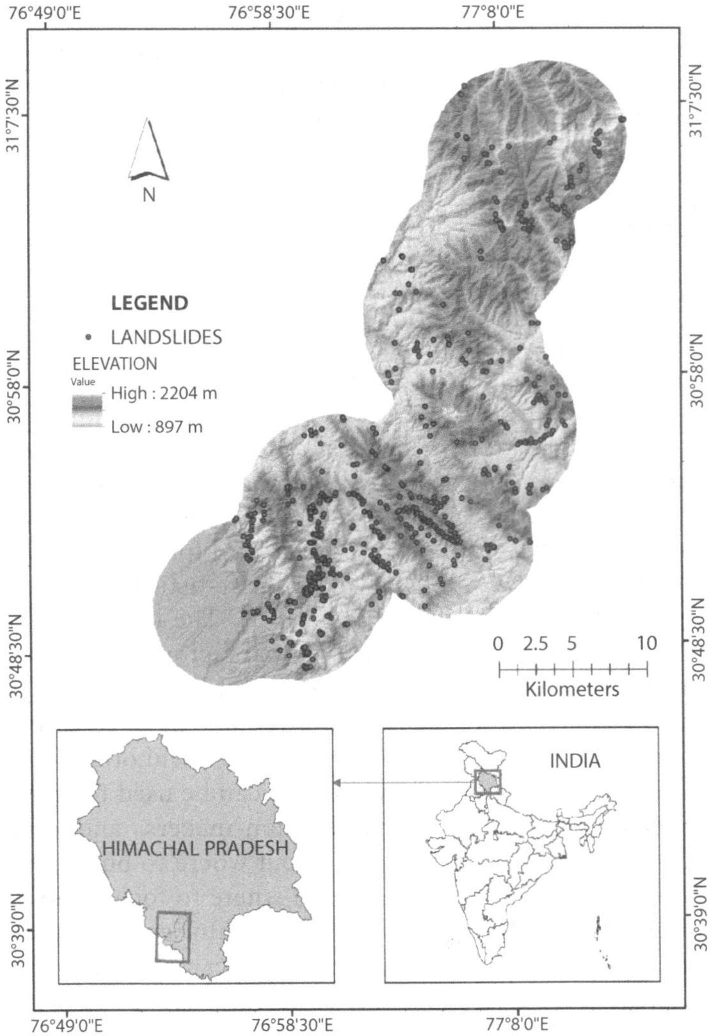

Figure 6.1 Geographical map of the study region showing the locations of landslide occurrences.

slope, leading to destabilization and an increased risk of slope failures. Moreover, the felling of trees along the highway alignment has further exacerbated the risk of landslide incidences. With increasing population growth and urban expansion in this hazard-prone hilly terrain, the impact of landslides has become a significant concern due to the increased exposure of both humans and the infrastructure. Thus, the selected study area serves as a suitable representative example of the unsustainable urban development problem and is an ideal candidate for demonstrating the efficacy of employing geographical information systems (GIS) to promote the sustainable development of mountainous regions.

6.3 Data

6.3.1 Landslide Inventory

It is a systematic and comprehensive collection of information related to landslides in a particular area. It involves the identification, mapping, and cataloging of all landslide events, including their location, type, size, frequency, and triggering factors. Landslide inventories are essential for understanding the spatial and temporal patterns of landslides, their causes, and their potential impacts on people, infrastructure, and the environment. These inventories are crucial for LS mapping (LSM) because they provide the necessary data for statistical analysis, spatial modeling, and validation of the LSMs [8, 9]. The landslide inventory data can be utilized to identify the principal factors that control landslide occurrences, such as geology, slope gradient, land use, and precipitation. By analyzing the landslide inventory data, researchers can develop empirical models that can predict the likelihood of landslides occurring in a given region. These models can be used to create LSMs that can help land-use planners, decision-makers, and emergency responders make informed decisions about where to build infrastructure, where to restrict development, and where to focus landslide risk reduction measures. Moreover, landslide inventories are important for early warning systems and monitoring. They can also be employed in the assessment of the effectiveness of landslide risk reduction measures and to track changes in landslide activity over time.

The landslide inventory for this investigation was compiled from multiple data sources, including the Geological Survey of India database, various government reports, and images from Google Earth [15]. Initially, the position of the landslides was mapped with the help of the GSI database and the government report as a reference. Then, additional landslide locations were also marked using high-resolution images from Google Earth to create an inventory database in GIS software. The final inventory used for LS assessment comprised a total of 592 landslide locations mapped as points (Figure 6.1).

6.3.2 Landslide Causal Factors

LSM is the process of predicting the likelihood of landslides occurring in a particular region based on the analysis of the parameters that impact slope stability. These factors can be broadly categorized into geological, topographical, hydrological, or environmental factors. Geological factors are the most significant contributors to LS. The presence of weak or poorly consolidated rocks, fractured bedrock, and geologic structures such as faults, joints, and bedding planes increase the potential for landslides. Geological factors also influence the type of landslide that may occur, including rockfall, rockslide, debris flow, and mudflow.

Topographical factors such as the slope gradient and the orientation of the slopes are crucial for LS evaluation as they influence landslide potential. Areas with steep slopes and concave upward landforms are more susceptible to landslides. Additionally, the slope gradient affects the rate of erosion and soil formation, which in turn affects the likelihood of landslides. Hydrological factors also contribute to LS including the presence of water, seepage, changes in groundwater levels, and rainfall intensity and duration. Areas with high rainfall intensity and duration, steep topography, and insufficient drainage systems are more susceptible to landslides. Conversely, areas with well-developed drainage systems that can remove excess water are less prone to landslides. Environmental factors, such as deforestation and land-use changes, can also influence LS. Deforestation reduces the ability of soil to absorb rainfall, leading to soil erosion and landslides. Land-use changes, such as urbanization, can increase impermeable surfaces, leading to increased runoff and erosion.

The choice of appropriate landslide causal factors is a critical component for the analysis of LS [36]. However, there are currently no established guidelines for their selection [13]. Typically, the selection of these factors for an investigation is based on data availability for the study region. In our analysis, we identified nine causal factors (Figure 6.2), including elevation, slope, aspect, curvature, terrain ruggedness index (TRI), distance to roads, distance to drainage, distance to faults, and land use/land cover (LULC), based on a review of the literature [13, 14, 32]. Each of these factors is briefly described below to provide clarity and context for the present study:

6.3.2.1 Elevation

The elevation is an important parameter for LS analysis and helps determine the occurrence of landslides [29]. Depending on the region, higher elevations typically have steeper slopes and can experience higher rainfall due to the orographic effects, which can increase the potential for landslides. Additionally, the types of rocks and soils found at higher elevations are often less stable and more prone to failure. The presence of steep slopes at higher elevations can increase the likelihood of landslides occurring. The angle of the slope affects the amount of stress and pressure on the soil, rocks, and other materials present on the slope. Steeper slopes are more susceptible to landslides as they have a higher destabilizing force acting on them. Furthermore, geological structures and soil characteristics at higher elevations can increase the LS. In mountainous regions, for example, rock formations may be more prone to fracturing and erosion due to the weathering and erosion processes occurring at higher elevations. The soils formed from these rocks may also be less stable and more prone to failure.

6.3.2.2 Slope

The slope gradient is one of the geomorphologic characteristics that are crucial for an LS analysis since it plays a vital role in gravitational motions. Because it regulates the shear pressures operating on hill slopes, the slope gradient is commonly regarded as one of the most crucial components in landslide modeling [25]. The steepness of a slope directly influences the stability of the ground surface and the potential for landslides to occur. The slope gradient is measured as the degree of inclination of the slope from the horizontal datum, and the greater its value, the higher the risk of slope failure. Other factors such as the nature of the soil and the presence of water can also contribute to the LS of slopes to landslides. For example, slopes composed of loose, poorly compacted soil are more prone to failure than those with well-compacted soils. Additionally, slopes that are saturated with water due to rainfall or water runoff can also become more susceptible to landslides. When assessing the LS, it is essential to consider the slope angle, soil type, and water saturation levels to determine the likelihood of landslide occurrence. By understanding the relationship between slope and other factors, appropriate measures can be put in place to minimize the risk of landslides in areas with high LS.

6.3.2.3 Aspect

Aspect is the spatial orientation of the slope faces and is an important parameter that influences landslide occurrence and LS in the Himalayas. The orientation of the slopes can significantly impact landslide initiation due to its influence on plant cover, moisture retention in the soil, and

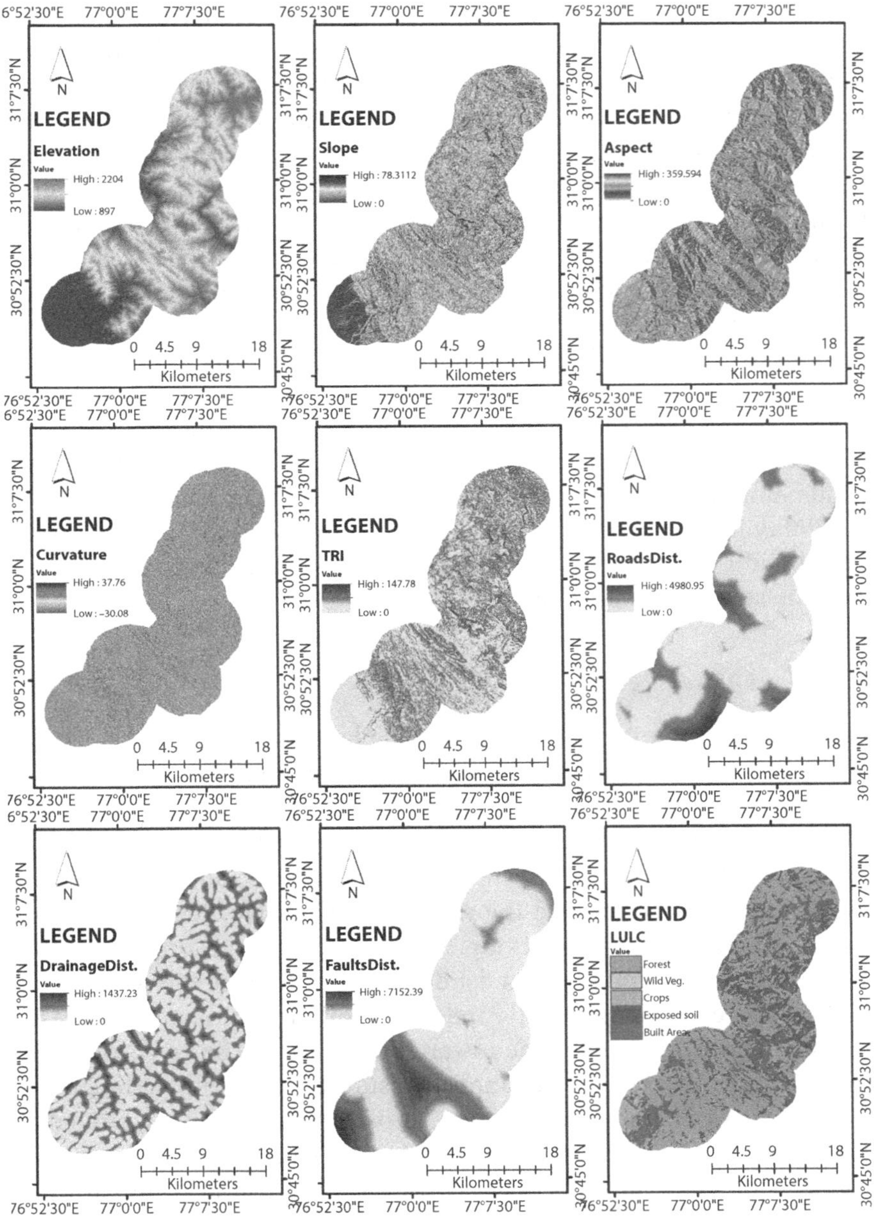

Figure 6.2 Digital maps of the landslide causal factors.

soil strength [7]. The Himalayan region exhibits complex topography, with steep slopes and high elevations, making it increasingly susceptible to landslides. The aspect of a slope can affect the amount of sunlight and water that it receives, which in turn can affect the stability of the ground surface. Slopes that face south and southwest are exposed to more direct sunlight and can experience more rapid thawing of ice and snow, leading to increased soil moisture and the potential for landslides. Additionally, slopes that face north and northeast receive less sunlight and may remain frozen for longer periods, which can also contribute to increased LS. The impact of the slope aspect on landslides in the Himalayas is evident from the prevalence of landslides on south and southwest-facing slopes during the summer monsoon season. During this time, heavy rainfall increases soil moisture, making these slopes more prone to failure. In contrast, landslides on the north and northeast-facing slopes tend to occur during the winter months when the slopes remain frozen for longer periods, leading to instability in the ground surface. Understanding the relationship between aspect and LS is crucial for developing effective measures to prevent and mitigate landslides in the Himalayan region.

6.3.2.4 Curvature

Curvature is a significant factor affecting landslide occurrence and LS. It is a measure of the changes in slope and direction of a terrain surface and can be classified as either concave (valley-like) or convex (ridge-like) curvature. The direction and degree of curvature can influence the distribution of surface water, as well as the erosion and weathering processes that occur on the slope [40]. Concave slopes, which are characterized by inward curving surfaces, tend to accumulate water, increasing the potential for landslides. This is because the water accumulates in depressions, which can lead to soil saturation and decreased soil stability. Additionally, the inward curvature of concave slopes can contribute to increased stress and pressure on the soil, leading to an increased risk of landslides. Convex slopes, which are characterized by outward curving surfaces, tend to drain water more efficiently, reducing the potential for landslides. However, they can also be susceptible to landslides due to the presence of overlying rock formations or soil layers that are prone to failure. The impact of curvature on LS can vary depending on the type of soil, vegetation cover, and other factors. In some cases, concave slopes may be more stable than convex slopes, depending on the local conditions.

6.3.2.5 Terrain Ruggedness Index (TRI)

The TRI represents the vertical variability of the terrain surface, and it can influence landslides. The TRI is calculated as the differences in elevation between adjacent cells in a DEM and indicates the roughness of the terrain surface. A higher value of TRI indicates more complex terrain with steeper slopes, which can increase the risk of landslides. Complex terrains can also lead to variations in soil type, vegetation cover, and drainage patterns, which can further contribute to the LS. Areas with higher TRI values generally exhibit higher LS due to the increased instability of the ground surface. The relationship between TRI and LS can vary depending on the local conditions. For example, in areas with dense vegetation cover or well-compacted soils, high TRI values may not necessarily correspond to high LS. Additionally, the influence of TRI on LS may be modified by other factors, such as the presence of water, seismic activity, or human activities. When assessing LS, it is important to consider the TRI along with other parameters including elevation, slope angle, and curvature. By understanding the relationship between terrain ruggedness and other factors, appropriate measures can be put in place to minimize the risk of landslides in areas with high LS.

6.3.2.6 Distance to Roads

Roads can act as important contributing factors to triggering or exacerbating landslides. First, the construction of roads often involves the removal of vegetation and the alteration of natural slopes, which can weaken the stability of the surrounding terrain. This disturbance can increase the susceptibility of the area to landslides. Additionally, roads can act as conduits for water, channeling rainfall, and groundwater directly into the slopes, leading to increased saturation and higher pore water pressure, which are key triggers for landslides. The presence of roads can also enhance human activity in landslide-prone areas, increasing the likelihood of slope modifications, improper drainage, and greater destabilization due to vehicular loading, all of which can exacerbate slope instability. Therefore, when assessing the LS of a region, the distance to roads and their associated impacts should be carefully considered to better understand and mitigate the potential risks posed by landslides.

6.3.2.7 Distance to Drainage

Drainage networks, such as rivers, streams, and channels, can erode and transport soil and rock materials, increasing the potential for landslides in adjacent areas [29]. The distance to drainage is often used as a proxy for the potential impact of water on slope stability. Slopes that are closer to drainage networks are more likely to be influenced by water, which can increase the risk of landslides. The presence of water can lead to soil saturation and decreased soil stability, which can increase the likelihood of landslides. Additionally, water can increase the weight and pressure on the slope, which can further contribute to instability. The relationship between distance to drainage and LS can vary depending on the local conditions. For example, in areas with well-drained soils or vegetation cover, the distance to drainage may not be a significant factor in LS. However, in areas with poorly drained soils or areas with high levels of precipitation, the distance to drainage can be a critical factor in LS. By understanding the relationship between distance to drainage and other factors, appropriate measures can be put in place to minimize the risk of landslides in areas with high LS.

6.3.2.8 *Distance to Faults*

The distance to faults is an important seismological parameter for LS assessment as it can directly influence the stability of the terrain. Faults are geological features where the Earth's crust has been fractured, and they can create zones of weakness in the surrounding rock and soil. When assessing LS, the distance to faults becomes crucial because the presence of faults can influence the likelihood of ground movement and trigger slope failures. If a landslide-prone area is located close to an active fault, the potential for ground shaking, ground displacement, and increased stress on the slopes significantly rises, making the area more susceptible to landslides. Therefore, in LS assessments, considering the distance to faults is essential for accurately identifying and evaluating landslide hazards in a given area.

6.3.2.9 *Land Use/Land Cover (LULC)*

LULC represents the way land is used and the type of vegetation or covers on the land. Changes in LULC can significantly influence slope stability and the potential for landslides. Generally, land use practices that disturb the natural vegetation cover or alter the slope gradient can increase the LS of an area to landslides [5, 28]. For example, deforestation, mining activities, and construction can all lead to changes in surface runoff, soil moisture content, and soil stability, which can increase the likelihood of landslides. On the other hand, the presence of natural vegetation cover can help to stabilize slopes and reduce the risk of landslides [21]. The roots of trees and other vegetation help bind soil particles together, increasing soil stability and reducing erosion. Vegetation cover also helps to regulate water infiltration, reducing the potential for saturation and soil liquefaction. The relationship between LULC and LS can vary depending on the local conditions. For example, in areas with steep slopes and dense vegetation cover, the LULC may not be a significant factor in LS. However, in regions where natural vegetation has been removed or the slope gradient has been altered, the LULC can be a critical factor in LS.

6.4 Methodology

Once the data on the landslide causal factors are collected as digital maps, all the maps are co-registered, preprocessed to remove any distortions, and re-projected to the same coordinate reference system before performing the FR analysis. The operation yields an LSM that shows the likelihood or probability of landslides occurring in different portions of the selected region. The LSM can be presented as a color-coded map, with different colors representing different levels of LS, or as a binary map, where regions with higher likelihood of landslides are classified as susceptible, and regions with lower likelihood of landslides are classified as non-susceptible. The preparation of an LSM requires careful consideration of the data, methods, and techniques used in the analysis, as well as validation and verification of the results. The following sections provide a concise overview of the approach employed in this study to create the LSZ map of the chosen area.

6.4.1 Frequency Ratio Method

The FR method is a popular statistical method commonly used for LS assessment that assumes a bivariate association between the landslide locations and the causal factors. It assumes that past landslide locations are a good indicator of future landslide occurrences and evaluates the correlation between the spatial distribution of landslide occurrences and the distribution of the causal factors. The FR method involves dividing the study area into spatial grids and calculating the landslide frequency in each of the grids. The landslide frequency is then compared with the frequency of each causal factor in the same grid to determine the correlation between the two. To calculate the FR, the number of landslides in a particular grid interval is divided by the number of all the grids in that interval that contain the causal factor of interest [1, 31]. This ratio is then compared with the overall landslide frequency in the study region to determine the relative significance of the causal factor in landslide occurrence (Equation 6.1).

$$FR = \frac{l_i / \Sigma(L_i)}{s_i / \Sigma(S_i)} \quad (6.1)$$

where l_i is the number of pixels affected by landslide in a certain interval, $\Sigma\ (L_i)$ is the number of landslide pixels in the whole study region, s_i represents the number of pixels associated with the *i*th causal factor in the corresponding interval, and $\Sigma\ (S_i)$ represents the total number of pixels associated with the *i*th causal factor within the study region.

The FR method is useful for LS analysis because it provides a simple and objective approach to evaluating the correlation between landslide occurrences and causal factors. It is a relatively straightforward method and can be easily applied using GIS software, which allows for easy integration of various datasets and simplifies the analysis process. However, the FR method has some limitations. The assumption that past landslide events are a good predictor of future events may not always be the case. Additionally, the FR method does not account for the interaction between different causal factors, and it may also overlook important factors that are not well represented in the available datasets. However, despite these limitations, the FR method remains a popular and widely used approach for LS analysis and provides valuable insights for landslide risk management [16, 22, 30, 37].

6.4.2 Weighted Linear Overlay

The weighted linear overlay (WLO) method is another commonly used technique in LS analysis that involves combining multiple thematic maps to generate an LSM. It involves several steps, including selecting the thematic maps, assigning a weight to them, and combining the various maps to generate a composite map showing the LS index for the study area. The method achieves this by using a mathematical formula (Equation 6.2) that considers the relative significance of all the causal factors when assigning the weights to generate an LSM:

$$LS\ index = \sum_{j=1}^{n} w_{ij} \times W_j \quad (6.2)$$

where w_{ij} is the weight assigned to the *i*th grid of the *j*th causal factor and W_j is the overall weight assigned to the *j*th causal factor.

The composite LS index map is then reclassified into different zones of LS based on the LS index values to generate the final LSZ map. The

WLO method is a flexible and widely used approach for LS analysis because enables the integration of multiple causal factors and datasets. This approach can incorporate both qualitative and quantitative data, making it suitable for an array of applications. However, the WLO method also has some limitations. It assumes that the correlation between various causal factors and the landslide occurrences is linear and additive, which may not always be the case. Additionally, the method can be sensitive to the selection of weights, which can be subjective and difficult to determine. Despite these limitations, the WLO method remains a popular and widely used approach for LS analysis due to its flexibility and ability to incorporate multiple datasets and causal factors.

6.5 Results and Discussion

The present study presents a simple geospatial framework for the sustainable development of the Indian Himalayas. We achieve this objective by generating an accurate LSZ map of a selected area within this region for formulating hazard mitigation strategies, planning sustainable infrastructure and urban development, and adopting sustainable development practices. The LSZ map of the research region is generated using a statistical-based method employing the use of the FR method of LS assessment. The following sections provide a concise discussion of the LS analysis and validation of the generated LSZ map.

6.5.1 FR Analysis and Relative Importance of Causal Factors

The FR method was used to derive the weights assigned to the spatial grids of each of the landslide causal factors by calculating the FR values. These values were calculated by dividing the fraction of landslide pixels within a grid of each of the landslide causal factors with the fraction of pixels occupied by the landslide causal factor in the same grid. Then, the relative frequencies (RF) for all the grids within a landslide causal factor were determined by dividing all the FR values with the smallest FR value for the particular landslide causal factor. Then, the weights for each of the spatial grids were determined by keeping the integer part of the corresponding RF values and ignoring the values after the decimal. Finally, the weights of all the causal factors themselves are assigned using their prediction rates. Table 6.1 presents the FR calculations for all the spatial grids within the study region.

It is seen that among all the landslide causal factors selected for this study, LULC receives the highest relative importance as visible from the prediction rate value. Particularly, the higher FR values observed for the "forest cover", "exposed rock/soil", and "wild vegetation" classes of LULC highlight the impact of anthropogenic interferences. These observations conform with the observations from the previous studies that show that anthropogenic interferences result in deforestation and urban expansion can increase the occurrence of landslide incidences [26, 33]. Observing the distribution of the FR values for the distance to roads parameter, which shows the highest FR value for the grids closest to the road network, further validates the conclusion that anthropogenic activities lead to increased incidences of landslide occurrences [11, 39].

It is also observed that the factors, namely, slope and TRI, receive some of the largest weights among all the landslide causal factors after LULC, indicating the importance of the topographical variables in determining the degree of slope instability. The trend analysis of their FR values also reveals that these values increase with the Slope and TRI values at a location. Similar results have been obtained by other studies and highlight the inherent instabilities of the mountainous topographies [24, 40]. Conversely, the parameters like the distance to drainage, curvature, and aspect received the lowest weights, as a weaker relationship was observed between these factors and landslide occurrences. Such observations have also been reported in other studies and validate the relative significance of the causal factors derived in this study [13].

Table 6.1 FR table showing the weights assigned to the various classes of the landslide causal factors.

Causal factor	Factor class	Fraction of study area pixels in class	Fraction of landslide pixels in class	FR	RF	Class weight	Prediction rate
Elevation	437–778 m	0.10	0.03	0.28	3.86	3	2.31
	778–1088 m	0.09	0.17	1.97	27.36	27	
	1088–1290 m	0.19	0.21	1.08	15.09	15	
	1290–1469 m	0.24	0.22	0.91	12.73	12	
	1469–1663 m	0.19	0.17	0.89	12.32	12	
	1663–1888 m	0.14	0.16	1.11	15.50	15	
	1888–2416 m	0.06	0.06	0.94	13.13	13	

(*Continued*)

Table 6.1 FR table showing the weights assigned to the various classes of the landslide causal factors. (*Continued*)

Causal factor	Factor class	Fraction of study area pixels in class	Fraction of landslide pixels in class	FR	RF	Class weight	Prediction rate
Slope	0°–11°	0.14	0.01	0.08	1.44	1	4.49
	11°–20°	0.25	0.12	0.47	8.15	8	
	20°–28°	0.30	0.32	1.04	17.87	17	
	28°–37°	0.23	0.34	1.48	25.47	25	
	37°–78°	0.08	0.22	2.74	47.07	47	
Aspect	0°–45°	0.13	0.08	0.64	8.11	8	1.02
	45°–90°	0.11	0.11	1.01	12.79	12	
	90°–135°	0.10	0.12	1.22	15.39	15	
	135°–180°	0.12	0.15	1.23	15.50	15	
	180°–225°	0.16	0.23	1.46	18.42	18	
	225°–270°	0.15	0.14	0.92	11.57	11	
	270°–315°	0.13	0.09	0.68	8.56	8	
	315°–360°	0.10	0.08	0.76	9.65	9	
Curvature	Negative	0.41	0.37	0.91	31.07	31	1.06
	Flat	0.19	0.16	0.85	29.06	29	
	Positive	0.41	0.47	1.16	39.87	39	

(*Continued*)

Table 6.1 FR table showing the weights assigned to the various classes of the landslide causal factors. (*Continued*)

Causal factor	Factor class	Fraction of study area pixels in class	Fraction of landslide pixels in class	FR	RF	Class weight	Prediction rate
TRI	0–6 m	0.16	0.01	0.09	0.41	0	5.21
	6–12 m	0.29	0.16	0.56	2.67	2	
	12–17 m	0.29	0.33	1.15	5.54	5	
	17–23 m	0.18	0.27	1.52	7.30	7	
	23–31 m	0.07	0.15	2.21	10.63	10	
	31–48 m	0.01	0.06	4.19	20.14	20	
	48–147 m	0.00	0.02	11.10	53.31	53	
Distance to roads	0–27 3 m	0.29	0.60	2.08	24.86	26	2.27
	273–644 m	0.22	0.08	0.36	4.70	4	
	644–1035 m	0.16	0.10	0.62	8.01	8	
	1035–1464 m	0.12	0.03	0.29	3.80	3	
	1464–1933 m	0.08	0.04	0.52	6.73	6	
	1933–2441 m	0.06	0.08	1.49	19.20	19	
	2441–3027 m	0.04	0.03	0.74	9.62	9	
	3027–3730 m	0.03	0.03	1.01	13.05	13	
	3730–4980 m	0.01	0.01	0.62	8.06	8	
Distance to drainage	0–90 m	0.17	0.08	0.49	5.19	5	1
	90–182 m	0.16	0.20	1.27	13.47	13	
	182–278 m	0.15	0.16	1.10	11.61	11	
	278–377 m	0.14	0.08	0.61	6.51	6	
	377–481 m	0.12	0.12	0.99	10.45	10	
	481–593 m	0.11	0.16	1.45	15.35	15	

(*Continued*)

Table 6.1 FR table showing the weights assigned to the various classes of the landslide causal factors. (*Continued*)

Causal factor	Factor class	Fraction of study area pixels in class	Fraction of landslide pixels in class	FR	RF	Class weight	Prediction rate
	593–718 m	0.08	0.12	1.40	14.85	14	
	718–878 m	0.06	0.05	0.94	9.97	9	
	878–1437 m	0.02	0.02	1.19	12.59	12	
Distance to faults	0–474 m	0.26	0.26	1.03	10.93	10	2.51
	474–1061 m	0.16	0.16	0.73	7.81	7	
	1061–1676 m	0.21	0.21	1.35	14.32	14	
	1676–2346 m	0.10	0.10	0.90	9.58	9	
	2346–3045 m	0.04	0.04	0.54	5.78	5	
	3045–3743 m	0.07	0.07	0.97	10.31	10	
	3743–4498 m	0.06	0.06	1.04	11.07	11	
	4498–5476 m	0.09	0.09	2.62	27.86	27	
	5476–7152 m	0.00	0.00	0.22	2.34	2	
LULC	Water bodies	0.00	0.00	0.00	0.00	0	5.28
	Forest cover	0.55	0.58	1.05	15.58	15	
	Wild vegetation	0.00	0.00	3.60	53.59	53	
	Crop fields	0.02	0.00	0.12	1.77	1	
	Exposed soil/ rock	0.22	0.23	1.04	15.51	15	
	Built area	0.20	0.18	0.91	13.54	13	
	Barren land	0.00	0.000	0.00	0.00	0	

6.5.2 LSZ Mapping and Its Validation

The evaluation of the generated LSZ map is done using the ROC curve (Figure 6.3) by utilizing the area under the curve value. According to the literature, an area under the ROC curve value greater than 70% is considered satisfactory for LSZ mapping, while higher values are even more desirable [17]. This value in the present case is determined to be 75.1% which validates the quality of the generated LSZ map. Hence, the LSZ map can be satisfactorily used for its intended purpose of sustainable urban planning in the mountainous terrain of the Indian Himalayas and may aid efficient disaster mitigation efforts by the concerned authorities.

The LSZ map (Figure 6.4) of the selected region is generated by dividing it into different classes of LS using Jenk's clustering algorithm, which minimizes the intra-class variance while maximizing the inter-class variance by identifying natural gaps within the data. The generated LSZ map divides the study area into five classes of LS including very high (8.13%), high (18.95%), medium (28.69%), low (26.06%), and very low (18.18%), respectively. The map effectively identifies and delineates areas within the study region that are more prone to landslides than the others. By categorizing the study area into five distinct LS zones, this map serves as a valuable resource for planners and decision-makers in assessing the terrain conditions and making informed choices for future development. By avoiding areas with a heightened risk of landslides, planners can minimize potential damages and ensure the safety of infrastructure and human lives [8].

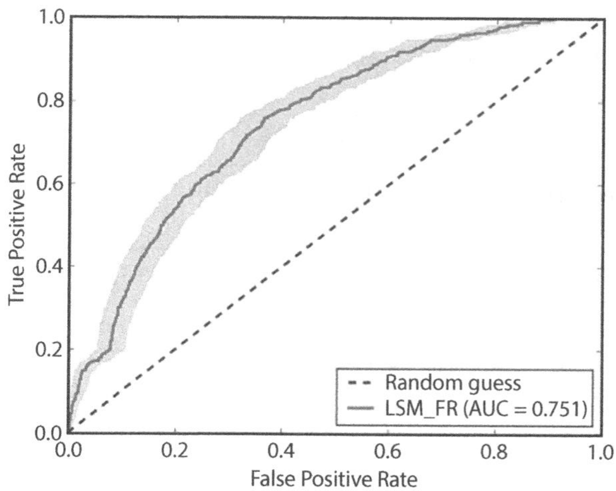

Figure 6.3 ROC curve for the generated LSZ map used for validation.

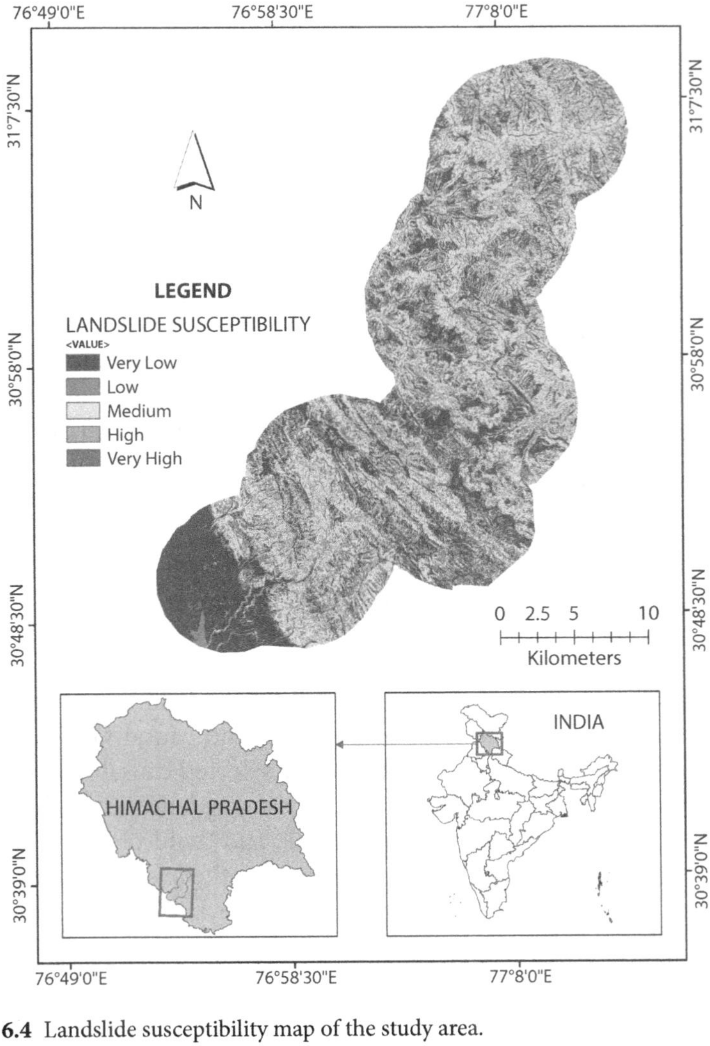

Figure 6.4 Landslide susceptibility map of the study area.

Interestingly, the identification of areas marked as unfavorable for future expansion presents opportunities for alternative land uses that align with ecological conservation and sustainable practices. These areas can be allocated for purposes such as biodiversity conservation, afforestation, soil conservation, and the enhancement of ecosystem services. Integrating intelligent land use planning with practices that support the ecosystem contributes to sustainable development in the challenging Himalayan terrain while simultaneously reducing the potential losses associated with landslide hazards. By leveraging this synergy, decision-makers can strike a balance between developmental aspirations and environmental preservation, ensuring long-term resilience and minimizing the socioeconomic impacts of landslides.

6.6 Conclusions

The surge in population has caused an urgent need for the urbanization of the Himalayas. Due to human interferences in the Anthropocene era and the increasing climate variability, the risk of natural disasters such as landslides has also increased. The rapid unplanned urbanization of the Himalayas has led to increased exposure of human life and infrastructure to adverse impacts of haphazard mountain development. LSZ mapping provides a convenient solution to this problem allowing for sustainable and planned development of important infrastructure while alleviating the risk of landslides.

The present study uses the FR method, which is a popular bivariate statistical technique, to generate an accurate regional scale LSZ map of a selected area in the Himalayas. The method follows a quantitative and statistical approach, which helps minimize subjectivity and bias in LS assessment. It provides an objective basis for decision-making by establishing a relationship between landslide occurrences and causal factors through FR values. The results reveal that among the selected causal factors, LULC receives the highest relative importance in causing landslides. The FR values of the "forest cover", "exposed rock/soil", and "wild vegetation" classes of LULC suggest anthropogenic interferences as the leading cause of landslide occurrences in the selected region. Moreover, the high prediction rate values for TRI and Slope parameters also highlight the susceptibility of the Himalayan topography to slope failures.

The accuracy of the generated LSZ map is assessed using the area under the ROC curve metric and it is found that the generated map exhibits satisfactory accuracy and provides a precise spatial distribution of LS within the study region. Furthermore, the outcome of the study shows that this approach can be easily utilized by planners and disaster management authorities for developing sustainable smart cities, critical infrastructure development, planning mitigation activities, geo-environmental conservations, and adopting sustainable development practices in the difficult Himalayan topography.

References

1. A. Aditian, T. Kubota, Y. Shinohara, Comparison of GIS-based landslide susceptibility models using frequency ratio, logistic regression, and artificial neural network in a tertiary region of Ambon, Indonesia, *Geomorphology.* 318 (2018) 101–111.
2. A. Arabameri, B. Pradhan, K. Rezaei, S. Lee, M. Sohrabi, An ensemble model for landslide susceptibility mapping in a forested area, *Geocarto Int.* 0 (2019) 000.
3. A. Arabameri, B. Pradhan, K. Rezaei, M. Sohrabi, Z. Kalantari, GIS-based landslide susceptibility mapping using numerical risk factor bivariate model and its ensemble with linear multivariate regression and boosted regression tree algorithms, *J. Mt. Sci.* 16 (2019) 595–618.
4. E. Arnone, A. Francipane, A. Scarbaci, C. Puglisi, L. v. Noto, Effect of raster resolution and polygon-conversion algorithm on landslide susceptibility mapping, *Environ. Modell. Software.* 84 (2016) 467–481.
5. L.P.H. Van Beek, T.W.J. Van Asch, Regional assessment of the effects of land-use change on landslide hazard by means of physically based modelling, *Nat. Hazard.* 31 (2004) 289–304.
6. S. Kumar, T. Kaur, M. K. Arora and S. Upadhyay. "Resource estimation and sizing optimization of PV/micro hydro-based hybrid energy system in rural area of Western Himalayan Himachal Pradesh in India." *Energy Sources Part A* 41 (2019) 2795-2807.
7. M. Capitani, A. Ribolini, M. Bini, The slope aspect: A predisposing factor for landsliding, *C.R. - Geosci.* 345 (2013) 427–438.
8. A. Chawla, S. Pasupuleti, S. Chawla, A.C.S. Rao, K. Sarkar, R. Dwivedi, Landslide Susceptibility Zonation Mapping: A Case Study from Darjeeling District, Eastern Himalayas, India, *J. Indian Soc. Remote Sens.* 47 (2019) 497–511.
9. Kumar S, Sandhu HA, Awasthi U, Singh SK, Agarwal P. Analysis of COVID-19 outbreak using GIS and SEIR model. In *Fractional Order Systems and Applications in Engineering* (2023) (pp. 215-225). Academic Press.
10. W. Chen, X. Xie, J. Wang, B. Pradhan, H. Hong, D.T. Bui, Z. Duan, J. Ma, A comparative study of logistic model tree, random forest, and classification and regression tree models for spatial prediction of landslide susceptibility, *Catena (Amst).* 151 (2017) 147–160.
11. R.K. Dahal, S. Hasegawa, T. Masuda, M. Yamanaka, Roadside Slope Failures in Nepal during Torrential Rainfall and their Mitigation Road construction practice in Nepal, *Proc. Interpaevent Int. Symp, Niijigata 2006, Disaster Mitigation of Debris Flow, Slope Failures and Landslides.* (2006) 503–514.
12. A. Dikshit, R. Sarkar, B. Pradhan, S. Segoni, A.M. Alamri, Rainfall induced landslide studies in Indian Himalayan region: A critical review, *Appl. Sci. (Switzerland).* 10(7) (2020).

13. J. Dou, A.P. Yunus, D. Tien Bui, A. Merghadi, M. Sahana, Z. Zhu, C.W. Chen, K. Khosravi, Y. Yang, B.T. Pham, Assessment of advanced random forest and decision tree algorithms for modeling rainfall-induced landslide susceptibility in the Izu-Oshima Volcanic Island, Japan, *Sci. Total Environ.* 662 (2019) 332–346.
14. A. Erener, A. Mutlu, H. Sebnem Düzgün, A comparative study for landslide susceptibility mapping using GIS-based multi-criteria decision analysis (MCDA), logistic regression (LR) and association rule mining (ARM), *Eng. Geol.* 203 (2016) 45–55.
15. S.K. Gupta, D.P. Shukla, Effect of scale and mapping unit on landslide susceptibility mapping of Mandakini River Basin, Uttarakhand, India, *Environ. Earth Sci.* 81(14) (2022).
16. S. Hidayat, H. Pachri, I. Alimuddin, Analysis of Landslide Susceptibility Zone using Frequency Ratio and Logistic Regression Method in Hambalang, Citeureup District, Bogor Regency, West Java Province, *IOP Conf. Ser. Earth Environ. Sci.* 280 (2019).
17. D.W. Hosmer, S. Lemeshow, R.X. Sturdivant, Assessing the Fit of the Model, in: *Applied Logistic Regression*, 3rd ed., Wiley & Sons, Inc., 2013: pp. 153–225.
18. F. Huang, Z. Cao, J. Guo, S.H. Jiang, S. Li, Z. Guo, Comparisons of heuristic, general statistical and machine learning models for landslide susceptibility prediction and mapping, *Catena (Amst).* 191 (2020).
19. F. Huang, K. Yin, J. Huang, L. Gui, P. Wang, Landslide susceptibility mapping based on self-organizing-map network and extreme learning machine, *Eng. Geol.* 223 (2017) 11–22.
20. M. Jaboyedoff, C. Michoud, M.H. Derron, J. Voumard, G. Leibundgut, K. Sudmeier-Rieux, C. Michoud, F. Nadim, E. Leroi, Human-Induced Landslides: Toward the analysis of anthropogenic changes of the slope environment, in: S. Vresa, L. Cascini, L. Picarelli, C. Scavia (Eds.), *Landslides and Engineered Slopes. Experience, Theory and Practice*, CRC Press, 2016: pp. 217–232.
21. B. Kalsnes, V. Capobianco, Use of Vegetation for Landslide Risk Mitigation, in: *Climate Adaptation Modelling*. Springer Climate, Springer Cham., 2022: pp. 77–85.
22. H. Khan, M. Shafique, M.A. Khan, M.A. Bacha, S.U. Shah, C. Calligaris, Landslide susceptibility assessment using Frequency Ratio, a case study of northern Pakistan, *Egypt. J. Remote Sens. Space. Sci.* 22 (2019) 11–24.
23. J.C. Kim, S. Lee, H.S. Jung, S. Lee, Landslide susceptibility mapping using random forest and boosted tree models in Pyeong-Chang, Korea, *Geocarto Int.* 33 (2018) 1000–1015.
24. D. Kumar, M. Thakur, C.S. Dubey, D.P. Shukla, Landslide susceptibility mapping & prediction using Support Vector Machine for Mandakini River Basin, Garhwal Himalaya, India, *Geomorphology.* 295 (2017) 115–125.

25. P. Lima, S. Steger, T. Glade, Counteracting flawed landslide data in statistically based landslide susceptibility modelling for very large areas: a national-scale assessment for Austria, *Landslides.* 18 (2021) 3531–3546.
26. M.K. Pandit, N.S. Sodhi, L.P. Koh, A. Bhaskar, B.W. Brook, Unreported yet massive deforestation driving loss of endemic biodiversity in Indian Himalaya, *Biodivers. Conserv.* 16 (2007) 153–163.
27. B.T. Pham, D. Tien Bui, I. Prakash, M.B. Dholakia, Hybrid integration of Multilayer Perceptron Neural Networks and machine learning ensembles for landslide susceptibility assessment at Himalayan area (India) using GIS, *Catena (Amst).* 149 (2017) 52–63.
28. L. Pisano, V. Zumpano, Malek, C.M. Rosskopf, M. Parise, Variations in the susceptibility to landslides, as a consequence of land cover changes: A look to the past, and another towards the future, *Sci. Total Environ.* 601–602 (2017) 1147–1159.
29. H.R. Pourghasemi, O. Rahmati, Prediction of the landslide susceptibility: Which algorithm, which precision?, *Catena (Amst).* 162 (2018) 177–192.
30. B. Pradhan, S. Lee, Landslide susceptibility assessment and factor effect analysis: backpropagation artificial neural networks and their comparison with frequency ratio and bivariate logistic regression modelling, *Environ. Modell. Software.* 25 (2010) 747–759.
31. B. Pradhan, S. Lee, Delineation of landslide hazard areas on Penang Island, Malaysia, by using frequency ratio, logistic regression, and artificial neural network models, *Environ. Earth Sci.* 60 (2010) 1037–1054.
32. Y.W. Rabby, A. Ishtiaque, M.S. Rahman, Evaluating the effects of digital elevation models in landslide susceptibility mapping in rangamati district, Bangladesh, *Remote Sens. (Basel).* 12(17) (2020).
33. Sangeeta, S.K. Singh, Influence of anthropogenic activities on landslide susceptibility: A case study in Solan district, Himachal Pradesh, India, *J. Mt. Sci.* 20 (2023) 429–447.
34. M. Schlögl, G. Richter, M. Avian, T. Thaler, G. Heiss, G. Lenz, S. Fuchs, On the nexus between landslide susceptibility and transport infrastructure - An agent-based approach, *Nat. Hazards Earth Syst. Sci.* 19 (2019) 201–219.
35. M. Sharma, R. Kumar, GIS-based landslide hazard zonation: A case study from the Parwanoo area, Lesser and Outer Himalaya, H.P., India, *Bull. Eng. Geol. Environ.* 67 (2008) 129–137.
36. A. Shirzadi, K. Solaimani, M.H. Roshan, A. Kavian, K. Chapi, H. Shahabi, S. Keesstra, B. Bin Ahmad, D.T. Bui, Uncertainties of prediction accuracy in shallow landslide modeling: Sample size and raster resolution, *Catena (Amst).* 178 (2019) 172–188.
37. F.E.S. Silalahi, Pamela, Y. Arifianti, F. Hidayat, Landslide susceptibility assessment using frequency ratio model in Bogor, West Java, Indonesia, *Geosci. Lett.* 6(1) (2019).

38. S. Wang, J. Zhuang, J. Zheng, H. Fan, J. Kong, J. Zhan, Application of Bayesian Hyperparameter Optimized Random Forest and XGBoost Model for Landslide Susceptibility Mapping, *Front. Earth Sci. (Lausanne).* 9 (2021).
39. M.G. Winter, B. Shearer, D. Palmer, D. Peeling, C. Harmer, J. Sharpe, The Economic Impact of Landslides and Floods on the Road Network, in: *Procedia Eng.*, Elsevier Ltd, 2016: pp. 1425–1434.
40. A.P. Yunus, X. Fan, S.S. Subramanian, D. Jie, Q. Xu, Unraveling the drivers of intensified landslide regimes in Western Ghats, India, *Sci. Total Environ.* 770 (2021).

7

Application of Geospatial Tools in Glacial Lake Outburst Floods: Mapping and Monitoring

Anita Sharma*, Vansheika and Chander Prakash

Civil Engineering Department, National Institute of Technology Hamirpur, Hamirpur, Himachal Pradesh, India

Abstract

Glaciers in the Himalayan area, which covers about 3 million hectares, are drastically receding. They are losing a mass of 10 to 15 approximately, causing the development of glacial lakes. The profound continuing effects on glaciers of climate change may impact the region's management of natural sources of freshwater and hydropower projects. Glacial lakes form and expand as glacier melt accelerates, increasing the likelihood of glacial lakes' outburst flooding. These floods present serious risks and may cause harm to settlements and infrastructure located downstream. Globally, the numbers, as well as the sizes of glacial lakes, have increased significantly in recent decades, threatening downstream populations. Fifteen million individuals on a global scale are at large risk from the consequences of GLOF, with 1 million of them living less than 10 km from a glacial lake in the high mountain Asia (HMA). Since they frequently reside near glacial lakes, populations in HMA are most susceptible to GLOF effects. The four nations with over half of the world's population, i.e., India, China, Pakistan, and the United States, are home to these glacial lakes. Consequently, comprehending the glacial lakes' creation, development, distribution patterns, and underlying mechanisms is crucial for GLOF hazard analysis and risk management. This chapter aims to understand glacial lakes, their types, spatiotemporal evolution, development, and growth using geospatial techniques like geographic information systems (GIS) and remote sensing (RS).

**Corresponding author*: Anita@nith.ac.in

Disha Thakur, Sanjay Kumar, Har Amrit Singh Sandhu and Chander Prakash (eds.) Sustainable Development Using Geospatial Techniques, (155–180) © 2024 Scrivener Publishing LLC

Keywords: Geospatial tools, climate change, glaciers, glacial lake outburst floods, mapping, monitoring

7.1 Introduction

The cryosphere comprises permafrost, marine ice, glaciers, ice sheets, and snow. This phrase is frequently used to explain the interplay between these frozen components and the Earth's climate system [1, 2]. Climate change significantly impacts the cryosphere; many of these frozen regions are melting swiftly and altering physically [3]. The impact of the cryosphere on the planet's energy balance and the global water cycle, according to the Intergovernmental Panel on Temperature Change (IPCC), is key in regulating the temperature of Earth [1]. Changes to the cryosphere, such as melting glaciers and sea ice, can have far-reaching impacts on sea level rise, ocean circulation, weather patterns, and biodiversity [2]. One example of the importance of the cryosphere is highlighted in the IPCC's Fifth Assessment Report, which states: "The reduction in snow cover is expected to exacerbate global warming, with the albedo effect (the reflectivity of the surface) decreasing as the amount of snow and ice decreases [1]. This will lead to more solar radiation absorbed by the Earth's surface, contributing to further warming." By permitting effective mapping and monitoring of these events, geospatial techniques have emerged as a critical factor in minimizing the disastrous results of glacial lake outburst floods (GLOFs) [4, 5].

Glacial lakes are increasingly identified and mapped in high-altitude locations sensitive to GLOFs, using remote sensing (RS) and geographic information system (GIS) technologies. Furthermore, these tools allow for identifying possible risks and constructing early warning systems to alert communities near affected areas. Continuous monitoring of glaciers, snow cover, and glacial lake water levels offers vital data for anticipating and mitigating the effects of GLOFs. Using geospatial tools in conjunction with advanced modeling techniques, such as hydrological models, slope stability models, and hazard mapping tools, improves the accuracy of GLOF predictions and the efficiency of mitigation actions. Using geospatial techniques in glacial lake mapping and monitoring constitutes an important step toward protecting communities living in high-altitude regions from the destructive effects of these natural calamities. To better forecast and prevent GLOFs, this chapter will examine how geospatial technologies are used in GLOF research, specifically in the mapping and monitoring of glacial lakes.

7.1.1 Climate Change and Glacial Recession

Glaciers are essential to the Earth's climate system and intimately related to climatic conditions. Global warming has caused glaciers to shrink and retreat, developing multiple glacial lakes worldwide. These lakes' outburst floods pose a serious hazard to residents downstream. The process of glacial recession has been attributed to various natural and man-made processes, including rising temperatures, changes in precipitation patterns, and a rise in greenhouse gasses in the atmosphere. Climate change is causing significant impacts on glaciers [6, 7]. In addition, climate change is also the cause of extreme weather events, such as heavy rainfall, that can initiate GLOFs. The region's glaciers are receding at an alarming rate, resulting in various ecological, hydrological, and socioeconomic challenges [8, 9]. Moreover, the temperature rise is impacting the Himalayan ecosystem, leading to the extinction of species, and reducing biodiversity [8]. Millions of people whose livelihoods depend on the rivers in the area are extremely concerned about the declining water supply. Additionally, the creation of new glacial lakes brought on by melting glaciers increases the risk of catastrophic flooding. The temperature rise also affects the Himalayan ecosystem, leading to species extinction and diminishing the region's biodiversity. These effects emphasize the significance of implementing effective policy measures and coordinating efforts to combat climate change and its implications for Himalayan glaciology. When a glacial lake created by retreating glaciers bursts because of too much water pressure or a weak glacial dam structure, it causes a GLOF. Globally, climate change influences the frequency and severity of GLOFs [9, 10]. These lakes' rapid water releases can result in disastrous floods that cause property destruction, human casualties, and community uprooting. Consequently, addressing climate change is crucial for lowering the probability of GLOFs and minimizing their effects on society.

7.1.2 Glacial Lakes

A glacial lake is an extensive waterbody that extends within or around a glacier and is a naturally occurring freshwater reservoir produced by glaciers and snow meltwater trapped by ice/debris located in glaciated regions and influenced by terrain, climate, and glacier movement as shown in Figure 7.1. Glacial lakes are vital ecosystems that provide freshwater, support aquatic and terrestrial life, and are crucial markers of climate change [11–13].

Figure 7.1 A rock and ice avalanche produced a 16-foot-high surge wave to flood a glacier lake called Dig Tsho in eastern Nepal in 1985. The naturally formed dam broke, releasing 177 million cubic feet of water and destroying a hydroelectric power station, multiple bridges, residences, and arable land in Thame Valley [14].

7.1.3 Formation of Glacial Lakes

Glaciers erode the terrain beneath them, leaving depressions, grooves, and debris ridges known as moraines [14]. Meltwater fills holes left by retreating glaciers to form glacial lakes, but ice and terminal moraines can also form natural dams that form glacial lakes [15, 16]. Moraine dams can be stable or leaky, capable of holding large lakes or slowly draining. Excessive melting, conversely, can cause them to overflow or burst, resulting in floods downstream and infrastructure damage [17–19]. Because they are porous, moraine walls are more likely to collapse, and a spike in glacier meltwater can swiftly produce flooding by undermining the moraine barriers. Glacial lakes vary from small water bodies to massive reservoirs holding tens of millions of cubic meters as largely depicted in Figure 7.2 [15]. To completely know glacier processes and activities such as glacier movement, moraine building, bed erosion, and depression formation, it is necessary first to understand glacier features and behaviors. A glacier is a compact and compressed mass of ice formed on land over time by the recrystallization of snow. This ice mass slides down mountain slopes because of its weight, gravity, and other considerations.

Glaciers can generate severe bed erosion and transport vast volumes of eroded material as they move downward. Glaciers' mass and movement operate as a natural bulldozer and conveyor belt, eroding the bed and transporting loose material along their course [18]. The process of bed erosion and material movement by glaciers causes depressions in the bed and

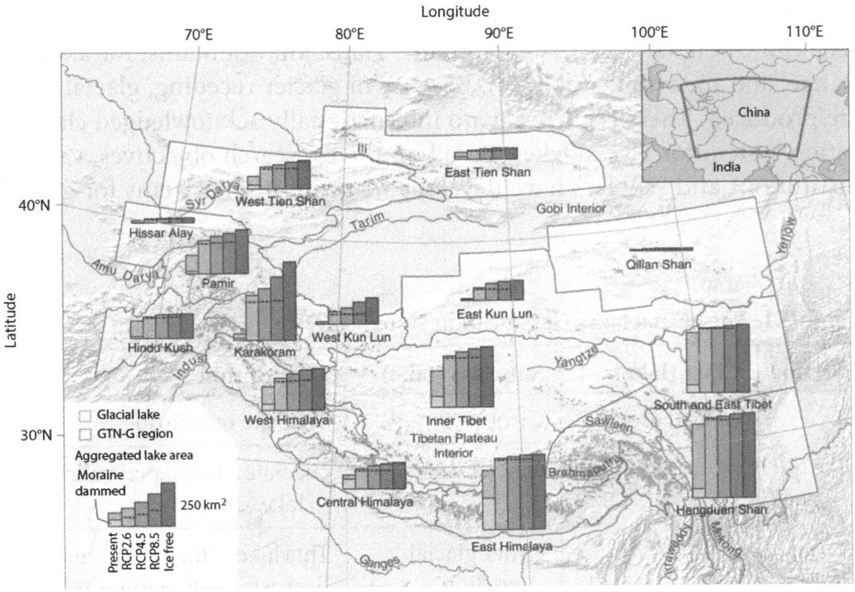

Figure 7.2 This map depicts the Himalayan mountain region, colloquially called the Third Pole, and its glacial lake locations. The accompanying bar charts use a variety of colors to depict current and prospective glacial lake areas based on three emission scenarios and an ice-free scenario. Gray indicates the proportion of glacial lakes that have been dammed by moraines to all glacial lakes. Additionally, dashed lines are given to show predicted changes in 2050 [15].

material deposition at the glacier's lateral and terminal margins [20]. The deposited material forms end or terminal moraines and lateral moraines. Moraines comprise various materials, ranging from small particles to large boulders, transported over great distances by glaciers and deposited along their margins [21–24]. Glacier advances during the "Little Ice Age" with subsequent retreats led to the creation of glacial lakes. Since then, glaciers retreated everywhere, causing the formation of newly formed glacial lakes [25]. These lakes are formed by meltwater build-up in the glacial bed, lateral/terminal moraines, or the front or edges of a glacier impounded by the glacier. They may also form on debris-covered glacier tongues as supraglacial lakes or ponds.

7.1.4 Kinds of Glacial Lakes

The majority of the glacial lakes are found in North America, South America, and Europe, specifically in the Andes, Alps, and Caucasus mountains, as well as the Rocky Mountain regions and Alaska's coastal ranges [26] and Asia (for example, the Altay Mountains, Tianshan Mountains, Karakorum Range, and the Himalayas [27]). Because of glacier receding, glacial lakes are produced. There is currently no internationally acknowledged classification system for glacial lakes. Based on their research objectives, various institutions and scholars have developed classification systems for glacial

Table 7.1 The glacial lakes classification system [29].

Sr. no.	Class (lake)	Subclass (lake)	Description
1	Glacial erosion	Cirque	The lake in a cirque.
		Glacial valley	Glaciated U-shaped valley lake.
		Other glacial erosion	This lake is formed by glacier erosion but belongs to other glacier lakes.
2	Moraine-dammed	End moraine-dammed	The lake between the end of the moraine ridge and the glacier's end.
		Lateral moraine-dammed	The lake beside the lateral moraine ridge
		Moraine thaw	The lake on the moraine ridge
3	Ice-dammed	Advancing glacier-dammed	The lake is blocked by a glacier that is advancing.
		Other glacier-dammed	The lake with existing glaciers.
4	Supraglacial		The lake over the surface of the glacier.
5	Subglacial		The lake inside the glacier or on top of the glacier bed.
6	Other glacial		Landslides, avalanches, debris flow, etc., block the lake.

lakes. According to ICIMOD, the Hindu Kush–Himalayan region has five main kinds of glacial lakes: lakes created by glacier erosion, lakes that moraines have blocked, lakes blocked by ice, supraglacial lakes, and subglacial lakes [28]. Yao *et al.* [28] established a comprehensive classification scheme for glacial lakes based on information such as the formation time, the lake's characteristics, the basin's geometry, the region affected by the lake's distance from its glacial source, and the level of risk. However, there is a discrepancy between the theoretical categorization scheme and how glacial lakes are categorized.

The glacial lake classification scheme includes six classes based on previous research and the principles of systematic, normalization, operability, and scalability: a glacial erosion lake, a moraine-dammed lake, an ice-blocked lake, a supraglacial lake, a subglacial lake, and the other glacial lakes, as shown in Table 7.1 with a description and examples of glacial lakes in Figure 7.3.

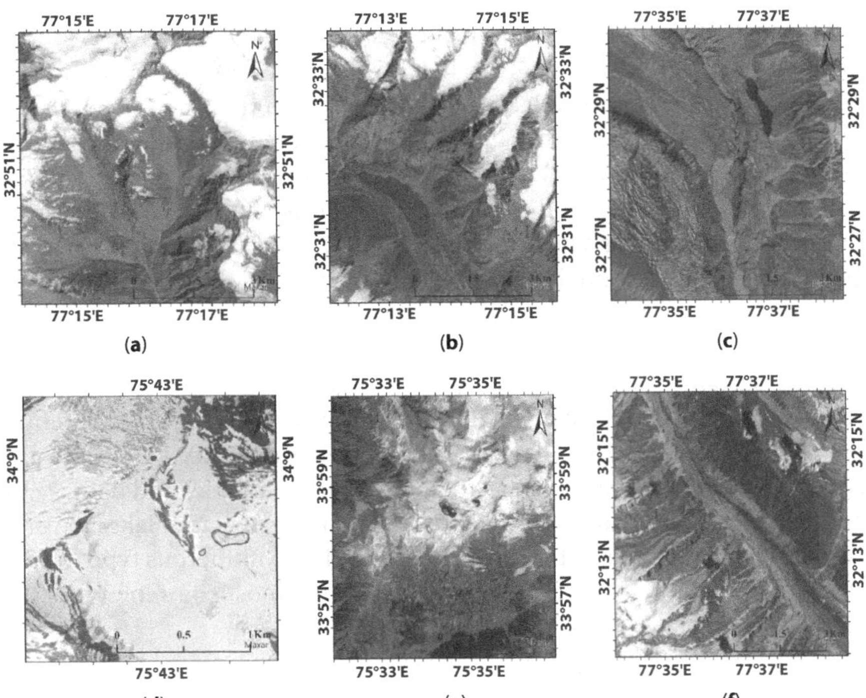

Figure 7.3 Maps illustrating several examples of glacial lake types, including (a) glacial erosion lake; (b) moraine-dammed glacial lake; (c) lateral moraine-dammed lake; (d) supraglacial lakes; (e) ice-dammed glacial lakes; and (f) other glacial lakes.

7.1.5 History of Glacial Lake Outburst Floods and Events in the Himalayas

Numerous moraine-dammed lakes had experienced rapid drainage, resulting in substantial flooding and debris surges. There are many causes for the collapse of the moraine dam, including piping problems, earthquakes, and melting glacier cores. High mountain regions have long experienced GLOFs, which are flash floods brought on by the sudden discharge of water from a glacial lake [10, 30]. GLOFs cause substantial water discharge and silt downstream, frequently resulting in catastrophic outcomes such as life casualty and property damage in downstream areas. The high mountain Asia (HMA) region has the highest GLOF hazard as of 2020, with a value of 0.313. This region contains 2211 lakes, spans 1256.09 km^2, and poses a threat to approximately 9.3 million persons. This region contains 2211 lakes, covers an area of 1256.09 km^2, and threatens around 9.3 million people [10, 31]. GLOF has grown to be a serious concern that has neither been restrained nor lessened in recent years as a conclusion of rising human occupancy, man-made activities, and unintentional infrastructure buildout in the delicate ecosystem of the high mountain regions [32]. A GLOF incident that affected 48 villages in the downstream areas occurred in 1929 because of the ice-dammed lake (Chog Kumdan) in the Shyok basin of the Ladakh region [33]. The Chorabari GLOF event, which resulted from the dam rupture on June 17, 2013, left the area in terrible shape [34]. Similar occurrences were seen in the Tibetan Himalayas Cirenamco (1981), the Karakoram Himalayas Ghulkin GLOF (1981), the Nepal Himalayas Dig Tsho (1985), and Tampokhari (1998), all of which resulted in a significant loss of lives as well as property downstream. The Andes, the European Alps, the Canadian Cordillera, and the Himalayan region, among other places, have all sustained significant damage because of GLOF events [18, 35]. Several serious GLOF incidents in the Himalayas have occurred recently [19], including the 2013 Kedarnath catastrophe and the 2021 Uttarakhand disaster, which resulted in a substantial number of fatalities and damage to infrastructure [36]. Numerous scientists have mapped glacial lakes in various Himalayan regions. GLOFs are categorized into numerous types based on their causes and features. Here are some of the most common types:

7.1.5.1 Dam-Break Floods

A dam-break flood happens when a glacial lake unexpectedly falls, unleashing tremendous water quickly. Dam-break floods are hazardous because they can travel quickly and carry significant amounts of silt and debris.

7.1.5.2 Subglacial Floods

Subglacial floods happen when water collects beneath a glacier and then rushes out, causing a rapid flood. These floods can be triggered by geothermal heat, melting ice, or upstream water pressure.

7.1.5.3 Ice-Dammed Lake Floods

A glacial lake sustained back by an ice dam reaches a critical level and then breaches the dam, releasing significant water. It can be caused by natural occurrences such as glacier melting or movement or human activity such as mining or construction.

7.1.5.4 Lake Floods Caused by Moraines

Moraines are accumulations of rock and silt deposited by glaciers as they travel. Moraine-dammed Lake floods occur when a glacier lake held back by a moraine bursts its natural dam, causing a sudden flood.

7.1.5.5 Rain-Triggered Floods

In glacial locations, rain-triggered floods can occur when heavy rainfall or rapid melting of snow causes a sudden increase in water volume in a glacier lake, resulting in a flood.

RS is an effective instrument for mapping [37], assessing, and detecting changes in water resources, particularly glacial lakes [38, 39]. With RS imagery from multiple satellites, it became possible to analyze vast glacial lakes. However, these glacial lakes have numerous inherent characteristics, such as lake-water color, dam structure, and nearby objects, which impede the rapid and accurate mapping of glacial lakes. Under these conditions, the most important step in glacial lake mapping over a significant region is to choose and develop an efficient and reliable method. In the following section, automated and semi-automated techniques are discussed.

7.2 Geospatial Techniques

Geospatial technologies such as RS and GIS are essential for analyzing GLOFs. These technologies provide a means to map and monitor glacial lakes, glaciers, and surrounding terrain, allowing for the identification of potential GLOF hazards [37]. The satellite imageries can calculate the proportions and the depth of glacial lakes as well as the changes in their volume over a period. The buildout that could be impacted due to GLOF, neighboring towns, and glacial lakes can all be represented using GIS. Geospatial technology has become an essential resource for mapping and predicting GLOFs. It integrates information from various sources to understand the hazard comprehensively [43]. This can enhance emergency preparedness and response to save lives and limit infrastructure damage by facilitating the development of early warning systems. Because of climate change-induced increases in the number and extent of glacial lakes, GLOFs are becoming a growing concern in glacier-containing regions. Downstream, these disasters can cause great damage to infrastructure, property, and even human life. Monitoring and forecasting GLOFs is essential for reducing the risk of such disasters [2]. Geospatial technology, such as satellite imagery and GIS, plays a substantial part in this endeavor by providing information on changes in glaciers and lakes [11, 40, 41]. For instance, a study by [42] monitored the growth and creation of glacial lakes using satellite imagery in the Himalayan region and identified areas at risk of GLOFs. Geospatial technology enables a better comprehension of GLOF hazards and aids in developing early warning systems by providing data on changes in glaciers and lakes.

According to research [43] by the International Centre for Integrated Mountain Development (ICIMOD), the Himalayan region has 3044 glacial lakes, 206 of which are potentially hazardous. To reduce the risk of GLOFs, early warning systems, glacial lakes monitoring, and community-based disaster risk reduction measures must be implemented [44, 45]. These floods occur when water trapped by glaciers is suddenly released, causing devastating floods downstream. With the increasing frequency and intensity of GLOFs due to climate change, monitoring and mapping glacial lakes to predict and prevent these disasters have become more critical, and geospatial tools, such as RS and GIS, are essential for mapping and monitoring GLOFs [40, 46, 47]. Himalayan glaciers lose mass and retreat at a rate of 10 to 15 m/year, creating new glacial lakes. These lakes' growth due to hastened glacier melting raises the possibility of flooding of glacial lakes with the potential to severely damage downstream areas [48]. As the size and number of glacial lakes increase worldwide, especially in the high mountain Asia, populations are becoming more susceptible to GLOF. According to satellite images taken between 1990 and 2018, globally, glacial lakes increased by 53% [49]. While Scandinavia, Iceland, and Russia have witnessed the most rapid growth in these lakes, their water volume remains petite compared to other places. While these lakes provide fresh water, they can also pose a considerable concern if suitable drainage systems are not in place, as they may burst, resulting in outbursts of floods. This phenomenon has profound implications since the abrupt surge of meltwater can harm infrastructure such as hydroelectric power plants, transit routes, and pipelines [55]. This might cause major environmental disasters, particularly in locations with large glacial lakes, such as the Himalayas and Alaska. Outburst floods are hazardous in the eastern Himalayas, where warming occurs faster than in the west [45]. Scientists have discovered that this region is twice as prone as other places to experience glacial outburst floods [10, 45]. As global warming continues, the risk in this area might triple by 2050, with new hotspots appearing further west [50], highlighting the critical need for effective mitigation strategies to reduce the danger of glacial outburst floods and their possible repercussions. Over the past few years, the global climate has significantly shifted due to climate change and unpredictable catastrophic GLOF events. These occurrences endanger both tourists and the residents of the downstream mountain region. Precise spatial and temporal inventories of glacier lakes are required to limit the risk of these catastrophes. It is feasible to construct flood hazard-prone zoning, apply risk control measures, and reduce the impact of floods on downstream populations by monitoring changes in their number and area coverage and related geo-hydrological features. As a result, it is critical to prioritize glacial lake monitoring to understand their behavior better and limit their influence on local communities.

Technologies like RS and GIS have permitted the mapping of glacial lakes at higher elevations that would otherwise have been outside the purview of traditional field research. The information gained from a systematic analysis of satellite images provides essential insights into the development and behavior of these bodies of water. Some glacial lake mapping and monitoring techniques are described in the below sections.

7.2.1 Techniques for Mapping and Monitoring of Glacial Lakes

Mapping and monitoring glacial lakes are required to analyze their spatiotemporal distribution, which is necessary to manage potentially hazardous glacier lake outburst risk. Several scholars have ventured to map glacial lakes in various glacial regions. Geospatial techniques have been recognized as an effective method for mapping, analyzing, and detecting changes in water resources, particularly glacial lakes. Using RS imagery from various satellites has aided in researching large-scale glacial lakes. The unique characteristics of individual glacial lakes, such as their water color, dam structures, and adjacent geology, hinder the mapping process. Consequently, selecting and establishing an efficient and accurate method for mapping glacial lakes across a vast area is essential. The subsequent section discusses techniques for mapping and monitoring glacier lakes.

7.2.2 Spectral Feature-Based Glacial Lake Mapping

In general, using spectral features is considered a simple and effective method for extracting information about glacial lakes from complex backgrounds. However, glacial lake mapping often relies on these features too heavily. Three main categories can be used to classify spectral features: Three types of features are 1) single-band reflectance [51], 2) two-band water spectral index [52], and 3) multiple-band transformation features [53]. There are several spectral bands available in primary satellite sensors. Primary satellite sensors use a single-band reflectance function like a near-infrared (NIR) band, to easily differentiate water bodies from their homogeneous surroundings [54]. For complicated backgrounds, combining the green/blue (G/B) bands and the NIR band is often used to create a water index for extracting glacial lake information. This is known as the Normalized Difference Water Index (NDWI). Equation 7.1 extracts glacial lake information [52].

$$\mathrm{NDWI} = (\mathrm{G} - \mathrm{NIR})/(\mathrm{G} + \mathrm{NIR}) \tag{7.1}$$

where G is the reflectance of green/blue bands, and NIR is the reflectance of the NIR band.

Similar to this, other water indices have been developed, such as the Modified NDWI (MNDWI) [55] which is provided in Equation 7.2 below.

$$MNDWI = \frac{pG - pSWIR}{pG + pSWIR} \tag{7.2}$$

where pG is the reflectance of green/blue bands and pSWIR is the reflectance of short-wave infrared bands.

These indices combine multiple spectral bands with improving water body extraction accuracy and reducing errors due to background effects. MNDWI uses green and mid-infrared bands, while AWEI uses green, red, and blue bands. The selection of the water indices is dependent on the study region's specific characteristics and the satellite sensor being used.

To map glacial lakes, water information has been extracted using various spectral features and picture modification approaches [56]. NIR bands,

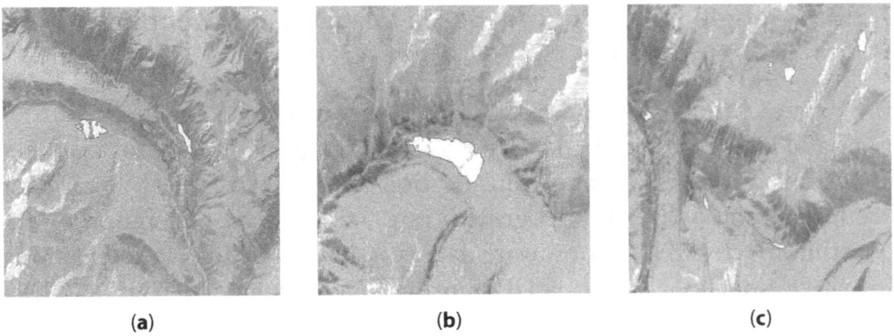

Figure 7.4 Represents some examples of Normalized Difference Water Index (NDWI).

green/blue (G/B) bands with improved water reflection, and NIR bands with reduced reflectance are only a few examples of simple threshold approaches based on the reflectance properties of a single band that are frequently used. These methods can be combined to create water indices like NDWI and MNDWI as depicted in Figure 7.4. Additionally, image transformation methods such as K-t transformation can be utilized by integrating multiple band features to obtain water information. In areas with complicated land cover, the Automated Water Extraction Index (AWEI) has been used in the direction of improving the accuracy of lake pixel extraction [53]. The misclassification of glacial lakes may result from uncertainty caused by meteorological disturbances such as snow cover, cloud cover, vegetation noise, soil moisture, and mountain shadow. To resolve these uncertainties, improved versions of the NDWI have been developed [55, 57]. It is essential to perform post-checking and editing for all automated glacial lake mapping methods and techniques. Consequently, combining computerized processes with manual digitization and overlay assessments using high-resolution Google Earth images is recommended. In addition, field observations of reference glacial lakes can help validate RS-based mapping results. Slope, aspect, and mountain shadow masks derived from Digital Elevation Models (DEMs) can correct topographically based water body misclassifications. Field investigations are needed to validate lake characteristics and topographic factors and create the final database [58, 59]. Spectral information methods are only relevant for small regions since the spectral features of glacial lakes and some terrestrial objects (such as mountain shadows and melted glaciers) are identical at larger scales. Glacial lakes can be mapped using cutting-edge technologies like Google Earth Engine (GEE) and artificial intelligence-based algorithms [60].

7.2.3 Google Earth Engine–Based Glacial Lake Mapping and Monitoring

Big data techniques have dramatically changed science and society [61]. Big data, including geospatial data, is a rapidly growing field that gathers enormous amounts of data from location-aware devices [72]. Such data accumulates at a 20% annual growth rate [62]. A daily generation of 2.5 quintillion bytes of location-aware data is estimated by the United Nations Initiative on Global Geospatial Information Management (UN-GGIM) [1 quintillion bytes = 1000 petabytes (PB); 1 PB = 1000 Terabytes (TB)]. Every single day, Google produces about 25 PB of data, a large percentage of which is spatiotemporal data [62]. Unmanned aerial systems (UAS) and satellite imagery are gathered more frequently plus at greater resolutions as the globe develops in technology; this trend is projected to accelerate even more [63]. With the exponential growth of vast amounts of geographic data, there is a growing demand for cloud-based computing and high-performance computing to facilitate the modeling, analysis, and simulation of geospatial data. Scholars and professionals in GIS and RS have recently given huge geospatial data a lot of attention [64]. The volume, spatial-temporal resolution, and complexity of RS data present formidable obstacles.

GEE is a practical and promising RS large data analysis solution. GEE is an accessible, cloud-based geospatial platform that offers a free cloud-computing platform for dealing with extensive geospatial data with various applications, such as glacial lake mapping and monitoring [65]. GEE can also be used with spectral indices and satellite images, such as Landsat and Sentinel, to extract, identify, and classify glacial lakes. Several stages are required to extract glacial lakes using GEE, including Image classification which is the process of finding and categorizing pixels in an image based on their spectral characteristics. Spectral indices, such as the NDWI, can classify water bodies in satellite images. Validation involves comparing the results classification with ground truth data to assess the preciseness of the glacial lake extraction. GEE can be a potent instrument for glacial lake extraction and monitoring, as it provides a quick and efficient method for analyzing large quantities of satellite imagery and geospatial data. This open-source platform utilizes Google Servers' computational resources to expand our earth observation capabilities. Glacial lakes in HMA were mapped using the MNDWI as well as a cloud computing strategy with GEE [57, 66].

7.2.4 Mapping and Monitoring of Glacial Lakes Using Deep Learning

Traditional algorithms for machine learning, such as decision trees and neural networks, have limited universal applicability because they require application-specific prior knowledge [67]. Deep learning (DL), which uses neural networks with more than two hidden layers, has become a successful technique for resolving challenging issues. For image analysis applications, including picture indexing, segmentation, and object detection, DL has grown in favor. By combining convolutional and pooling layers, CNNs (convolutional neural networks), a form of the neural network, have been successful at extracting mid and high-level abstract properties from unprocessed images. Recent research has demonstrated the effectiveness of CNNs for large-scale picture segmentation tasks, object recognition, and semantic segmentation [68–70]. With more data and computer resources readily available, deep learning is becoming increasingly popular in RS. Based on picture attributes, semantic image segmentation techniques such as distributed iteration and object-oriented segmentation can extract glacier lake boundaries precisely. With more data and computer resources readily available, deep learning is becoming increasingly popular in RS. Based on picture attributes, semantic image segmentation techniques such as distributed iteration and object-oriented segmentation can extract glacier lake boundaries precisely [67]. However, post-processing operations are still required, and the segmentation scale and lake size variations limit the wide application of these methods [71]. Due to variations in pixel spectra, existing glacier lake extraction algorithms have difficulty evaluating the spectrum, form, texture, and other glacial lake properties. To increase classification accuracy, deep learning, a multilayer neural network learning system, can automatically extract intrinsic and deep characteristics [72]. Deep learning algorithms have outperformed traditional approaches in classifying remotely sensed images, handling complicated data better, and having higher autonomous learning capabilities [73]. With their code-decoding structure being employed to segment ground objects, full convolutional neural networks (FCN) have demonstrated advantages in image semantic segmentation [74, 75]. Based on FCN, U-shaped symmetric networks (U-Net) have been extensively used in the categorization of remotely sensed pictures because they incorporate low- and high-dimensional data to improve segmentation accuracy [76, 77]. Many modeling techniques and software programs are available for modeling dam failures, the spread of flood water, and sediment flow. For better hazard assessment, forecasting the danger of future outbursts, and improving knowledge of diverse GLOF processes, outburst flood modeling is crucial [40]. The components of GLOF modeling differ based on their location and chain process. The two primary categories of breaching that can be modeled are 1) breaching of dams and 2) flood outburst routing. Dam breaching can be caused by the sudden overtopping of a water wave on a moraine dam or the internal collapse of the moraine due to piping and seepage. When the propelling forces exerted by displacement waves or channelized flow exceed the material strength of the reservoir, erosion and failure occur. Due to the formation of these breaches, the discharge from the lake increases, which leads to an increase in hydrodynamic forces, which further progressively enlarge the breach, resulting in a partial or complete draining of lakes and inundation downstream. Detailed modeling and simulation of breach formation and development, which are necessary for hazard assessment [67], can be used to estimate the geometry of the flood hydrograph and peak discharge at the flood source. Numerous approaches and models are available for modeling and simulating the earthen dam breach process and breach discharge, including empirical, analytical, and parametric models and physical numerical models.

In an outburst flood routing, GLOFs are highly unsteady flows that alter dramatically as they propagate downstream. The type and rate of fracture formation initially govern the abrupt discharge of a failing dam. However, the inundation is governed by the flow's topography, vegetation, and sediment ingestion in the region downstream of the outburst. Typically, many non-cohesive sediments are entrained in the flow, transforming it into a hyper-concentrated or granular detritus flow. Particle properties and particle interactions with water govern flow conditions. Consequently, water sediment content, sediment size, and sediment type are crucial flood modeling parameters [58, 68, 69]. Various empirical, mathematical, and physical models are used to model and simulate the propagation and transport of floods and sediments downriver. GIS-built flow routing and physically founded numerical flow models are examples. One-dimensional numerical models, such as HEC-RAS [70], MIKE 11 [71], NWS-FLDWAV [72], and BOSS-DAMBRK [73], are utilized to predict the behavior of weather systems.

7.3 Discussion and Conclusion

GLOFs seriously threaten mountain populations. Floods brought on by the abrupt release of water from lakes trapped by moraines have resulted in significant property destruction and fatalities. The vulnerability of the reservoir to failure and the lake's exposure to potential trigger mechanisms are two crucial aspects that affect the possibility of catastrophic flooding from unstable moraine reservoirs. The dam's stability is affected by its shape, material properties, internal structure, and stability. Dynamic slope motions that threaten the dam or the lake, such as ice avalanches, rock falls, landslides, or debris flows, can cause GLOFs. Rapid water buildup from glaciers or snowmelt, downpours, earthquakes, or sediment pipes in the dam is among more potential reasons. The possibility of GLOF due to the moraine reservoir's failure is a complicated subject impacted by several factors, including the region's topography, climate, and seismic characteristics.

The identification of lakes that have the potential to cause hazards and the modeling of outburst floods are crucial steps in pinpointing vulnerable areas and mitigating the risks associated with GLOFs. Regional glacial lake inventories with medium to high-resolution geographical and temporal data are needed for efficient risk management and mitigation. Only the Himalayas in Sikkim and Uttarakhand have undergone a thorough assessment. To better understand the impacts of climate change on these lakes, it is urgently necessary to perform a thorough and methodical assessment of Indian Himalayan glacier lakes. A comprehensive inventory should evaluate glacial lakes' spatiotemporal evolution and development patterns and their geographical sensitivity to outburst floods. Prioritizing potentially vulnerable places for risk management and mitigation may be made easier by identifying the regional GLOFs' probable source, location, and effect area. The complex problem of GLOFs in the Himalayas requires a multidisciplinary approach that brings together knowledge in climate science, glaciology, geology, hydrology, and risk management. These experts can work together to develop practical ways to control and lessen the risks brought on by GLOFs. RS satellite data and approaches are less expensive and expeditious than conventional field studies for the extended surveillance of glacial lakes, which are difficult to reach due to their enormous size, remote locations, and perilous mountainous terrain conditions. Remotely sensed data products, such as DEMs and satellite imagery, can extract critical terrain and other information about surface features, lake location and size, moraine dam geometry, and surrounding physical conditions such as slope, elevation, and steepness to evaluate the likelihood of GLOFs.

As a result, glacier lake monitoring and potential outburst threats are increasingly being assessed using satellite pictures. Satellite data can be classified according to its spectral, geographical, radiometric, and temporal resolution and features. The spectral resolution and properties are employed mainly for recognizing and distinguishing various spatial features. The spatial resolution of RS determines the mapping size and corresponding accuracy. Landsat multispectral satellite data with 30-m ground resolution are extensively utilized for glacier lake identification and mapping. The resolution was increased to 24-m IRS LISS-III [4], 5-m IRS-LISS-IV multispectral data, 2- to 4-m panchromatic CORONA aerial photographs, and about 1m resolution IKONOS and Quickbird satellite data. Radiometric resolution is the number of gray values/digital numbers used to differentiate the amount of radiometric information in a specific spectral band. The amount of radiometric information required for surface object discrimination is critical. Because illumination and surface reflectance are extremely changeable in such areas, radiometric features are critical in high mountain RS applications. Landsat-TM/ETM+ and Landsat 7 optical sensor systems typically feature an 8-bit radiometric resolution (256-gray values). IRS Resourcesat-1 and Resourcesat-2 sensors have a 10-bit radiometric resolution (1024-gray values), but IKONOS and QuickBird have an 11-bit radiometric resolution (2048-gray values), indicating their importance and potential for investigations in high mountain locations.

Finally, the risk of GLOFs in the Himalayas is a significant issue that requires immediate attention. Effective risk management and mitigation strategies require a thorough regional inventory of glacial lakes and their vulnerability to GLOFs. This will help to lessen the possible threats posed by GLOFs and safeguard the safety of the region's communities. Experts from diverse professions must work together to address this complicated issue and protect vulnerable populations.

References

1. IPCC, *The Ocean and Cryosphere in a Changing Climate.* Cambridge University Press, 2022. doi: 10.1017/9781009157964.
2. B. R. Arora *et al.*, 'The Himalayan Cryosphere: Appraisal of Climate-Glacier Inter-linkages', *Proc. Indian Natl. Sci. Acad.*, vol. 99, no. 0, Oct. 2018, doi: 10.16943/ptinsa/2018/49509.
3. B. D. Vishwakarma *et al.*, 'Challenges in Understanding the Variability of the Cryosphere in the Himalaya and Its Impact on Regional Water Resources', *Front. Water*, vol. 4. Frontiers Media S.A., Jul. 28, 2022. doi: 10.3389/frwa.2022.909246.
4. B. K. Govindha Raj, V. K. Kumar and S. N. Remya, 'Remote sensing-based inventory of glacial lakes in Sikkim Himalaya: Semi-automated approach using satellite data', *Geomatics Nat. Hazards Risk*, vol. 4, no. 3, pp. 241–253, 2013, doi: 10.1080/19475705.2012.707153.
5. Y. Nie *et al.*, 'A regional-scale assessment of Himalayan glacial lake changes using satellite observations from 1990 to 2015', *Remote Sens. Environ.*, vol. 189, pp. 1–13, Feb. 2017, doi: 10.1016/j.rse.2016.11.008.
6. P. D. A. Kraaijenbrink, M. F. P. Bierkens, A. F. Lutz and W. W. Immerzeel, 'Impact of a global temperature rise of 1.5 degrees Celsius on Asia's glaciers', *Nature*, vol. 549, no. 7671, pp. 257–260, Sep. 2017, doi: 10.1038/nature23878.
7. O. King, S. Ghuffar, A. Bhattacharya, R. Yao, T. Yao and T. Bolch, 'Glaciological and climatological drivers of heterogeneous glacier mass loss in the Tanggula Shan (Central-Eastern Tibetan Plateau), since the 1960s', *J. Glaciol.*, pp. 1–18, Apr. 2023, doi: 10.1017/jog.2023.5.
8. B. Wouters, A. S. Gardner and G. Moholdt, 'Global glacier mass loss during the GRACE satellite mission (2002-2016)', *Front. Earth Sci. (Lausanne)*, vol. 7, Apr. 2019, doi: 10.3389/feart.2019.00096.
9. Intergovernmental Panel on Climate Change, Climate change 2014: Synthesis Report: longer report.
10. C. Taylor, T. R. Robinson, S. Dunning, J. Rachel Carr and M. Westoby, 'Glacial lake outburst floods threaten millions globally', *Nat. Commun.*, vol. 14, no. 1, Dec. 2023, doi: 10.1038/s41467-023-36033-x.
11. Y. Nie, Q. Liu, J. Wang, Y. Zhang, Y. Sheng and S. Liu, 'An inventory of historical glacial lake outburst floods in the Himalayas based on remote sensing observations and geomorphological analysis', *Geomorphology*, vol. 308, pp. 91–106, May 2018, doi: 10.1016/j.geomorph.2018.02.002.
12. K. B. G. Raj and K. V. Kumar, 'Inventory of Glacial Lakes and its Evolution in Uttarakhand Himalaya Using Time Series Satellite Data', *J. Indian Soc. Remote Sens.*, vol. 44, no. 6, pp. 959–976, Dec. 2016, doi: 10.1007/s12524-016-0560-y.

13. A. Emmer, J. Klimeš, M. Mergili, V. Vilímek and A. Cochachin, '882 lakes of the Cordillera Blanca: An inventory, classification, evolution and assessment of susceptibility to outburst floods', *Catena (Amst)*, vol. 147, pp. 269–279, Dec. 2016, doi: 10.1016/j.catena.2016.07.032.
14. A. Emmer and V. Vilímek, 'Review article: Lake and breach hazard assessment for moraine-dammed lakes: An example from the Cordillera Blanca', *Nat. Hazards Earth Syst. Sci.*, vol. 13, no. 6. Copernicus GmbH, pp. 1551–1565, Jun. 18, 2013. doi: 10.5194/nhess-13-1551-2013.
15. A. Emmer and A. Cochachin, 'The causes and mechanisms of moraine-dammed lake failures in the cordillera blanca, North American Cordillera, and Himalayas', *Acta Univ. Carol. Geogr.*, vol. 48, no. 2, pp. 5–15, 2013, doi: 10.14712/23361980.2014.23.
16. R. J. McKillop and J. J. Clague, 'A procedure for making objective preliminary assessments of outburst flood hazard from moraine-dammed lakes in southwestern British Columbia', *Nat. Hazards*, vol. 41, no. 1, pp. 131–157, Apr. 2007, doi: 10.1007/s11069-006-9028-7.
17. R. J. McKillop and J. J. Clague, 'Statistical, remote sensing-based approach for estimating the probability of catastrophic drainage from moraine-dammed lakes in southwestern British Columbia', *Global Planet. Change*, vol. 56, no. 1–2, pp. 153–171, Mar. 2007, doi: 10.1016/j.gloplacha.2006.07.004.
18. R. Osti, T. N. Bhattarai and K. Miyake, 'Causes of catastrophic failure of Tam Pokhari moraine dam in the Mt. Everest region', *Nat. Hazards*, vol. 58, no. 3, pp. 1209–1223, Sep. 2011, doi: 10.1007/s11069-011-9723-x.
19. G. Zheng *et al.*, 'Increasing risk of glacial lake outburst floods from future Third Pole deglaciation', *Nat. Clim. Change*, vol. 11, no. 5, pp. 411–417, May 2021, doi: 10.1038/s41558-021-01028-3.
20. O. Korup and F. Tweed, 'Ice, moraine, and landslide dams in mountainous terrain', *Quat. Sci. Rev.*, vol. 26, no. 25–28, pp. 3406–3422, Dec. 2007, doi: 10.1016/j.quascirev.2007.10.012.
21. P. Iribarren Anacona, K. P. Norton and A. Mackintosh, 'Moraine-dammed lake failures in Patagonia and assessment of outburst susceptibility in the Baker Basin', *Nat. Hazards Earth Syst. Sci.*, vol. 14, no. 12, pp. 3243–3259, Dec. 2014, doi: 10.5194/nhess-14-3243-2014.
22. A. Mandal *et al.*, 'An 11-year record of wintertime snow-surface energy balance and sublimation at 4863ma.s.l. on the Chhota Shigri Glacier moraine (western Himalaya, India)', *Cryosphere*, vol. 16, no. 9, pp. 3775–3799, Sep. 2022, doi: 10.5194/tc-16-3775-2022.
23. A. Mandal *et al.*, '1-year record of wintertime snow surface energy balance and sublimation at 4863 m a.s.l. on Chhota Shigri Glacier moraine (western Himalaya, India)', doi: 10.5194/tc-2021-386.
24. J. J. Clague and S. G. Evans, 'A review of catastrophic drainage of moraine-dammed lakes in British Columbia', 2000.

25. A. Emmer, S. Merkl and M. Mergili, 'Spatiotemporal patterns of high-mountain lakes and related hazards in western Austria', *Geomorphology*, vol. 246, pp. 602–616, Oct. 2015, doi: 10.1016/j.geomorph.2015.06.032.
26. M. Zhang, F. Chen, H. Zhao, J. Wang and N. Wang, 'Recent changes of glacial lakes in the high mountain A sia and its potential controlling factors analysis', *Remote Sens. (Basel)*, vol. 13, no. 18, Sep. 2021, doi: 10.3390/rs13183757.
27. P. K. Mool, S. R. Bajracharya and S. P. Joshi, 'Review Reviewed Work (s): Inventory of Glaciers, Glacial Lakes and Glacial Lake Outburst Floods, Monitoring and Early Warning Systems in the Hindu Kush-Himalaya Region: Nepal by P. K. Mool, S. R. Bajracharya and S. P. Joshi; Inventory of Glac', *Mt. Res. Dev.*, vol. 24, no. 3, pp. 20–23, 2004.
28. X. Yao, S. Liu, L. Han and M. Sun, 'Definition and classification systems of glacial lake for inventory and hazards study', *Dili Xuebao/Acta Geogr. Sin.*, vol. 72, no. 7, pp. 1173–1183, 2017, doi: 10.11821/dlxb201707004.
29. P. I. Anacona, A. Mackintosh and K. Norton, 'Reconstruction of a glacial lake outburst flood (GLOF) in the Engaño valley, chilean patagonia: Lessons for GLOF risk management', *Sci. Total Environ.*, vol. 527–528, pp. 1–11, Sep. 2015, doi: 10.1016/j.scitotenv.2015.04.096.
30. D. A. Cenderelli and E. E. Wohl, 'Peak discharge estimates of glacial-lake outburst floods and AnormalB climatic floods in the Mount Everest region, Nepal', 2001. [Online]. Available: www.elsevier.comrlocatergeomorph
31. N. R. Khanal *et al.*, 'A comprehensive approach and methods for glacial lake outburst flood risk assessment, with examples from Nepal and the transboundary area', *Int. J. Water Resour. Dev.*, vol. 31, no. 2, pp. 219–237, Apr. 2015, doi: 10.1080/07900627.2014.994116.
32. 'Inglis2016-Massive1929HimalayanFloodisaCautionaryTale'.
33. S. K. Allen, A. Linsbauer, S. S. Randhawa, C. Huggel, P. Rana and A. Kumari, 'Glacial lake outburst flood risk in Himachal Pradesh, India: an integrative and anticipatory approach considering current and future threats', *Nat. Hazards*, vol. 84, no. 3, pp. 1741–1763, Dec. 2016, doi: 10.1007/s11069-016-2511-x.
34. A. Emmer and V. Vilímek, 'New method for assessing the susceptibility of glacial lakes to outburst floods in the Cordillera Blanca, Peru', *Hydrol. Earth Syst. Sci.*, vol. 18, no. 9, pp. 3461–3479, Sep. 2014, doi: 10.5194/hess-18-3461-2014.
35. S. K. Allen, P. Rastner, M. Arora, C. Huggel and M. Stoffel, 'Lake outburst and debris flow disaster at Kedarnath, June 2013: hydrometeorological triggering and topographic predisposition', *Landslides*, vol. 13, no. 6, pp. 1479–1491, Dec. 2016, doi: 10.1007/s10346-015-0584-3.
36. D. J. Quincey *et al.*, 'Early recognition of glacial lake hazards in the Himalaya using remote sensing datasets', *Global Planet. Change*, vol. 56, no. 1–2, pp. 137–152, Mar. 2007, doi: 10.1016/j.gloplacha.2006.07.013.

37. A. Aggarwal, S. K. Jain, A. K. Lohani and N. Jain, 'Glacial lake outburst flood risk assessment using combined approaches of remote sensing, GIS and dam-break modelling', *Geomatics Nat. Hazards Risk*, vol. 7, no. 1, pp. 18–36, Jan. 2016, doi: 10.1080/19475705.2013.862573.
38. T. Bolch *et al.*, 'Earth Observation to Investigate Occurrence, Characteristics and Changes of Glaciers, Glacial Lakes and Rock Glaciers in the Poiqu River Basin (Central Himalaya)', *Remote Sens. (Basel)*, vol. 14, no. 8, Apr. 2022, doi: 10.3390/rs14081927.
39. S. Passang, S. Schmidt and M. Nüsser, 'Topographical Impact on Snow Cover Distribution in the Trans-Himalayan Region of Ladakh, India', *Geosciences (Switzerland)*, vol. 12, no. 8, Aug. 2022, doi: 10.3390/geosciences12080311.
40. B. Bajracharya, A. B. Shrestha and L. Rajbhandari, 'Glacial lake outburst floods in the Sagarmatha region: Hazard assessment using GIS and hydrodynamic modelling', *Mt. Res. Dev.*, vol. 27, no. 4, pp. 336–344, Nov. 2007, doi: 10.1659/mrd.0783.
41. A. Kääb, D. Treichler, C. Nuth and E. Berthier, 'Brief Communication: Contending estimates of 2003-2008 glacier mass balance over the Pamir-Karakoram-Himalaya', *Cryosphere*, vol. 9, no. 2, pp. 557–564, Mar. 2015, doi: 10.5194/tc-9-557-2015.
42. ICIMOD, '*Glacial Lakes and Glacial Lake Outburst Floods in Nepal* THE WORLD BANK', 2011.
43. V. Ruiz-Villanueva, S. Allen, M. Arora, N. K. Goel and M. Stoffel, 'Recent catastrophic landslide lake outburst floods in the Himalayan mountain range', *Prog. Phys. Geogr.*, vol. 41, no. 1, pp. 3–28, Feb. 2017, doi: 10.1177/0309133316658614.
44. A. Emmer *et al.*, 'Progress and challenges in glacial lake outburst flood research (2017-2021): a research community perspective', *Nat. Hazards Earth Syst. Sci.*, vol. 22, no. 9, pp. 3041–3061, Sep. 2022, doi: 10.5194/nhess-22-3041-2022.
45. S. K. Jain, A. K. Lohani, R. D. Singh, A. Chaudhary and L. N. Thakural, 'Glacial lakes and glacial lake outburst flood in a Himalayan basin using remote sensing and GIS', *Nat. Hazards*, vol. 62, no. 3, pp. 887–899, Jul. 2012, doi: 10.1007/s11069-012-0120-x.
46. M. Amin *et al.*, 'Mapping and monitoring of glacier lake outburst floods using geospatial modelling approach for Darkut valley, Pakistan', *Meteorol. Appl.*, vol. 27, no. 1, Jan. 2020, doi: 10.1002/met.1877.
47. S. Dubey and M. K. Goyal, 'Glacial Lake Outburst Flood Hazard, Downstream Impact, and Risk Over the Indian Himalayas', *Water Resour. Res.*, vol. 56, no. 4, Apr. 2020, doi: 10.1029/2019WR026533.
48. D. H. Shugar *et al.*, 'Rapid worldwide growth of glacial lakes since 1990', *Nat. Clim. Change*, vol. 10, no. 10, pp. 939–945, Oct. 2020, doi: 10.1038/s41558-020-0855-4.

49. C. Prakash and R. Nagarajan, 'Glacial lake changes and outburst flood hazard in Chandra basin, North-Western Indian Himalaya', *Geomatics Nat. Hazards Risk*, vol. 9, no. 1, pp. 337–355, Mar. 2018, doi: 10.1080/19475705.2018.1445663.
50. S. K. Jain, R. D. Singh, M. K. Jain and A. K. Lohani, 'Delineation of flood-prone areas using remote sensing techniques', *Water Resour. Manage.*, vol. 19, no. 4, pp. 333–347, 2005, doi: 10.1007/s11269-005-3281-5.
51. S. K. McFeeters, 'NDWI BY McFEETERS', *Remote Sens. Environ.*, vol. 25, no. 3, pp. 687–711, 1996.
52. G. L. Feyisa, H. Meilby, R. Fensholt and S. R. Proud, 'Automated Water Extraction Index: A new technique for surface water mapping using Landsat imagery', *Remote Sens. Environ.*, vol. 140, pp. 23–35, 2014, doi: 10.1016/j.rse.2013.08.029.
53. P. S. Frazier and K. J. Page, 'Water body detection and delineation with Landsat TM data', *Photogramm. Eng. Remote Sens.*, vol. 66, no. 12, pp. 1461–1467, 2000.
54. H. Xu, 'Modification of normalized difference water index (NDWI) to enhance open water features in remotely sensed imagery', *Int. J. Remote Sens.*, vol. 27, no. 14, pp. 3025–3033, 2006, doi: 10.1080/01431160600589179.
55. S. Wang *et al.*, 'A simple enhanced water index (EWI) for percent surface water estimation using Landsat data', *IEEE J. Sel. Top. Appl. Earth Obs. Remote Sens.*, vol. 8, no. 1, pp. 90–97, 2015, doi: 10.1109/JSTARS.2014.2387196.
56. V. K. Gautam, P. K. Gaurav, P. Murugan and M. Annadurai, 'Assessment of Surface Water Dynamics in Bangalore Using WRI, NDWI, MNDWI, Supervised Classification and K-T Transformation', *Aquat. Procedia*, vol. 4, no. Icwrcoe, pp. 739–746, 2015, doi: 10.1016/j.aqpro.2015.02.095.
57. W. Chen, T. Yao, G. Zhang and S. Li, 'Accelerated glacier mass loss in the largest river and lake source regions of the Tibetan Plateau and its links with local water balance over 1976 - 2017', 2021.
58. M. Mergili and J. F. Schneider, 'Regional-scale analysis of lake outburst hazards in the southwestern Pamir, Tajikistan based on remote sensing and GIS', pp. 1447–1462, 2011, doi: 10.5194/nhess-11-1447-2011.
59. L. Wu, X. Li, S. Liu, P. K. Mool, S. Joshi and S. Bajracharya, 'Remote sensing based glacial lake inventory in the Hindu Kush-Himalaya region', *2011 International Conference on Remote Sensing, Environment and Transportation Engineering, RSETE 2011 - Proceedings*, pp. 1451–1454, 2011, doi: 10.1109/RSETE.2011.5964557.
60. J. Li, T. A. Warner, Y. Wang, J. Bai and A. Bao, 'Mapping glacial lakes partially obscured by mountain shadows for time series and regional mapping applications', *Int. J. Remote Sens.*, vol. 40, no. 2, pp. 615–641, 2019, doi: 10.1080/01431161.2018.1516314.
61. L. Yang, A. M. Maceachren, P. Mitra and T. Onorati, 'Visually-enabled active deep learning for (geo) text and image classification: A review', *ISPRS Int. J. Geoinf.*, vol. 7, no. 2, pp. 1–38, 2018, doi: 10.3390/ijgi7020065.

62. J. G. Lee and M. Kang, 'Geospatial Big Data: Challenges and Opportunities', *Big Data Res.*, vol. 2, no. 2, pp. 74–81, 2015, doi: 10.1016/j.bdr.2015.01.003.
63. C. D. Lippitt and S. Zhang, 'The impact of small unmanned airborne platforms on passive optical remote sensing: a conceptual perspective', *Int. J. Remote Sens.*, vol. 39, no. 15–16, pp. 4852–4868, 2018, doi: 10.1080/01431161.2018.1490504.
64. Z. Liu, H. Guo and C. Wang, 'Considerations on Geospatial Big Data', *IOP Conf. Ser. Earth Environ. Sci.*, vol. 46, no. 1, 2016, doi: 10.1088/1755-1315/46/1/012058.
65. X. Sun, B. Cao, B. Pan, K. Li, X. Zhao and W. Guan, 'Identification of Hazardous Glacial Lakes in the Yarlung Zangbo River Basin Based on Lakes Changes Determined Using Google Earth Engine', *Front. Earth Sci. (Lausanne)*, vol. 10, no. February, pp. 1–15, 2022, doi: 10.3389/feart.2022.825482.
66. M. Mei Zhang, F. Chen and B. Sen Tian, 'An automated method for glacial lake mapping in High Mountain Asia using Landsat 8 imagery', *J. Mt. Sci.*, vol. 15, no. 1, pp. 13–24, 2018, doi: 10.1007/s11629-017-4518-5.
67. J. Luo, Y. Sheng, Z. Shen and J. Li, 'Automatic and High-Precise Extraction of Water Information using a Hierarchical Iterative Method' Institute of Remote Sensing Applications, Chinese Academy of Sciences, Beijing, 1001012. Department of Geography, University of California, Los Ang', *Int. J.*, pp. 3–4.
68. H. Zhao, F. Chen and M. Zhang, 'A Systematic Extraction Approach for Mapping Glacial Lakes in High Mountain Regions of Asia', *IEEE J. Sel. Top. Appl. Earth Obs. Remote Sens.*, vol. 11, no. 8, pp. 2788–2799, 2018, doi: 10.1109/JSTARS.2018.2846551.
69. L. Ma, Y. Liu, X. Zhang, Y. Ye, G. Yin and B. A. Johnson, 'Deep learning in remote sensing applications: A meta-analysis and review', *ISPRS J. Photogramm. Remote Sens.*, vol. 152, no. March, pp. 166–177, 2019, doi: 10.1016/j.isprsjprs.2019.04.015.
70. R. Kemker, R. Luu and C. Kanan, 'Low-shot learning for the semantic segmentation of remote sensing imagery', *IEEE Trans. Geosci. Remote Sens.*, vol. 56, no. 10, pp. 6214–6223, 2018, doi: 10.1109/TGRS.2018.2833808.
71. L. Zhu *et al.*, 'Landslide susceptibility prediction modeling based on remote sensing and a novel deep learning algorithm of a cascade-parallel recurrent neural network', *Sensors (Switzerland)*, vol. 20, no. 6, 2020, doi: 10.3390/s20061576.
72. C. Pelletier, G. I. Webb and F. Petitjean, 'Temporal convolutional neural network for the classification of satellite image time series', *Remote Sens. (Basel)*, vol. 11, no. 5, pp. 1–25, 2019, doi: 10.3390/rs11050523.
73. V. Badrinarayanan, A. Handa and R. Cipolla, 'SegNet: A Deep Convolutional Encoder-Decoder Architecture for Robust Semantic Pixel-Wise Labelling', 2015.
74. S. Rajat, B. Rajeshwar Singh, C. Prakash and S. Anita, 'Glacier retreat in Himachal from 1994 to 2021 using deep learning', *Remote Sens. Appl.*, vol. 28, no. November, p. 100870, 2022, doi: 10.1016/j.rsase.2022.100870.

75. S. Guo *et al.*, 'Rapid large-area mapping of ice flow using Landsat 8', *J. Hydrol. (Amst)*, vol. 7, no. May, pp. 95–102, 2020, doi: 10.1016/j.epsl.2017.02.008.
76. E. Shelhamer, J. Long and T. Darrell, 'Fully Convolutional Networks for Semantic Segmentation', *IEEE Trans. Pattern. Anal. Mach. Intell.*, vol. 39, no. 4, pp. 640–651, 2017, doi: 10.1109/TPAMI.2016.2572683.
77. O. Ronneberger, P. Fischer and T. Brox, 'U-Net: Convolutional Networks for Biomedical Image Segmentation', in *MICCAI*, 2015.

8

Dynamic Coastal Flood Risk Assessment of a Coastal Island in West Bengal, India

Praneta, N., Aishwarya, N. and Bharath, H. A.*

Energy and Urban Research Group, Ranbir and Chitra Gupta School of Infrastructure Design and Management, Indian Institute of Technology Kharagpur, Kharagpur, West Bengal, India

Abstract

Coastal ecosystems are extremely productive and dynamic due to their continuous interaction with the sea, making them highly vulnerable to coastal flooding. By 2050, India's 3461 coastal villages will expose 36 million people to flooding due to rising sea levels. Local governments and stakeholders' coastal flood mitigation strategies typically rely on short-term solutions, but the advent of remote sensing data has made it easier for resource-strapped developing nations to formulate viable solutions. This study presents a dynamic flood risk model to analyze the business-as-usual (BAU) scenario of Sagar Island, which has been continuously imperiled by tropical cyclones, including Yaas (2021). After being assessed with a multi-criteria decision model, the chosen vulnerability and exposure factors are combined to estimate the future flood risk, using inundation height data from the Kolkata Port Trust and high-resolution data from an Indian satellite. A multi-level logistic patch growth algorithm model: FUTURES reveals an increase of 10.53% in the built-up over the next three decades (2020–2050), resulting in increased flood vulnerability. Core areas of the island have very low to moderate risk when compared to the coast, indicating an urgent need for sustainable solutions for circumventing prolonged flooding.

***Keywords*:** Coastal floods, risk modeling, exposure, land-use analysis, FUTURES, Sagar Island

**Corresponding author*: bharath@infra.iitkgp.ac.in

Disha Thakur, Sanjay Kumar, Har Amrit Singh Sandhu and Chander Prakash (eds.) Sustainable Development Using Geospatial Techniques, (181–216) © 2024 Scrivener Publishing LLC

8.1 Introduction

Coastal ecosystems are home to many while providing crucial services to the world [1]. These zones are frequently prone to various risks ranging from floods to deadly cyclones, causing heavy damage to human lives and infrastructure. The risks are often derived from anthropogenic interference [2]. A significant percentage (~10) of the world's populace resides in the 10-m low-elevation coastal zone, which equals 2% of the Earth's land area [3]. In India, more than 25% of the population lives within 50 km of the coast, while around 40% of the world's people reside within 100 km of the coast worldwide [3]. This increases the vulnerability of the coasts, demanding an immediate risk management framework to combat the situation. The rising need for efficient mitigation measures for coastal floods is one of the primary research focuses in this domain. Coastal flooding is caused by several factors, including tidal activity-induced high tides, wind-generated waves, high sea levels combined with high river flows, and sea level rise as a result of global warming [4]. Annual increases in global warming are also contributing to an increase in the frequency of coastal floods. Seawater expands as a result, increasing its likelihood of flooding coastal areas and moving farther toward the mainland [5]. Most typically, cyclonic occurrences result in coastal floods. India is particularly susceptible to disasters because it is a tropical nation that experiences cyclones on a regular basis. It is bound by the sea on three sides and is quite prone, especially during the high tides. The west coast of India, due to its shape and bathymetry, attracts these floods frequently [6, 7].

Researchers around the world have demonstrated the risk of floods at the coast using various statistical regression and spatial techniques [6]. These risk calculations help build policies and solutions for prevention. The hazard exposure is determined by taking the intersection of the hazard with the communities and assets that could be impacted [8]. Exposure refers to the susceptibility of natural hazards in the area under consideration, therefore having a direct relation to the topographical features of the area. The area that gets affected is directly dependent upon the physiographical conditions of the area [9]. Each community has a diverse reaction to the coastal hazard based on their behavioral patterns, community size, social position, etc. [10]. Some may show extreme resilience, while others may very well adapt to the consequences. Hence, considering the socioeconomic aspect becomes inescapable.

Several methods and approaches, such as a coastal vulnerability index (CVI) [6, 11, 12], data envelope analysis [13, 14], and a socioeconomic vulnerability index (SeVI) [10], have been used. For predicting coastal vulnerability through the calculation of sediment dynamics, erosion rate assessments, historical erosion, flood assessments, etc., CVI has been one of the most valued approaches [11]. Though a subjective term, CVI is considered to be a function of coastal slope, tidal range, displacement, bathymetry, and other physiographical features. Kumar *et al.* [6] made the novel addition of including the coastal regional elevation and tsunami run-up rate. Pantusa *et al.* [12] took into account the geomorphology, shoreline erosion, slope, new beach width, relative sea-level change, dune width, mean significant wave height, the width of vegetation behind the beach, and the mean tide range. This has also been used for calculating the risk of natural hazards along the coastal line [6]. Hence, even after the new additions and tweaks in the formula of CVI, it is still an ineffective method, as the method has always concentrated on the risk and damage that would be caused to the coastal strip and not the entire coastal area and the community.

As mentioned earlier, the community's reaction to the floods and their resistance to floods also play a major role in vulnerability and risk calculations. Hence, the SeVI is more specific, capturing the pure essence of the socioeconomic parameters in the assessment of climate change–based vulnerability by scrutinizing the community sensitivity, exposure, and adaptive capacity [15]. Coasts being the hub for multiple activities and various communities, acquiring a deep understanding of the impact of the flood on an individual community as well as the overall impact on all the existing communities becomes important. McLaughlin *et al.* [10] calculated SeVI using the following indicators: roads, population, railways, cultural heritage, land use, and conservation status. Ahsan and Warner [15] considered demographics (male-female ratio, age groups, and density), social (literacy, pakka houses, and household type), and economic (job type and property or land ownership). Narendr *et al.* [9] used socioeconomic parameters along with a new addition of accessibility parameters (proximity to the shoreline, amenities, roads, etc.).

There can be numerous reasons leading to increased flood risk in a region, one of which can be unplanned or unforeseen developmental growth. This leads to a hefty flood risk in the future, causing monumental damages and economic losses. Land-use planning plays a crucial role in hazard risk reduction. A good land-use strategy will focus on minimizing the damage by correctly analyzing the hazard-sensitive parts. It is a well-known fact that huge economic and life losses will occur in the event of any damage to the infrastructure and the built-up. Coastal cities like Mumbai and Kolkata in India, due to their exponential unplanned growth and poor city planning strategies, have failed to ensure safety from the coastal flood risk [16]. These cities still face enormous life and infrastructural losses due to the flood events. Thus, planning these effectively to reduce their exposure to floods will lead to resilience and minimal damage.

Calculating the present risk is necessary, but analyzing and estimating the risk in the future is even more necessary [16, 17]. With the continual risk of global warming and connected events such as sea-level rise (SLR), the flood risk in coastal areas is increasing simultaneously. Future risk calculations require indicators with the projected values of the future, such as the expected population, changed land-use patterns, and development growth. Thus, growth models have been used to predict future development trends, characterize and evaluate the effects of future development, and investigate the possible effects of various policy options existing or proposed in the area [18]. Numerous growth models have been built to predict the land-use patterns of the area using various parameters as inputs and trend calculations [19].

Land change models such as SLUETH [20], MOLUSCE (Modules for Land Use Change Simulations), and recently FUTURES (FUTure Urban-Regional Environment Simulation) have been used for determining the possible land-use change patterns for the flood risk calculations [19]. FUTURES is provided by GRASS GIS to generate a predicted development using the concept of a patch-growing algorithm, showing realistic results [21]. The model carries out the parameter scoring based on temporal land use, which removes the bias of traditional MCDA scoring. Furthermore, it also takes into account the historical population trend and builds upon it to give the results, making them more accurate [22]. The advantage of the FUTURES model is that it does not apply one deterministic condition throughout the area but works heterogeneously and provides land change based on contextual data [19].

Therefore, this research aims to project the risk for the year 2050 using an MPFR (multidimensional parametric flood risk model). Dynamic from the aspect of the temporal nature of the analysis, i.e., the changing land-use pattern over the years, is the key to this research. Overall, this paper demonstrates a temporal risk analysis in a BAU situation, considering the socioeconomic effect on land-use development or modification in the region based on which appropriate land-use zoning guidelines can be recommended. High-resolution data have been used to capture even the minutest change in the pattern to provide highly realistic results and calculate the risk accurately.

8.2 Study Area

Sagar Island, a part of the South 24 Parganas in the state of West Bengal and the largest island in the Indian Sundarbans, is chosen as the research area [23]. Due to its natural heritage, it is recognized as a UNESCO World Heritage Site and is home to the biggest mangrove forest in the world [24]. Low-lying salt marshes, mudflats, sand flats, and marshy plains make up the area in the deltaic zone [25].

It is the largest and most populated island with a population of around 212,037 (2011 census) of the Sundarbans delta regions, with a north-south length of approximately 30 km and a maximum approximate width of 12 km [26].

Sagar, with a total area of around 282 km^2, is divided into 47 administrative zones (Figure 8.1), the majority of people being below the poverty line [27]. A tropical climate prevails throughout the area. The island has had about 40 tropical cyclones and 13 super-cyclones between 1882 and 2020. Sagar Island's sea levels have been demonstrated to be rising over a long period [28], with an annual increase rate closer to 4.4 mm. The livelihood of coastal residents has been impeded by rising sea levels and more severe cyclonic events, both of which are caused by rising global temperatures. The island has been subject to extreme cyclonic events such as Aila

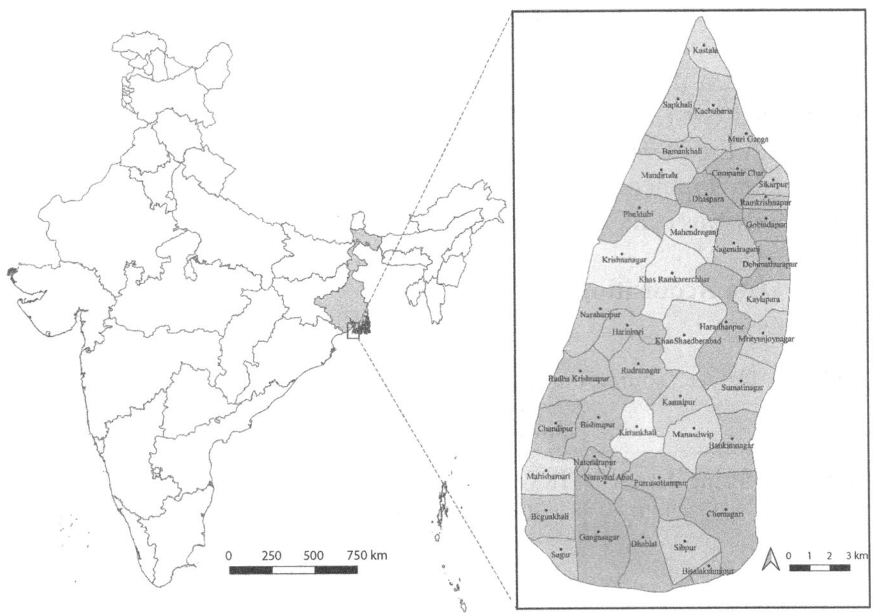

Figure 8.1 Study area: Sagar Island.

(2009) and Yaas (2021), which have severely affected the island. The island has about 70% semi-permanent housing, making it even more vulnerable to damage from floods [9]. Due to the accessibility and amenities on the island, it has been a constant refuge for the villagers living around the island. Over the years, the island has seen a reduction in the mangrove belt due to development, exposing it more to the risk of inundation.

8.3 Method

8.3.1 Multidimensional Parametric Flood Risk Model (MPFR - Model)

The method comprises an MPFR model that works on the principle of flood risk as a function of hazard exposure and vulnerability elements at risk. The rapid rate of unplanned development makes it essential to maintain a dynamic outlook on risk perception. Hence, the proposed method aims to estimate the change in the flood risk profile of Sagar Island concerning changes in land use in BAU conditions over 30 years. The model has been developed considering the available data and the scope of open-source software (QGIS and GRASS GIS). Table 8.1 consists of the dataset required for the MPFR model, and Figure 8.2 explains the model structure, components, and flow.

Data collection through satellite imagery and other valuable sources is done as explained in Table 8.1. Secondly, the growth parameters have been obtained from the processed data. Following this, land-use classification was carried out (2012, 2014, 2018, and 2020), which was taken as input along with the identified growth parameters for the simulation model to obtain a development map for 2050. Using the map of 2050, the risk factor in the study was computed, generating a risk zone map.

Step 1: Data Acquisition

Table 8.1 Dataset, source, and description required for the analysis.

Data	Source	Description
Remote sensing data at 5.8 m - 2012–2020	IRS LISS IV & ResourceSat 2A	Flood exposure mapping, delineation of coastline, land-cover dynamics, and creeks

(Continued)

Table 8.1 Dataset, source, and description required for the analysis. (*Continued*)

Data	Source	Description
Administration boundaries 'mouza (village)'	Census of India - GoI, land and land data - government of West Bengal	Area of interest, sensitivity/ vulnerability assessment
Stereo pair satellite data at 2.5 m	IRS Cartosat-1–NRSC, government of India	Flood inundation, topography, and exposure mapping
Flood extent data (5 m) 26th May 2021	Sentinel 1-A - Copernicus program	Yaas cyclone flood mapping
Social and economic data, infrastructure, and demographic data	Census of India - GoI	Vulnerability assessment
Police stations, road networks, Banks, bus stops, schools, hospitals and docks	Virtual Earth Database, Google Earth	Vulnerability to infrastructure accessibility and also, for land-use modeling.
GPS-based field data	NRSC Bhuvan, field investigations	Land-use analysis, understanding damage due to climate extremes
Historic topographic data	1:50,000 scale (SOI)	Base data for terrain mapping

Step 2: Data Processing:

a) Layer Extraction: Exposure and Vulnerability Parameters

The flood exposure is calculated based on the physiographic parameters (Table 8.2) and the land-use map, while the vulnerability factors (Table 8.3) are calculated based on the socioeconomic parameters.

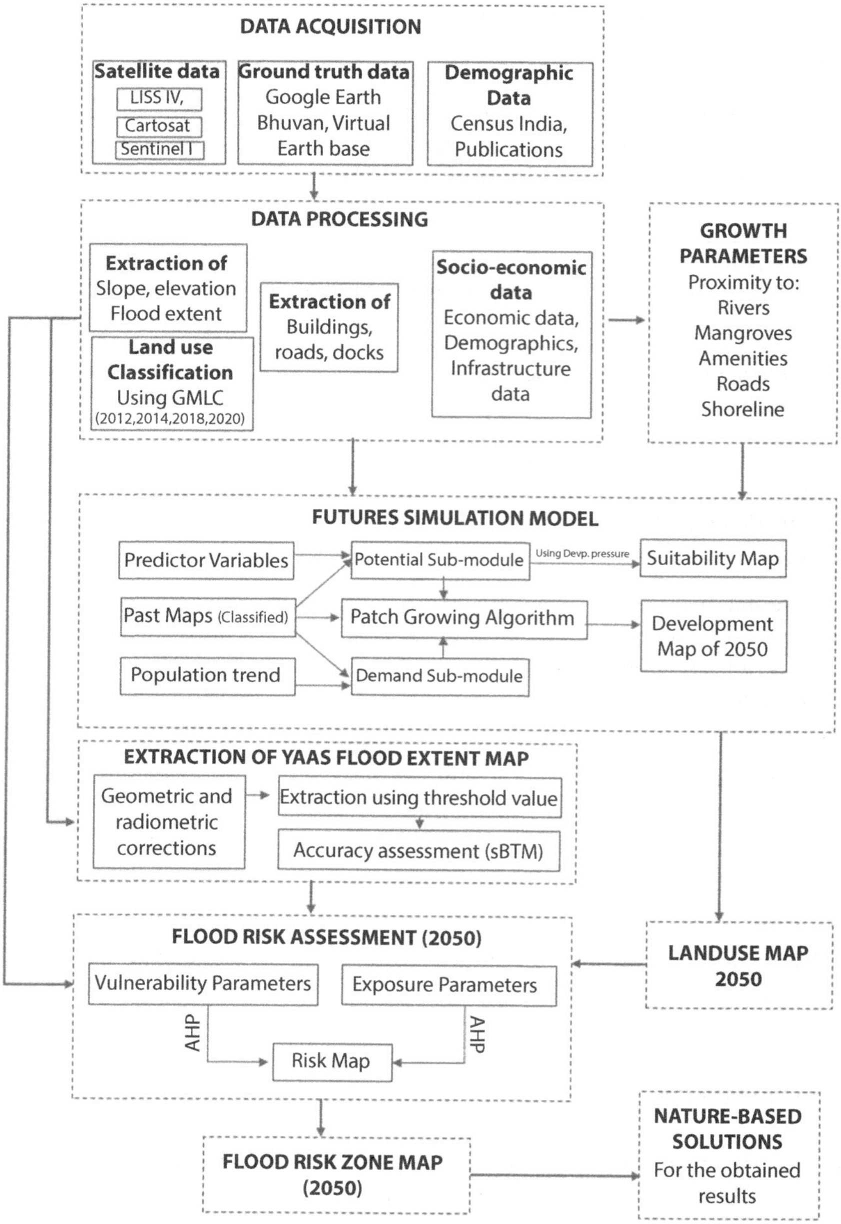

Figure 8.2 Multidimensional parametric flood risk model (MPFR - model) method.

Table 8.2 Exposure parameters.

Physiographic/ exposure parameters	Description
Slope (in degrees)	Slope regulates the velocity of stormwater. It influences the extent and duration of coastal inundation extracted from DEM
Elevation (in meters)	Elevation is used for predicting coastal vulnerability due to inundation. Extracted from DEM
Distance from the coast (in meters)	Responsible for flood spread and increasing sensitivity to flooding. Areas close to the shoreline show high exposure to coastal flooding.
Distance from the stream (in meters)	Tidal water affects the adjacent ground and inundates the low-lying areas across the streams
Land use	A classified map for 2012, 2014, 2018, and 2020 is used for the analysis. Housing typology is highly vulnerable due to flooding. Kutcha houses can be destroyed because of flood action. Saltwater due to coastal flooding affects agricultural fields
Flood extent map	An inundation map for Yaas cyclone. The map shows the areas prone to flooding when a similar situation occurs in the future.

b) Parameter Classification

All the chosen parameters are classified into five categories using the Likert scale, from very low to very high. The factors considered in the classification are mentioned in Tables 8.4, 8.5, 8.6, 8.7, and 8.8.

i. Dependent Population - The percentage of the population that is dependent on the household's wage earners is included in this parameter. This represents the non-working population, which comprises female household members who are not employed, individuals over 65, and children under the age of 15.

Table 8.3 Vulnerability parameters.

Vulnerability indicators		Description
Sociodemographic	Dependent population	Non-earning population
	Illiterate population	Population with no or primary level education
	Deprived population	Scheduled caste (SCs) and Scheduled Tribe (STs) population.
Economic	Job typology	No job or low-income job
	Housing typology	Number of Kutcha houses
	Landholding	No land ownership; Marginal worker population
Infrastructure services	Health care facility	The presence or absence of infrastructural
	Treated water supply facilities	
	Drainage system	
Accessibility to infrastructure	Accessibility to infrastructures such as banks, schools, hospitals, docks, main roads, metal roads, and police station	Approachability to critical infrastructures such as banks, schools, hospitals, and police stations.

ii. Illiterate Population - Percentage of residents in the study area with only an elementary education or no education at all. Lack of education contributes to disaster-related ignorance.

iii. Backward or Deprived Population - Due to societal injustice, scheduled castes (SCs) and scheduled tribes (STs) are disadvantaged [9]. As a result, the underprivileged population experiences disparities in access to education, equal rights, and opportunities for resilience.

Table 8.4 Exposure indicator classification.

Parameter (%)	Exposure				
	Very high (1)	High (2)	Moderate (3)	Low (4)	Very low (5)
Elevation (m)	<2	2 to 3	3 to 4	4 to 6	>6
Slope (degree)	<1	1 to 2	2 to 3	3 to 5	>5
Proximity to coast (m)	<500	500 to 750	750 to 1000	1000 to 1250	>1250
Proximity to creek (m)	<200	200 to 300	300 to 400	400 to 500	>500
Landscape features	Built-up	Agriculture	Horticulture	Others	Water and inland vegetation

Table 8.5 Sociodemographic indicator classification.

Parameter (%)	Sociodemographic				
	Very high (1)	High (2)	Moderate (3)	Low (4)	Very low (5)
Dependent	>35	30 to 35	25 to 30	20 to 25	<20
Illiterate	>40	35 to 40	30 to 35	25 to 30	<25
Backward/ deprived	>60	40 to 60	20 to 40	10 to 20	<10

Table 8.6 Economic indicator classification.

Parameter (%)	Economic				
	Very high (1)	High (2)	Moderate (3)	Low (4)	Very low (5)
Housing typology	>40	30 to 40	20 to 30	10 to 20	<10
Landholding	>60	45 to 60	30 to 45	15 to 30	<15
Job type (low/no income)	>4	3 to 4	2 to 3	1 to 2	<1

Table 8.7 Infrastructure indicator classification.

Parameter (%)	Infrastructure				
	Very high (1)	**High (2)**	**Moderate (3)**	**Low (4)**	**Very low (5)**
Health care	Absence	-	-	-	Presence
Treated water supply	Absence	-	-	-	Presence
Drainage	Absence	-	-	-	Presence

Table 8.8 Accessibility indicator classification.

Parameter (m)	Accessibility				
	Very high (1)	**High (2)**	**Moderate (3)**	**Low (4)**	**Very low (5)**
Bank	>4000	3000–4000	2000–3000	1000–2000	<1000
Bus stop	>1000	800–1000	500–800	300–500	<300
Docks	>4000	3000–4000	2000–3000	1000–2000	<1000
Hospital	>4000	3000–4000	2000–3000	1000–2000	<1000
School	>1000	800–1000	500–800	300–500	<300
Main road	>2000	1500–2000	1000–1500	500–1000	<500
Metal road	>1000	800–1000	500–800	300–500	<300
Police station	>4000	3000–4000	2000–3000	1000–2000	<1000

i. Housing typology - It is based on the concept of temporary houses, which are easily susceptible to floods [9].

ii. Landholding - The population's percentage of non-landowners is depicted. It includes the marginal workforce, which is defined as those employed for fewer than six months (Census of India).

iii. Infrastructure accessibility denotes resource mobility, which involves the flow of people, products, and services. Reducing risk during flood disasters aids in boosting resilience. Better accessibility may make it easier for people to flee or receive aid.

c) Land-Use Classification for 2012, 2014, 2018, and 2020

From the derived satellite images from LISS IV, land-use classification is done for 2012, 2014, 2018, and 2020. A supervised technique of land-use classification using a maximum likelihood classifier was applied, which considers training input for each class to be normally distributed. Furthermore, it computes the probability, or likelihood, of a pixel belonging to a class. Eight classes, namely, agriculture (sown and fallow), inland vegetation, water, mangroves, built-up, others, and Casaurina (a type of deciduous tree often found on the island), were identified based on the surveys and the secondary data.

d) Extraction of Growth Parameters

The growth parameters were identified based on the visual analysis of temporal land-use maps. The development of built-up areas near these factors (as per Table 8.9) has demonstrated their essentiality for being selected. These parameters serve as the base layers for the growth model.

e) Flood Extent Map for Cyclone Yaas

SAR (Synthetic Aperture Radar) imagery from Sentinel-1 was used for Yaas flood event mapping (May 26, 2021). To produce images of higher quality, speckle filtering was used to minimize the noise in the image. The water and non-water pixels were distinguished using a histogram of the filtered backscatter coefficient and applying thresholds over it. Thresholds (0.0434) are applied by the user, which is considered an unsupervised classification method. Once the flood mask was obtained, it was assessed for accuracy by comparing it with the developed bathtub model flood map. A connected bathtub model using the DEM and the water level data from the Kolkata port trust was created which, when compared to the obtained Yaas flood mask, gave an accuracy of 87% [29].

This helps in understanding the potential risk that can be faced by the community if a similar event, such as the Yaas cyclonic floods, occurs in

Table 8.9 Growth influencing parameters.

Category	Growth parameters	Description
Physiographic	Slope	Construction and living on the gentler slopes are viable and easier than settling on the steeper slopes.
Accessibility	Distance to amenities	Amenities such as hospitals, schools, police stations, shops, etc., attract people and are helpful during times of coastal disasters.
	Distance to rivers	Rivers are great communication channels and sources of economic activities like fishing.
	Distance to shoreline	These are great tourist attraction points and sources of economic activity.
	Distance to mangroves	They are natural buffers and help protect people from floods, tides, and surges.
	Distance to roads	Roads are the most important internal connectivity channels leading to development.

the region. It gives a clear picture of the areas that are likely to be inundated; thus, it is considered an exposure parameter.

Step 3: Future Built-Up Prediction Using the FUTURES Model:
The patch-growing algorithm-based model takes land-use classified maps, slope values, demographic data, and socioeconomic data as inputs to derive a development map for 2050. An open-source urban growth model called FUTURES (Figure 8.3) works on the following three main sub-models: demand, potential, and patch growing algorithm (PGA). The whole ecosystem is based on the interaction of these sub-models.

a) Potential sub-model:
The potential sub-model is derived from socioeconomic conditions, using which a land-use suitability map is produced. It uses the influence of

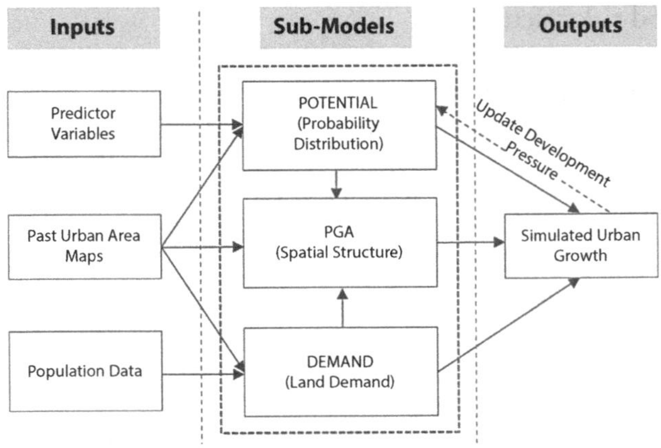

Figure 8.3 FUTURES model conceptual framework [19].

growth parameters (proximity to roads, rivers, shorelines, amenities, etc.) to determine the chance that a region will be developed [22]. The parameters were ranked using logistic regression rather than the standard CA model method.

b) Demand sub-model:
The per-capita demand is derived from the projected population using the population trend of the past years via the demand module.

c) Patch growing algorithm:
The PGA module is a stochastic algorithm used to generate patches of various sizes and shapes based on the surrounding conditions [21]. Using the computation of these patches, the model predicts the possible size and location of the development of the area.

Once the binary development map of 2050 is generated, the classified map of 2020 will serve as the foundation for the development map, which will then be patched to it and utilized to calculate risk.

Step 4: Ranking Using AHP and Risk Modeling
The parameters in each category are ranked based on the MCDM technique, analytical hierarchical process (AHP), which generates separate vulnerability and exposure maps for the island. In this study, every value aij in the matrix depicts the relative sensitivity of the flood vulnerability parameters. According to Saaty (2008), relative importance is assigned on a scale of 1 to 9 based on the scale of preference.

Coastal flood risk calculations:

i. Using the product of vulnerability and exposure maps, coastal risk zoning is carried out. The main Equation 8.1 for calculating the risk is:

$$R_f = H_f * V_f \tag{8.1}$$

where R_f is the risk factor, H_f is the exposure, and V_f represents the overall vulnerability.

ii. Flood exposure calculations:
Six parameters [slope, elevation, flood inundation, land use (2050), and proximity to streams and coast] are taken as functions of exposure.

a. Slope: The ratio of the elevation change to the horizontal distance between any two points on the coast is known as the "coastal slope." The steepness or flatness of the coastal region's slope affects how easily the region is inundated [6]. Areas with slopes less than 1° are considered highly vulnerable.
b. Elevation: Minimum 2-m tidal amplitude variations were typical resulting in frequent inundation of the land of 2 m above the MSL. Furthermore, rare yet exceedingly damaging, very high tides with an amplitude of 4 m or more are frequently related to cyclones.
c. Proximity to streams: According to Coastal Regulatory Zone (CRZ) standards, a 200-m natural buffer must be kept between a site and streams and creeks. The 200-m buffer zone from the streams was therefore thought to be particularly vulnerable to flooding.
d. Proximity to the shoreline: A buffer zone 500 m away from the high tide zone was deemed to be particularly vulnerable to inundations.
e. Land use: In the case of floods, the damage to infrastructure and lives will be most valuable; hence, the built-up class has been considered to have maximum exposure when compared to other classes such as mangroves, forests, and inland vegetation.

Exposure maps are derived spatially using Equation 8.2

$$H_f = \sum_{i=1}^{i=6} W_i * S_i \tag{8.2}$$

where H_f denotes food-induced sensitivity, W_i denotes the parameter i's weight, and S_i denotes the parameter i's spatial sensitivity, with values between 1 and 6 (using AHP calculations) denoting very high to very low sensitivity.

iii. The vulnerability parameter has been categorized into four sub-parameters, which are even further divided. The four sub-parameters, namely, sociodemographic, economic, infrastructural, and accessibility, have been explained in Section 8.4.3.2.

Vulnerability maps are derived using Equation 8.3

$$V_f = \sum_{\substack{i=1 \\ j=1}}^{\substack{i=2 \\ j=n}} W_i * W_j * V_j = W_1 \sum_{j=1}^{j=n} W_j * V_j + W_2 \sum_{j=1}^{j=n} W_j * V_j +$$
$$W_3 \sum_{j=1}^{j=n} W_j * V_j + W_4 \sum_{j=1}^{j=n} W_j * V_j$$
$$\sum_{j=1}^{j=n} W_j = 1 \tag{8.3}$$

where V_f stands for overall food vulnerability, W_i stands for the weight of vulnerability parameter i, and W_j stands for the weight of vulnerability sub-parameter j. 'i' ranging from 1 to 4 indicates (1) sociodemographic, (2) economic, (3) infrastructure, (4) accessibility parameters.

8.4 Results and Discussion

8.4.1 Land-Use Mapping

As explained in the previous section (8.3.1), supervised land-use classification was conducted for the years 2012, 2014, 2016, and 2020. With observations from Google satellite imagery and ground surveys, eight land-use classes were chosen, namely, agricultural land (sown and fallow), water, casuarina, inland vegetation, mangroves, others, and built-up. Figure 8.4 shows maps of 2012 and 2020, and changes in the pattern of land use can be observed. The change observed in the inland vegetation over the eight years is negligible, and the mangroves have decreased by 0.2%, indicating the risk of seawater penetration due to the decreased overall water

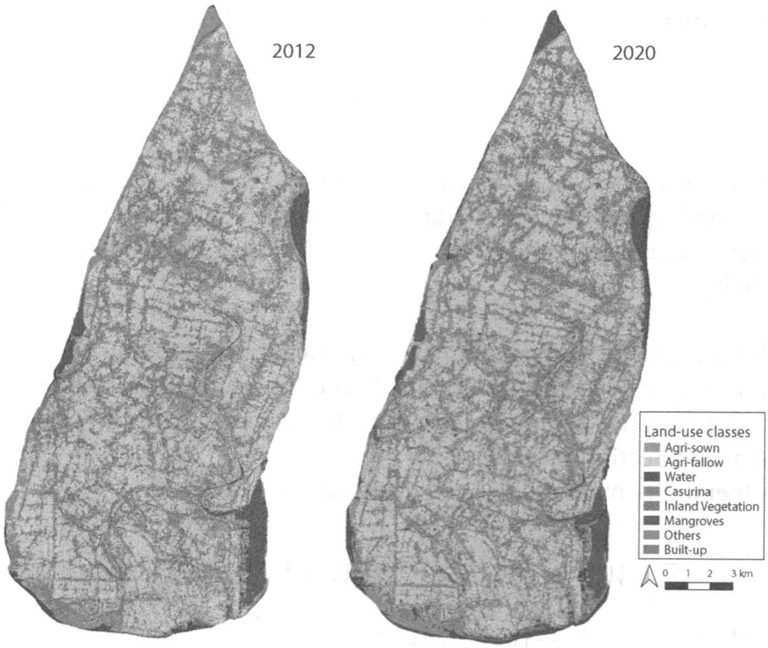

Figure 8.4 Land use for 2012 and 2020.

retention capacity of the mangroves. From Table 8.10, we can see the variation in terms of percentage over eight years. We observe an increase in water class (by 1.98%), indicating the penetration of water into the island through seawater inundation. Built-up has also increased by 2.02%, indicating considerable development happening across the island. This will exponentially grow over time, demanding a more intense need for flood mitigation strategies to deal with this increase. Thus, indicating a dire need for accurate land-use planning techniques and methods to minimize the risk and damage that would be caused to the newly increased development in the future.

Figure 8.5 shows the growth pattern (from 2012 to 2020) in detail for two areas as depicted in the keymap. The first set of images shows the increase in built-up (depicted in red) along the road network, and the second set of images shows the growth along the creeks. In 8 years, built-up areas have increased 2.5 to 3 times in Mahishamari, Bakimnagar, Gobindpur, and Ramkrishnapur and 1.5 times in Chandipur, Mrityunjoynagar, Dhablat, Sumatinagar, Naraharipur, and Mandirtala. In the villages of Bishalakshmipur, Muri Ganga, Khan Saheber Abad, and Campanir Char, the minimum growth was less than 0.5 times. Based on similar

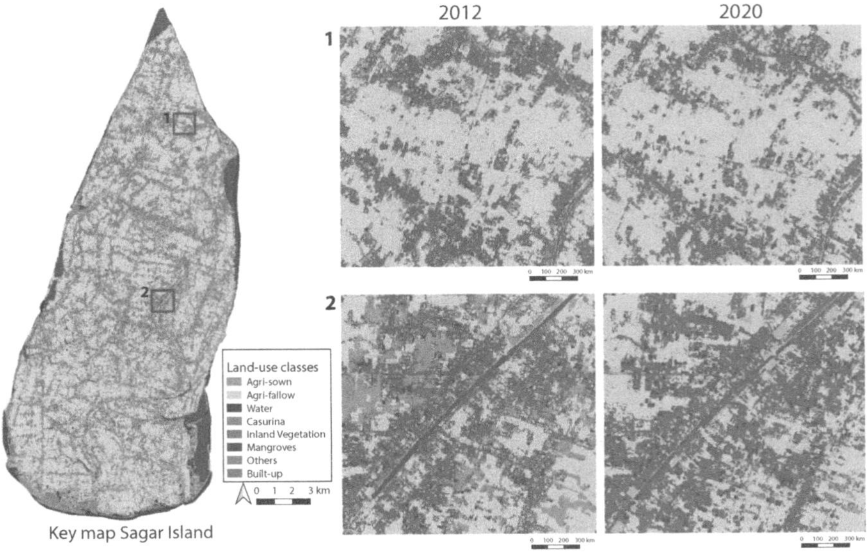

Figure 8.5 Land use for 2012 and 2020.

observations, the growth modeling parameters were selected. This signifies the role of socioeconomic parameters in transforming the region, as built-up originates near the road, amenities, creeks, etc. The built-up growth in the region indicates an obvious future development, leading to the next step in built-up growth prediction.

8.4.2 Future Built-Up Prediction

8.4.2.1 Mapping Growth Parameters

Growth parameters are the parameters used as input parameters for simulating the predicted built-up for 2050. The parameters in Figures 8.6 and 8.7 show the proximity to the respective features. Red has been used to represent a nearness to the growth parameter, whereas the color changes to green and then to blue as the distance increases. Features like roads, amenities, and creeks have a high possibility of attracting development, whereas areas near the shore and high slopes are less desirable to any development. When there is the development of a settlement, it happens due to some influencing condition. Generally, a lot of development is seen around the marketplaces or near some business areas; these conditions are termed the "growth parameters." A multi-level logistic regression is carried out on these six identified parameters by the FUTURES model, which assigns the

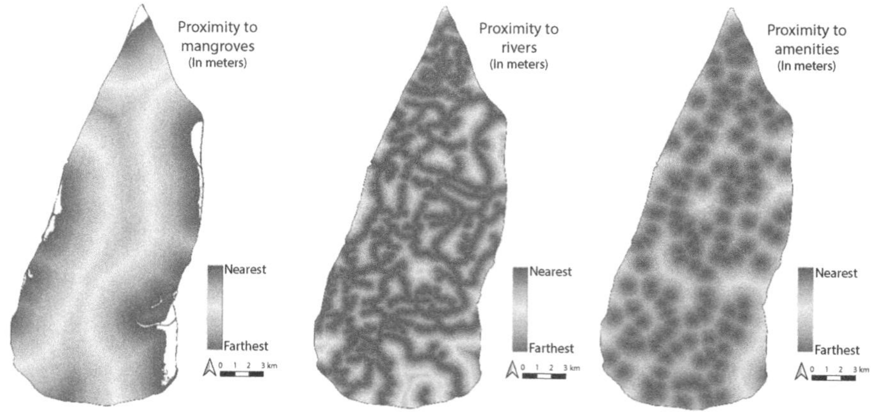

Figure 8.6 Proximity analysis of mangroves, rivers, and amenities.

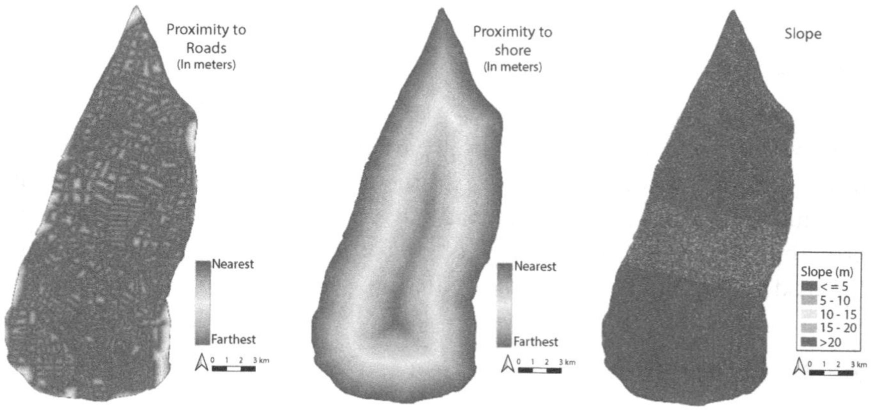

Figure 8.7 Proximity analysis of roads, shoreline, and slope.

weights based on the existing development trends from the provided land-use maps.

8.4.2.2 Land Suitability Analysis

As an outcome of the potential sub-model, a development pressure map was generated as the intermediate layer from the FUTURES model to determine "development suitability" for the areas. The model uses the past development pressure map of 2020 for Sagar Island to predict a series of future suitability until 2050. Suitability is the function of proximity

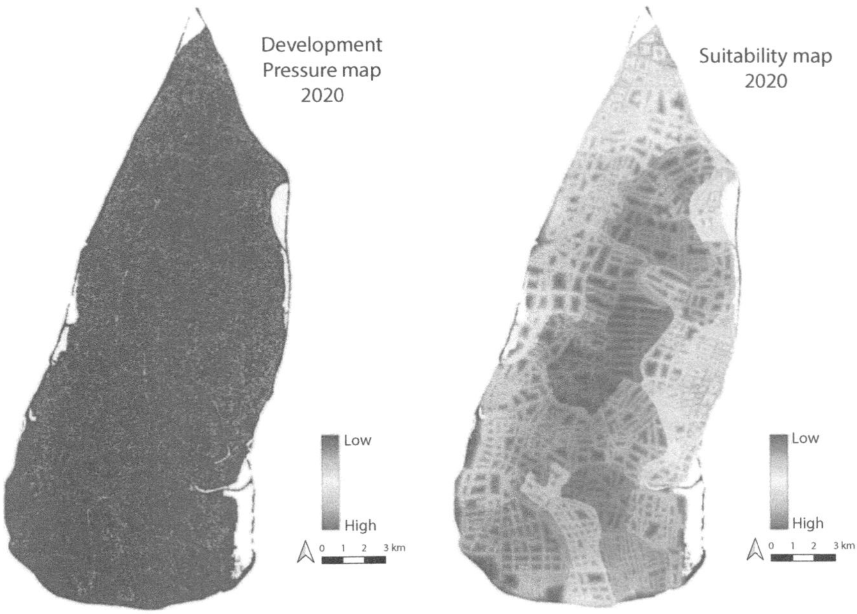

Figure 8.8 Development pressure and Suitability map.

parameters along with the slope and past development pressure, generating future suitability for 2050 [21].

In Figure 8.8(a), the bright yellow regions have a high chance of getting developed. The suitability map in Figure 8.8(b) is produced as an output from the potential sub-module. The green areas in the suitability map have a high development probability compared to the red regions. It can be observed from the map that, the land area in the buffers of the roads shows up yellow, indicating excellent suitability along the road networks. This can be similarly explained in the case of abundant amenities present in the south and jetties and docks on the eastern coast, which are green, showing even higher suitability. The villages on the east coast, Muriganga, Sumatinagar, Mrityunjoynagar, and Kastala, show promising suitability for development due to the presence of influencing factors such as roads and coasts.

8.4.2.3 Predicted Land-Use Map of 2050

Figure 8.9 shows the land-use map for 2020 and 2050. An obvious increase in the built-up can be seen in the predicted map of 2050. Dense development in the north (Kastala, Sapkhali, and Kachubaria) and east coast

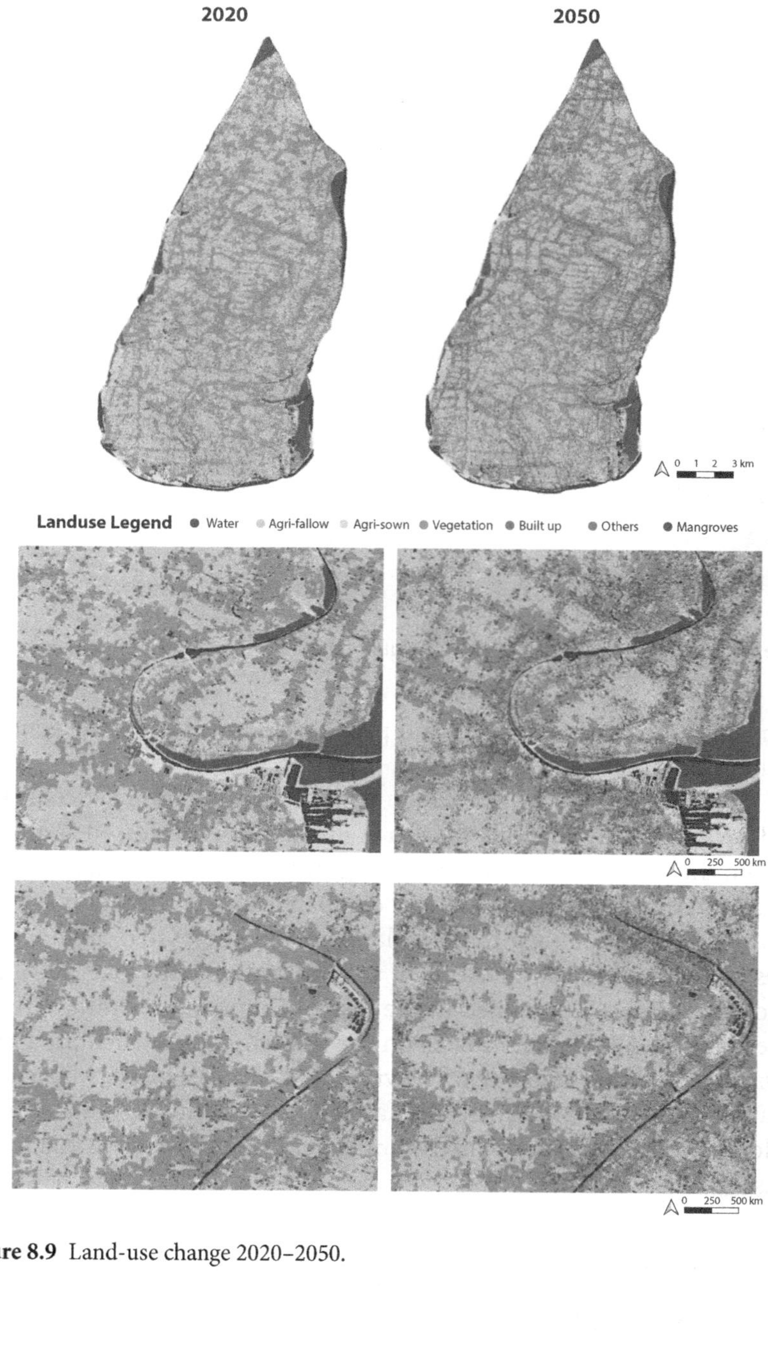

Figure 8.9 Land-use change 2020–2050.

Table 8.10 Comparison of land-use features in 2012, 2020, and 2050.

Features	2012	2020	2050
Agriculture-Sown	9.90%	5.03%	4.38%
Agriculture-Fallow	46.50%	47.83%	43.5%
Water	2.75%	4.73%	4.35%
Others	7.15%	6.81%	5.92%
Inland vegetation	28.32%	28.40%	24.13%
Mangroves	3.28%	3.07%	3.07%
Casuarina	0.27%	0.20%	0.20%
Built-up	1.84%	3.92%	14.45%

(Sumannagar, Mrityunjoynagar, and Haradhanpur) is observed due to the presence of jetties and connectivity to mainland India. Moderate to low development can be observed in the core areas of the region due to the presence of required and well-functioning amenities and infrastructure that support the community. As expected, the majority of development is around roads, creeks, and amenities. Table 8.10 shows the change in the land-use classes for 2012, 2020, and 2050. There is an evident change of approximately 10.5% in the built-up class from 2020 to 2050. The historical population trend had shown an exponential trend, which is also reflected in the developmental growth (built-up and infrastructure), thus leading to real results for our approach. Since there is considerable development, the vulnerability also increases, leading to more flood risk.

8.4.3 Parameter Ranking Using AHP

8.4.3.1 Yaas Flooding as an Exposure Parameter

The majority of the blue parts can be seen along the coast and in the northern parts of the island. As per the observations (Figure 8.10), the villages of Muri Ganga, Kachubaria, Chemgari, Bilaslakshmipur, Sibpur, Bishnupura, and Sumannagar were the most affected areas. Lower elevations and gentler slopes result in more inundation in these areas.

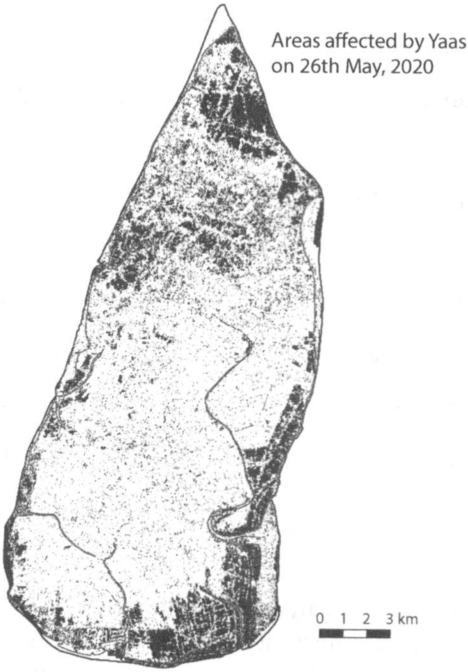

Figure 8.10 Flood inundation map (Yaas).

8.4.3.2 Ranking Using AHP

i. Exposure Parameters
According to Table 8.11 and Figure 8.11, the distance to the shoreline (weight = 0.35) and the distance to the creek (weight = 0.2) have higher weights, indicating relatively higher importance in the overall exposure calculations. These indicators are followed by the slope (weight = 0.16), elevation (weight = 0.13), land use (weight = 0.09), and the flood case of Yaas (weight = 0.07), showing that higher elevations and steeper slopes are good for resisting floods (Narendr *et al.*, 2022; Kumar *et al.*, 2010). Cyclone Yaas represents one in a 1000-year return period event. Hence, the probability of the occurrence of this event is less in the current scenario; however, given the climate change predictions, the probability of such occurrences may be more frequent in the future.

ii. Economic Parameters (Vulnerability Parameter)
The sociodemographic, economic, infrastructure, and accessibility characteristics have weights of 0.06, 0.52, 0.15, and 0.27, respectively (Figure 8.12).

Table 8.11 Indicator score based on AHP.

Component	Weight	Parameter	Weight	Sub-parameters
Flood exposure	0.13	Elevation		
	0.16	Slope		
	0.2	Proximity to creek		
	0.35	Proximity to shore		
	0.09	Land use		
	0.07	Yaas flood		
Flood vulnerability	0.06	Sociodemographic	0.2	Population: Dependent
			0.31	Population: Illiterate
			0.49	Population: Deprived
	0.52	Economic	0.49	Housing typology
			0.2	Landholding
			0.31	Job type
	0.15	Infrastructure	0.56	Health care
			0.2	Water supply
			0.14	Drainage
	0.27	Accessibility	0.08	Banks
			0.14	Docks
			0.06	Bus stops
			0.19	Hospital
			0.16	Roads: Major
			0.06	Roads: Metal
			0.23	Schools
			0.08	Police station

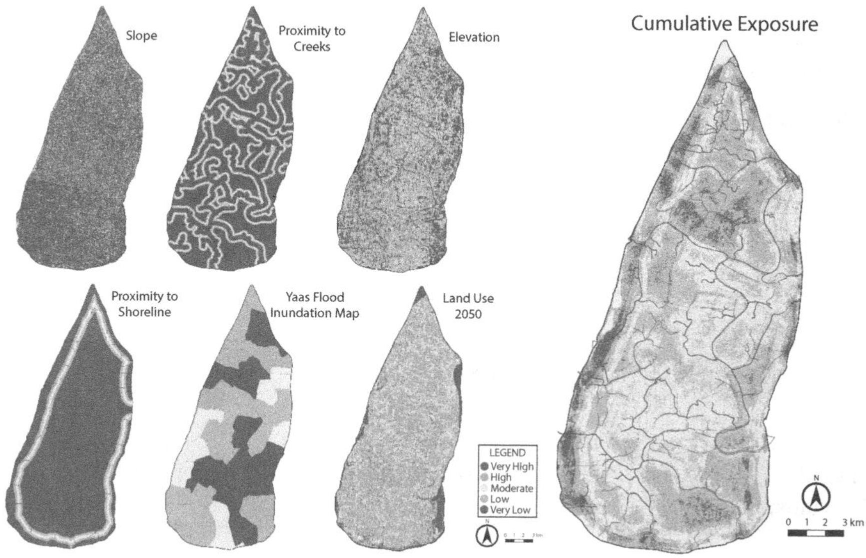

Figure 8.11 Exposure mapping.

Researchers have drawn attention to the importance of economic factors in ensuring a community's well-being by enabling speedy recovery after disasters. Economically strong regions tend to recover faster as they can restore resources. Due to immigration from other islands, seawater intrusion, and small or no landholdings, Sagar's agriculture-based economy has lost both its financial stability and overall flood security. The economic factor is indicated by three sub-factors: job type (permanent-temporary), housing typology (kutcha-pakka), and land-holding ownership. Residents with a permanent job or permanent land ownership are bound to stay despite the high risk of flooding, making them more prone to such events.

iii. Accessibility Parameters (Vulnerability Parameter)

The economic parameter is followed by the accessibility indicator (Figure 8.13), with a weight of 0.27, owing to the explanation that it is quite important to have the right facilities, such as hospitals, police stations, and roads at an accessible distance to build an efficient and well-functioning disaster management system. Schools have been considered as they function as evacuation centers during cyclones and other natural disasters in addition to their regular operations; hence, a growing commute time could compromise mitigation efforts [9, 30]. People have been unable to access essential

Figure 8.12 Vulnerability mapping (sociodemographic, economic, and infrastructural).

medical needs due to the lack of all-weather routes, which has resulted in serious impairment or even casualties. Thus, these reasons make the accessibility indicator fairly important. This is followed by the infrastructure indicator, with a score of 0.15.

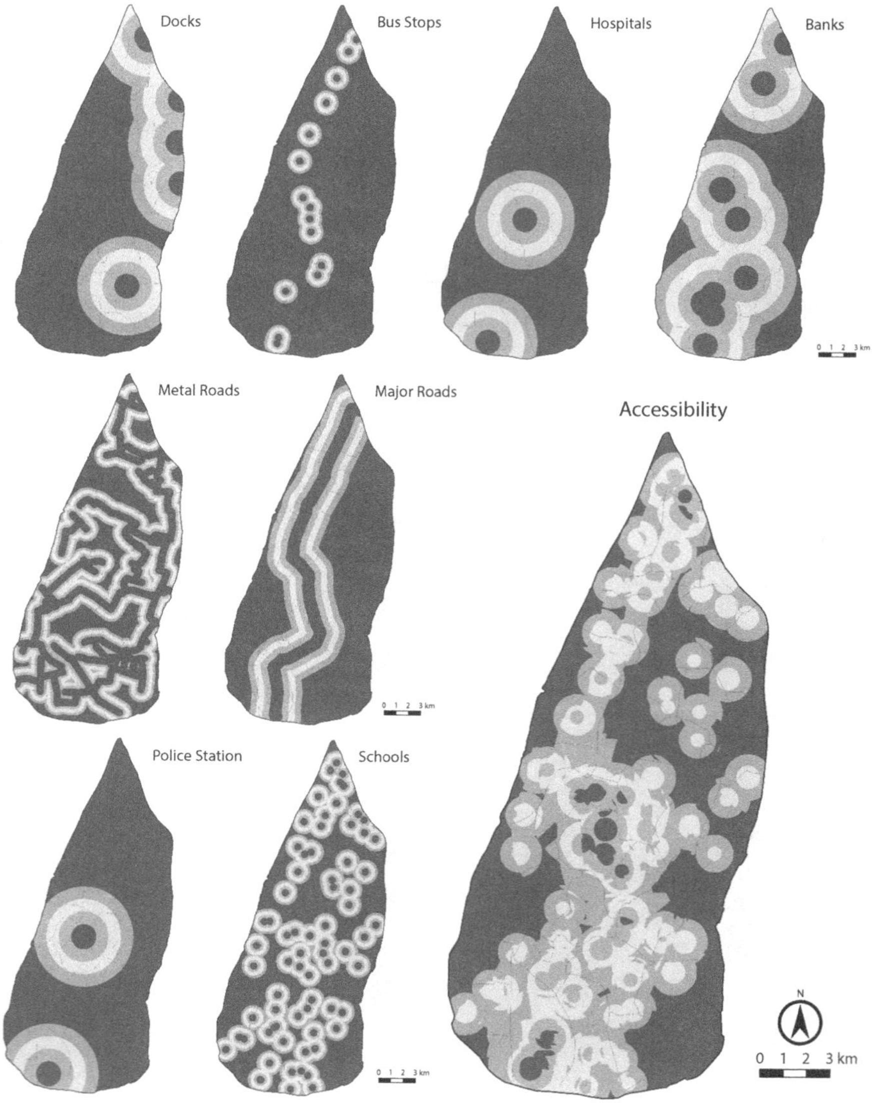

Figure 8.13 Vulnerability mapping (accessibility).

iv. Infrastructural Parameters (Vulnerability Parameter)

It has been determined by checking the presence of basic amenities and facilities such as water supply, health care (clinics and hospitals), and drainage. The absence of these amenities would lead to severe degradation in the lifestyle of the residents and attract numerous diseases and health issues [9]. Ramkrishnapur and Sumatinagar would be extremely vulnerable without these amenities.

v. Sociodemographic Parameters (Vulnerability Parameter)
Due to the remote location of the island, awareness regarding floods and societal norms is substandard and poor; hence, the sociodemographic indicator has been considered to add more accuracy to the vulnerability calculations. As a result, the dependent population is given a weight of 0.2, followed by the illiterate population groups (0.31), who lack information and awareness of meteorological risk updates, and the socially backward and deprived populations (0.49). Ancient customs such as child marriages, caste discrimination, female feticide, etc. are still practiced, contributing to the vulnerability.

8.4.4 Flood Exposure Maps

The map of coastal flood exposure examines the region's relative susceptibility to coastal flooding, its physiography, land use, and Yaas flood inundation. We see a high exposure on the west coast due to the continuous, gently sloped shoreline allowing easy penetration of water. The eastern coast has lower elevations and moderately dense settlements. Low-lying coastal areas are the first to be affected by sea pressures, including recurring flooding, coastal erosion, submergence, and saltwater intrusion [31].

As the distance from the coasts has the highest weightage (Table 18.1), the southern or coastal region in the exposure map shows the highest exposure. Furthermore, creeks are the first to be affected in case of floods. Hence, the coastal villages of Mrityunjonagar, Sumatinagar, Muriganaga, Debimaturapur, Krishnanagar, Radhakrishnapur, and Gangasagar are severely prone to exposure to floods. Mangrove forests, however, help the area's communities, particularly those in Gobindapur, Inamnagar, Chemagari, Naraharipur, Shikarpur, and Kastala, be more resilient to flooding. Hazard exposure mapping shows that 33% of habited villages in the core would have moderate exposure levels, 20% along the shore would have high to extremely high exposure levels, and 47% would be in the low to the very low zone.

8.4.5 Flood Vulnerability Maps

Flood vulnerability is directly related to the community's behavior toward floods and the related demographic and socioeconomic trends [10, 32]. The uninformed and unaware are the first to be affected. The island is rural,

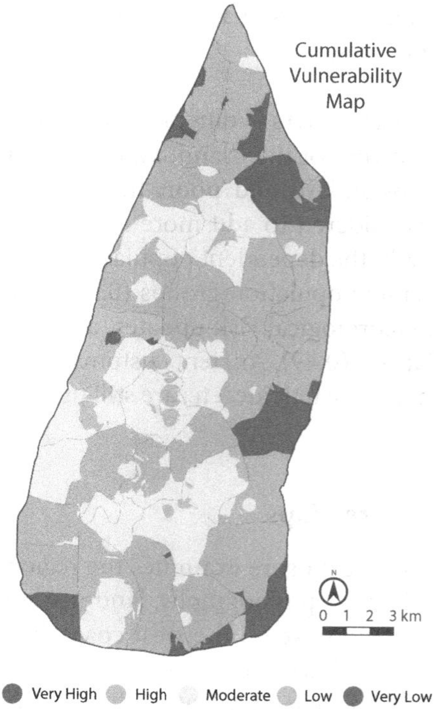

Figure 8.14 Cumulative flood vulnerability.

remains unaware of the threats of the floods, and is unprepared. Thus, to capture this very scenario, vulnerability is included in the study. All of the selected vulnerability variables (Table 8.11) were taken into account when creating the cumulative vulnerability map in Figure 8.14 (economic, sociodemographic, and accessibility infrastructure). It shows that around 30 mouzas, or 69% of the island, are classified as high or very high vulnerability zones, whereas 10 villages, or 32%, are classified as moderate vulnerability zones, and Naraharipur village is classified as medium vulnerability. The deserted villages of Sagar and Bishalakshmipur have the lowest socioeconomic vulnerability ratings. As the economic and accessibility parameters were given importance, it can be observed that the eastern coast shows a very high to very high vulnerability all along, depicting the need for extra care and concentration while planning flood mitigation solutions.

8.4.6 Flood Risk Zoning

Both the exposure to and vulnerability due to local flood hazards have been taken into account during the evaluation procedure. The island core is classified as having a moderate to low risk of flooding, whereas the whole shoreline is deemed to be in the high to very high-risk group. The central zone of the island has good facilities and easy accessibility, hence catering well to the population living there; thus, this zone falls under the low to moderate risk zone. The northern part of the island is dominated by kutcha housing and a vulnerable population; hence, it has a comparatively higher risk potential. Along the shoreline (especially the east coast), the development by 2050 will be abundant, as can be seen in Figure 8.9. Furthermore, the elevation is also lower along the shoreline, resulting in extreme risk in those areas. Figure 8.15 is a representation of the risk in 2020 and the risk in 2050 under the Yaas flood scenario. There is an apparent increase in the red zones, indicating increased flood risk shortly. An increased risk in the core areas along the creeks is observed as a result of the increase in water levels due to SLR. There is also a considerable increase in the risk along the shoreline, indicating the depletion of the mangroves and the growth of settlement along the shorelines. Furthermore, a high- to very high–risk zone covers 14% (2020) and 25% (2050) of the entire island; a moderate

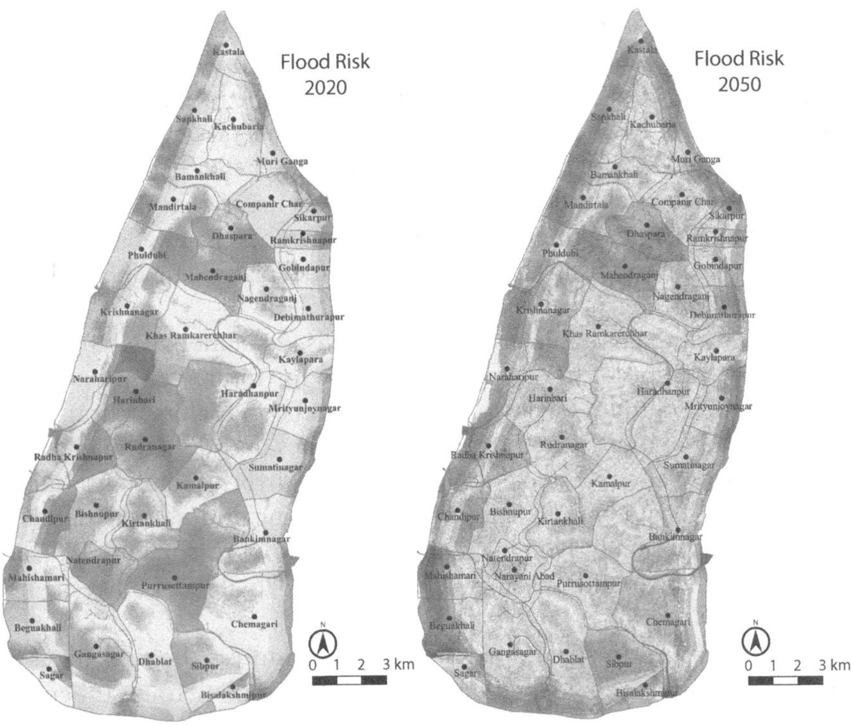

Figure 8.15 Coastal flood risk: (a) 2020 and (b) 2050.

zone covers 37% (2020) and 30% (2050); and a low- to very low–risk zone covers 49% (2020) and 45% (2050). Very high– to high-risk zones show an increase of 11% in 30 years, suggesting a heavy threat to the island. It is significant to highlight [from Figure 8.15(b)] that the island's high flood risk is mostly caused by the vulnerability parameters' poor performance.

8.5 Conclusion

With the rising demand for efficient coastal flood mitigation plans, this research formulates a risk calculation method for Sagar Island, taking into consideration all the necessary exposure and vulnerability parameters. Calculating and analyzing the future risk is crucial to mitigating the damage that could be caused. The framework for assessing flood risk includes the probability density of flood hazards and socioeconomic indicators. This research strives to present the risk situation from the perspective of the Yaas cyclonic floods and determine the possible vulnerability and exposure.

From the study, we conclude that there is a possibility of 25% of high-risk zones and 45% of low-risk zones on the island by 2050 due to the considerable rise in built-up by 10.53% as predicted by the FUTURES growth model. Furthermore, the study has also presented the increase of risk from 2020 to 2050 under the considered scenario of the Yaas cyclone, demonstrating the necessity of taking quick action to lower the chance of loss and damage from coastal flooding. Though many structural measures like the construction of embankments and dike walls can be suggested, something more long-lasting and sustainable, such as spatial planning combined with accurate risk zoning, is demanded.

Spatial planning combined with nature-based solutions (NbS) can go a long way to support flood mitigation plans for future generations. For instance, NbS solutions (such as mangroves, reefs, and salt marshes), managed natural solutions (such as restored beaches and dunes, mangroves, artificial reefs, and planted saltmarshes), and hybrid solutions (dune-dike systems) that combine structural engineering and natural elements in an environmentally benign manner (bamboo construction) [33] are both cost-effective and sustainable alternatives to flood risk resilience applications.

NbS, compared to structural solutions, can address the difficulties of climate mitigation and adaptation while providing numerous co-benefits for both people and the environment. For instance, reestablishing natural mangrove forests in upper catchments can improve carbon sequestration, save biodiversity, preserve hydrological processes and behavior, and keep residents downstream from flooding [34, 35]. Sagar Island can support the mangrove ecosystem due pre-existing base. Hence, strategically replanting the mangroves in a way that water penetration from the sea is substantially reduced can immensely help the community's battle against flood risk.

Acknowledgments

We are grateful to the Indian Institute of Technology Kharagpur and Ranbir and Chitra Gupta School of Infrastructure Design and Management for the financial and infrastructure support. We thank NRSC, USGS, and Kolkata Port Trusts for providing temporal remote sensing data and Tidal data.

References

1. L. Burke, Y. Kura, K. Kassem, C. Revenga, M. Spalding and D. McAllister, "*Pilot analysis of global ecosystems: Coastal ecosystems.*" WRI: World Resources Institute, Apr. 01, 2000.
2. N. N. V. Sudha Rani, A. N. V. Satyanarayana and P. K. Bhaskaran, "Coastal vulnerability assessment studies over India: A review," *Nat. Hazards*, vol. 77, no. 1, pp. 405–428, May 2015, doi: 10.1007/s11069-015-1597-x.
3. United Nations Environment Programme (UNEP), "Physical Alteration and Destruction of Habitats", 2007. www.unep.org (accessed February 27, 2023).
4. M. A. Marfai and L. King, "Monitoring land subsidence in Semarang, Indonesia," *Environ. Geol.*, vol. 53, no. 3, pp. 651–659, Nov. 2007, doi: 10.1007/S00254-007-0680-3/TABLES/3.
5. S. Vitousek, P. L. Barnard, C. H. Fletcher, N. Frazer, L. Erikson and C. D. Storlazzi, "Doubling of coastal flooding frequency within decades due to sea-level rise," *Sci. Rep.*, vol. 7, no. 1, pp. 1–9, Dec. 2017, doi: 10.1038/S41598-017-01362-7/TABLES/1.
6. T. S. Kumar, R. S. Mahendra, S. Nayak, K. Radhakrishnan and K. C. Sahu, "Coastal Vulnerability Assessment for Orissa State, East Coast of India," *J. Coast. Res.*, vol. 26, no. 3 (263), pp. 523–534, May 2010, doi: 10.2112/09-1186.1.
7. A. Mukhopadhyay, S. Mukherjee, S. Mukherjee, S. Ghosh, S. Hazra and D. Mitra, "European Journal of Remote Sensing Automatic shoreline detection and future prediction: A case study on Puri Coast, Bay of Bengal, India," *Eur. J. Remote Sens.*, vol. 45, no. 1, pp. 201–213, 2017, doi: 10.5721/EuJRS20124519.
8. K. Phongsapan *et al.*, "Operational Flood Risk Index Mapping for Disaster Risk Reduction Using Earth Observations and Cloud Computing Technologies: A Case Study on Myanmar," *Front. Environ. Sci.*, vol. 7, p. 191, Dec. 2019, doi: 10.3389/FENVS.2019.00191/BIBTEX.

9. A. Narendr, S. Vinay, B. H. Aithal and S. Das, "Multi-dimensional parametric coastal flood risk assessment at a regional scale using GIS," *Environ. Dev. Sustain.*, vol. 24, no. 7, pp. 9569–9597, Jul. 2022, doi: 10.1007/S10668-021-01839-6/FIGURES/7.
10. S. McLaughlin, J. McKenna and J. A. G. Cooper, "Socio-economic data in coastal vulnerability indices: Constraints and opportunities," *J. Coast. Res.*, vol. 36, no. 36 (10036), pp. 487–497, Mar. 2002, doi: 10.2112/1551-5036-36.
11. E. Doukakis, "Coastal Vulnerability and Risk Parameters," *Eur. Water*, vol. 11, pp. 3–7, 2005.
12. D. Pantusa, F. D'Alessandro, L. Riefolo, F. Principato and G. R. Tomasicchio, "Application of a Coastal Vulnerability Index. A Case Study along the Apulian Coastline, Italy," *Water 2018, Vol. 10, Page 1218*, vol. 10, no. 9, p. 1218, Sep. 2018, doi: 10.3390/W10091218.
13. Y. M. Wei, Y. Fan, C. Lu and H. T. Tsai, "The assessment of vulnerability to natural disasters in China by using the DEA method," *Environ. Impact Assess. Rev.*, vol. 24, no. 4, pp. 427–439, May 2004, doi: 10.1016/j.eiar.2003.12.003.
14. D. Huang, R. Zhang, Z. Huo, • Fei, M. • Youhao and W. Zheng, "An assessment of multidimensional flood vulnerability at the provincial scale in China based on the DEA method", doi: 10.1007/s11069-012-0323-1.
15. M. N. Ahsan and J. Warner, "The socioeconomic vulnerability index: A pragmatic approach for assessing climate change led risks–A case study in the south-western coastal Bangladesh," *Int. J. Disaster Risk Reduct.*, vol. 8, pp. 32–49, Jun. 2014, doi: 10.1016/J.IJDRR.2013.12.009.
16. L. Weinstein, A. Rumbach and S. Sinha, "Resilient Growth: Fantasy Plans and Unplanned Developments in India's Flood-Prone Coastal Cities," 2019, doi: 10.1111/1468-2427.12743.
17. J. A. Pollard, T. Spencer and S. Jude, "Big Data Approaches for coastal flood risk assessment and emergency response," Wiley Interdiscip. Rev. Clim. Change, vol. 9, no. 5, p. e543, Sep. 2018, doi: 10.1002/WCC.543.
18. M. Al-Shalabi •, L. Billa, B. Pradhan, S. Mansor, • Abubakr and A. A. Al-Sharif, "Modelling urban growth evolution and land-use changes using GIS based cellular automata and SLEUTH models: the case of Sana'a metropolitan city, Yemen", doi: 10.1007/s12665-012-2137-6.
19. P. Jayasinghe and L. N. Kantakumar Bharati Vidyapeeth, "Comparative Evaluation of Open Source Urban Simulation Models Applied to Colombo City and Environs in Sri Lanka," *Artic. Int. J. Geoinformatics*, vol. 17, no. 3, 2021, doi: 10.52939/ijg.v17i3.1897.
20. A. Ilyassova, L. N. Kantakumar and D. Boyd, "Urban growth analysis and simulations using cellular automata and geo-informatics: comparison between Almaty and Astana in Kazakhstan," 2019, doi: 10.1080/10106049.2019.1618923.

21. R. K. Meentemeyer, W. Tang, M. A. Dorning, J. B. Vogler, N. J. Cunniffe and D. A. Shoemaker, "FUTURES: Multilevel Simulations of Emerging Urban-Rural Landscape Structure Using a Stochastic Patch-Growing Algorithm," *Ann. Assoc. Am. Geogr.*, vol. 103, no. 4, pp. 785–807, 2013, doi: 10.1080/00045608.2012.707591.
22. A. Petrasova *et al.*, "Open source approach to urban growth simulation Open Source Approach To Urban Growth Simulation Commission VII, SpS10-FOSS4G: FOSS4G Session (co-organized with OSGeo)," 2016, doi: 10.5194/isprs-archives-XLI-B7-953-2016.
23. A. Khan and S. Chatterjee, "SPRINGER BRIEFS IN OCEANOGRAPHY *Coastal Risk Assessment A Comprehensive Framework for the Bay of Bengal*" 2018, doi: 10.1007/978-3-319-69992-9. Available: http://www.springer.com/series/11754.
24. A. A. Danda, "Resolution of Public-good Dilemma through Induced Collective Action," *J. Infrastruct. Dev.*, vol. 3, no. 1, pp. 97–105, 2011, doi: 10.1177/097493061100300105.
25. M. Sahana and H. Sajjad, "Vulnerability to storm surge flood using remote sensing and GIS techniques: A study on Sundarban Biosphere Reserve, India," *Remote Sens. Appl. Soc. Environ.*, vol. 13, pp. 106–120, Jan. 2019, doi: 10.1016/J.RSASE.2018.10.008.
26. S. Bandyopadhyay, "Natural Environmental Hazards and Their Management: A Case Study of Sagar Island, India," *Singap. J. Trop. Geogr.*, vol. 18, no. 1, pp. 20–45, Jun. 1997, doi: 10.1111/1467-9493.00003.
27. A. Narendr, P. Nadupalli, S. Das and B.H. Aithal, "Flood resilient scenario modelling (FReSMo): Dynamic assessment of climate-induced effects on land use" in *Proceedings of 58th International Society of City and Regional Planners (ISOCARP)*, Brussels, Belgium, Oct, 2022.
28. National Oceanography Centre, "Permanent Service for Main Sea level," 2020. [Online]. Available: https://psmsl.org/.
29. S. Vinay, N. Aishwarya and H. A. Bharath, "Role of Local Sea-level Rise on Disaster Exposure in Coastal Island," *5th World Congr. Disaster Manag.*, pp. 319–327, Dec. 2022, doi: 10.4324/9781003341932-36.
30. B. Yan, S. Li, J. Wang, Z. Ge and L. Zhang, "Socio-economic vulnerability of the megacity of Shanghai (China) to sea-level rise and associated storm surges", doi: 10.1007/s10113-015-0878-y.
31. N. H. Ae, R. Nicholls and R. Nicholls, "Global sea-level rise and coastal vulnerability," *Sustain. Sci.*, vol. 3, pp. 5–7, 2008, doi: 10.1007/s11625-008-0049-x.
32. S. F. Balica, N. G. Wright and F. van der Meulen, "A flood vulnerability index for coastal cities and its use in assessing climate change impacts," *Nat. Hazards*, vol. 64, no. 1, pp. 73–105, Sep. 2012, doi: 10.1007/s11069-012-0234-1.
33. N. Pontee, S. Narayan, M. W. Beck and A. H. Hosking, "Nature-based solutions: lessons from around the world," 2016, doi: 10.1680/JMAEN.15.00027.

34. N. Seddon, A. Chausson, P. Berry, C. A. J. Girardin, A. Smith and B. Turner, "Understanding the value and limits of nature-based solutions to climate change and other global challenges," *Philos. Trans. R. Soc. B*, vol. 375, no. 1794, Mar. 2020, doi: 10.1098/RSTB.2019.0120.
35. Y. Huang *et al.*, "Nature-based solutions for urban pluvial flood risk management," *Wiley Interdiscip. Rev. Water*, vol. 7, no. 3, p. e1421, May 2020, doi: 10.1002/WAT2.1421.

9

A Review of Methods for Studying Glacier Dynamics Due to Climate Change in the Himalayas

Shanta Kumar[1,2*] **and Anurag Linda**[1,3]

[1]*Department of Environmental Sciences, Central University of Himachal Pradesh, Dharamshala, Himachal Pradesh, India*
[2]*Department of Environmental Studies, Indraprastha College for Women, University of Delhi*
[3]*Department of Environmental Sciences, Central University Jharkhand, Ranchi, India*

Abstract
Mountain glaciers are the primary source of most of the perennial rivers. Glaciers are widely regarded as reliable indicators of climate fluctuations in response to various factors such as temperature, precipitation, and other environmental variables. Changes in the regional energy budget induced by severe climatic conditions in mountain regions have altered the local hydrological cycle over the past few decades. Therefore, monitoring glaciers is crucial to comprehend mountain-related hazards and ensure sustainable management of water resources. As a result, analyzing the gap, limitations, and glacier monitoring in the Himalayas is essential for developing long-term adaptation strategies. The present study presents a review of suitable methods for glacier monitoring. This study provides a compilation of different approaches to analyzing the mass balance of Himalayan glaciers over different periods using geodetic, direct, and remote sensing methods. It extends a tactic to enhance diverse mass balance measurements over various periods with one of the most satisfactory glacier mass balance methods in the Mountainous region.

Keywords: Climate change, Himalayan, glacier, remote sensing, mass balance methods

**Corresponding author*: shantakumar.cuhp@gmail.com

Disha Thakur, Sanjay Kumar, Har Amrit Singh Sandhu and Chander Prakash (eds.) Sustainable Development Using Geospatial Techniques, (217–240) © 2024 Scrivener Publishing LLC

9.1 Introduction

Glaciers in high mountains provide indispensable freshwater sources for people living in the valleys of the mountains worldwide [1]. The region of Hindu Kush–Himalaya (HKH) encompasses and is recognized as the most extensive mountain range globally and a large storehouse of freshwater outside the Polar Region [2, 3]. Glaciers supply life-sustaining and ecosystem-maintaining freshwater for people worldwide, so they must be preserved [4–6]. Cryosphere, especially glaciers, are considered a dependable indicator to assess climate change. The fluctuations in temperature and precipitation, the main climatic parameters, that influence glacier's movement and dynamics can be easily traced and correlated [7]. The increasing trend in temperature and dwindling trends in precipitation stimulate climate change and affect the regional and global climate [8, 9]. The mean global temperature has been raised up to 1.5°C in 2017 since the pre-industrial era at the rate of 0.1°C to 0.3°C increment per decade (reported by IPCC) [4]. In the past decade, the local hydrological cycles have experienced disruptions caused by alterations in the regional glacier energy budget [7]. Many glaciers in the Himalayan region are retreating rapidly, which is evident from the temporal glacier inventory [9–11] (Figure 9.1).

The melting response of snow cover and Himalayan glaciers is strongly influenced by climate change [12, 13]. Most of the Himalayan glaciers have

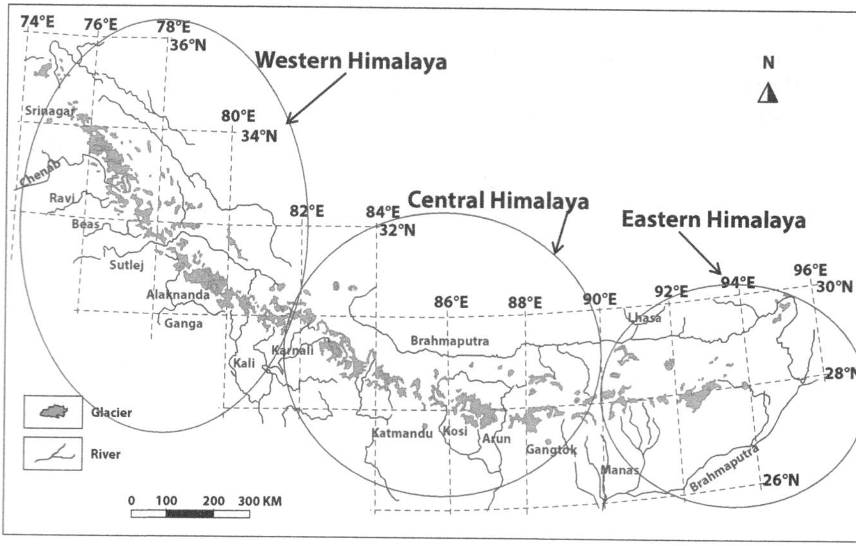

Figure 9.1 Distributions of glaciers in the Himalaya. *(Source: Linda; 2007).*

undergone a retreat process due to an escalated rate of global warming over the past century [14–17]. The glacier area, length, snout position, debris cover, surface elevation, snowline altitude, and surface mass balance (SMB) are among the essential features of a glacier that are most important for climate change impact studies. These features are dominated by growth and melting processes at the glacier's surface, which are influenced mainly by climatic and solar insolation conditions [18–20].

The glacier growth/melting can be studied through *ex situ* and *in situ* methods. *In situ* glacial study includes estimating glacier mass balance measurements, snout monitoring, surface elevation change, glacier thickness, snowline, etc. [21, 22]. Himalayan glaciers are challenging to study due to their rugged terrain and inaccessibility, which tend to be expensive and labor-intensive [23, 24]. In such conditions, the remote sensing data and technique prove to be a helpful tool that can be used to acquire information about different glaciological features [25]. Numerous *ex situ* methods have been used to study Himalayan glacier dynamics using geospatial tools, but more comprehensive mass balance analyses are available only for a few glaciers, such as Chota Shigri, Nardu, Shaune Garang, and Samudra Tapu, in the Himalayas due to topographical constraints [14, 26–29]. Using geospatial tools in conjunction with advanced modeling techniques for the analysis of glacier accumulation area ratio, equilibrium line of altitude, snowline, moraine-dammed lake, and snow cover mapping tools can improve the accuracy of predictions of glacier health [9, 29–31].

Therefore, satellite- and aerial data–based studies pose feasible and rational approaches for obtaining numerically illustrative data of glacier mass balance estimation. Consequently, an all-data set needs some validation for analysis; it needs field-based studies to generate data for validating remotely sensed data.

This review focuses on the *in situ* and *ex situ* techniques available for indirectly or directly determining the mass balance of Himalayan glaciers. *In situ* glaciological techniques include measurement of change in mass balance, snout, surface elevation, and glacier thickness [7, 32]. These methods include stakes and pits, area change, surface GPR surveys, hypsometry, scanning, and geodetic, while remotely sensed methods include air, space, and ground measurements. In this work, an effort has been carried out to review the various techniques for studying glacier mass change in the Himalaya under changing climate. The advantages and disadvantages of methods used to study glacier dynamics and the most appropriate method for studying Himalayan glaciers are discussed.

9.2 Monitoring of Himalayan Glaciers by Different Methods

9.2.1 Mass Balance Through Remote Sensing Methods

Remote sensing is a scientific field that includes collecting information about various Earth surface features devoid of compelling with direct physical contact. It is successful in mapping studies; estimating the surface velocity; estimating mass balance, snow line, and snowmelt runoff/change; and predicting the spatial-temporal change of glaciers [33]. Due to the extensive geographic area and the difficult terrain, remote sensing-based data, tools, and approaches are one of the most useful approaches for the collection of evidence of snow and glacier changes throughout the Himalayan region [18].

It is impossible to quantify accumulation and ablation zone directly from space [34], but factors such as glacier boundary, terminal location, snowlines, surface velocity, surface elevations and glacier lake can be utilized to estimate glacier mass balances with the introduction of new remote sensing satellite sensors such as Landsat (TM, MSS, and ETM+), ASTER, IRS, SPOT, Sentinel-1 and -2 and more recently developed sensor ALOS. The above available sensors (medium resolution: 10 to 90 m) dataset has become freely accessible for cryosphere studies. For in-depth glacier research at the regional level, different sensors with high spatial resolution and altitudinal resolution, such as Quick-bird, InSAR, GeoEye-1, IKONOS, and other optical sensors equivalent to aerial photography, help in the studies. However, their employment at wide geographical scales is limited by their high prices, small sweep sizes (resolution of IKONOS and Quick bird is 11 and 16 km), and lengthy repetition intervals in months [35–37].

These technologies provide a means to analyze and monitor glaciers, surface velocity, glacial lakes, and surrounding terrain, allowing for the identification of potential lake outburst floods for forecasting [3–5]. Nevertheless, the proficiency to obtain numerical facts from remotely sensed satellite imagery information for use in computations of the yearly net mass increase or decrease of the glaciers continues to be an ongoing task [6]. Geospatial tools, such as satellite imageries and Geographic Information Systems (GIS), play a significant part in this endeavor by providing information on changes in glaciers and glacier lakes, debris, moraines, and moraine dams [38–40]. To illustrate and monitor the change of a glacier along with their associated features, like glacier lakes, in the Himalayan region, satellite imagery is used [34]. Assessments of glacier characteristics may now be developed across larger regions and longer periods. Nowadays, the use of remotely sensed satellite imagery along with geospatial tools has increased due to their accessibility and improved appropriate spatial and temporal resolution with minimal financial expenses. In unexplored glacierized regions like the Himalaya, regular evaluation of the distinctive features of mountain glaciers is necessary.

9.2.2 Mass Balance Through AAR/ELA

Mostly remote sensing method is used for calculating glacier mass balance due to regional constrain in the Himalaya. Remote sensing–based techniques/approaches are being used to conduct glacier surveys and monitoring through space- and aerial-based platforms and sensors. The net glacier mass balance (growth/melting) cannot be assessed directly from satellite or aerial datasets [41]. However, glacier-wide or cumulative mass balances may be estimated using data collected from aerial and satellite scans (glacier boundary, snouts location, and snowlines), etc.

9.2.2.1 Mass Balance Through Area Accumulation Ratio (AAR) Methods

The area accumulation ratio (AAR) is the estimate of the total seasonal snow (winter) accumulation and the seasonal ablation of snow and ice in summer. AAR is a value between 0 and 1, which is the ratio of the total accumulation zone area to the total glacier area [10]. This includes the accumulation of solid precipitation in the winter and melting in the summer through insolation, calving, evaporation, etc. This method has been applied to single glaciers based on their location at the end of the melting/ablation period [42]. For example, the AAR method was applied in Himachal Pradesh on the Gara and Gor-Garang glaciers. In this method, correlation coefficients (r) were estimated at 0.88 and 0.96 m w.e. a^{-1}, using 6 to 7 years of data, respectively. The data from all the glaciers were combined, and a composite relationship between AAR and mass balance was established. The regression analysis was further applied in association with r, equal to the value of 0.75. Furthermore, the AAR value was found to be 0 and the amount of mass balance was calculated as 0.44 [42]. In Tipra and Rataban glaciers, the calculated values were 0.47 and 0.49 during the studied period of 1962–2008 [13]. Also, in the mass balance studied in the Alaknanda and Bhagirathi basins, the estimated specific mean mass balance through the AAR method was found to be 0.14 ±0.02 and 0.11±0.27 and mean mass balance loss was estimated at -1.1 ± 0.03 m w.e. a^{-1} and -1.01 ±0.07 m w.e a^{-1} during 2001 to 2013 [43].

9.2.2.2 *Equilibrium Line Altitude (ELA) Method*

The equilibrium line altitude indicates the glacier portion in which annual ablation and accumulation are equal. The ELA separates the ablation zone from the accumulation zone. It is a crucial indicator for studying and assessing the impact of climate change on the dynamics of a glacier. The ELA is traditionally studied, and it is seen at the end of the melting period [15, 16]. Despite this, the unusual snowfall and cloud cover on the satellite images make it difficult to identify the ELA. This has led to an approximation of ELA for the ablation season using the highest snowline [16]. In the Himalayan glaciers, ELA and field mass balance data were also studied and the correlation between them was observed. A high correlation coefficient was seen from field data of the two glaciers, namely, Gara and Gor-Garang, i.e., −0.92 and −0.94 during the years 1986–1987 and 1987–1988, respectively [42]. In this study, they found a positive mass balance by analyzing satellite data, topographic maps, and earlier available field-based data. From the field-based study of Nardu glacier, ELA has been studied in the field and ELA was reconstructed using mass balance data from 1979 to 2013, and the results show a favorable trend with 96.73 m of change in ELA. The magnitudes of ELA were found to be varying in the winter, pre-monsoon, post-monsoon, and monsoon seasons having values of 2.84, 3.24, 4.02, and 2.04 m/year, respectively [17]. The ELA is an effective indicator for assessing a glacier's health (mass balance). It is well adapted to characterize glacier's response to climate change.

9.2.3 Normalized Difference Snow Index (NDSI) Methods

The NDSI has proven to be particularly useful for mapping areas with and without snow cover. It is the difference between high visible light reflection and mid-infrared absorption for the NDSI calculation [21]. This approach makes use of snow reflectance data; snow often has high reflectivity in the electromagnetic spectrum of the visible range and low reflectivity in SWIR. The snow in the visible spectrum band is white and black in the SWIR spectrum [21]. Initially, NDSI was used to calculate the automatic snow mapping of surface reflectance of satellite images in Moderate Resolution Imaging Spectroradiometer (MODIS) bands 4 and 6 [21].

$$NDSI = \frac{Band\ Green - Band\ SWIR}{Band\ Green + Band\ SWIR} \tag{9.1}$$

Although NDSI is a reliable approach for identifying glacial borders, it performs a poor task of separating snow from glaciers, which led to bigger glacier areas than what was obtained using the other two methods.

9.2.4 Mass Balance Through Ice Surface Velocity

The glacier surface velocity is defined as the amount of ice that moves downward due to deformation and basal slide in response to gravitational force and the glacier itself [23, 44, 45]. Also, the glacier can move downhill because of its plastic property nature and the high pressure exerted by the accumulation of snow/ice [44]. The speed with which the surface ice moves in the direction of a glacier terminal is known as "surface ice velocity". As to the amount of ice that melted, the flow may vary considerably or not at all. Moving glaciers move more ice to the terminus where it melts which is one of the most important factors that affects the mass balance of a glacier. The surface velocity of the Chota Shigri glacier has been estimated for the years 2003–2009 using image correlation techniques and was found that surface velocity varied from approximately 20 to 40 m/year having maximum velocity at the mid-ablation and accumulation zones [44].

In the Gangotri glacier, surface velocity in the accumulation region was found to be between 14 and 85 ma^{-1} and near the snout 20 to 30 ma^{-1}. The glaciers of the Dhauliganga basin showed a displacement of 28 m/year [45, 46]. Thus, measurements show that the glacier's geomorphology and geomorphometry significantly influence the glacier's surface ice velocity. Some of the remote sensing-based methods with their merit and demerits are discussed in Table 9.1.

9.2.5 Mass Balance Through Glacier Delineation

It is necessary and of paramount significance that world glacier dynamics are accurately mapped, observed, and recorded. Assessing how glaciers around the world will be altered by changing climate [47]. An overview of mapping glacier (delineation) improvements in the Himalaya, with a focus on recent developments and breakthroughs of manual, semiautomatic, and automated AI-based algorithm approaches for the identification of the glaciers, debris-cover, glacier lakes, moraines, moraines dams, etc. is urgently required.

Table 9.1 List of surface velocity calculations by remote sensing.

Sl. no.	Broad category	Subcategory	Merits	Demerits
Field-based method	Field-based method	----	Accurate up to centimeter level	Time taking; difficult due to rugged glacier terrain; only a few points could be collected, so do not give an overall picture of ice dynamics.
Remote sensing–based methods	1. Using SAR data	(i) SAR intensity tracking	Not affected by cloud	Due to the oblique viewing angle of SAR images, data presents high shadowing and layover in mountainous terrain, a high cost of data.
		(ii) SAR interferometry	Very accurate; not affected by cloud	Due to the oblique viewing angle of SAR images, data presents high shadowing and layover in mountainous terrain and fails in high-speed glaciers due to the de-correlation of data. High cost of data, which limits its application
	2. Using optical data	(i) Manual Feature tracking	Quickly to implement and cost-effective; large time series data available	Affected by cloud and shadow; not very accurate; user-dependent
		(ii) Correlation of time series data	Quickly implement and cost-effective	Affected by cloud and shadow, accuracy depends on the data resolution, but the results are very accurate.

The glacier monitoring and mapping analyses help us to identify the spatiotemporal changes in the glaciers. The focus is on harnessing the benefits of thermal and optical datasets, as optical data helps to distinguish primarily over-glacial cover types (fresh snow, ice, water, glacier lake, mixed ice, rock, and glacier debris). The use of thermal data to differentiate between debris from glaciers and glacial lakes allows for a differentiation in the glacier mass waste.

9.2.6 Manual Digitization

The manual digitization of glacier boundaries began in GIS software using satellite imageries by using false color composites (FCC) [47–49]. This procedure is time-consuming and effort-demanding. The introduction of human error is intrinsic to the method as the assessment of various glacier features on remotely sensed satellite images may differ from analyst to analyst depending on the visual perception of the analyst [50, 51]. The glacier boundaries are delineated using different software, which requires high-resolution and high-quality datasets for manual generation. These data have been obtained by visual interpretation of satellite imagery with RGB band (red, green, and blue band) [52, 53]. Therefore, it has been found that the integration of optical data with thermal data is very important for glacier studies because the impossibility of deciphering ice extensions under a debris cover and enabling glacier boundary mapping in areas covered by rubble can play an essential role in resolving ambiguities [47, 54]. The approach has been positive and shows that the combination of optical and thermal data obtained through satellite imagery can be used efficiently to delineate and map the debris on glaciers. The manual delineation process is good for studying the glacier boundary, but it is laborious and time-consuming. But sometimes, it is not accurate in area calculations [55, 56].

9.2.7 Automatic Digitization

Nowadays, glacier mapping is done using an integrated approach for delineation by semiautomatic and automatic/algorithm approaches [57, 58]. A very useful way to map the glacier's covered area is through classification based on morphometric parameters and thermal information [59–61]. Different platforms developed for the analysis of glaciers are machine learning, deep learning, and the Google Earth engine for mapping and monitoring glaciers on a regional or global scale [58, 62, 63]. These methods were developed for the spectral and hyperspectral indices data analysis through algorithms and modeling. It helps us understand and develop the accuracy to differentiate between glacier, snow, and debris-covered areas. These data require the iteration of preprocessing and postprocessing operations for segmentation, pixel-based, and object-oriented–based imagery analysis [64–71]. These introduced new computation techniques for rigorous sampling-resampling and orthorectification of satellite imagery. These techniques have the ability of self-learning, accurate, fast, and unbiased for classification [36, 72].

9.2.8 Field-Based Methods

The manual field–based measurement of glacier mass balance in Himalayan glaciers has been conducted using glaciological, hydrological, and direct-geodetic methods.

9.2.8.1 Mass Balance Through Glaciological Approach

The technique of glaciological method for mass balance study is based on *in situ* field measurements. At various places on the surface of glaciers, mass changes are calculated from two dates in different hydrological years. When multiplied by the density of the glacier surface, the resulting mass computed is the mass balance at that point. At two different dates and depths of snow, the exposed stakes can measure changes in glacier mass as compared to recognized stratigraphic surfaces during the preceding summers or winter seasons. In the accumulation area, snow density shall be measured by digging a snow pit to its reference surface [26, 73, 74].

The glaciological approach involves the insertion of bamboo stakes to form a network and pith holes on the glacier surface and assessing the change in the surface level of the glacier while accounting for the density of snow/firn/ice. A seasonal mass balance is produced by comparing data taken after the ablation and accumulation seasons, as opposed to measurements taken between two specified dates, which provide an annual mass balance [75].

The Geological Survey of India initiated a study on Himalayan glaciers in 1974 using glaciologically based methods to determine their mass balances. Understanding the significance and interactions with climate and water systems of mass inputs in different areas of development, a detailed study was done on the Gara glacier situated in Himachal Pradesh [10, 29, 73]. A comprehensive investigation was conducted on the Gara glacier located in Himachal Pradesh to comprehend the significance of mass balance inputs in diverse domains of advancement and their interplay with

Table 9.2 List of glaciological methods used in Himalayan glaciers.

S. no.	Glacier	Observation year	Area (km^2)	Elevation (masl)	Region	Orientation	References
1	Gara Garang, Shaune Garang	1974/1975–1981/1982	5.2	4710–5600	Himachal	NE	[78, 79]
2	Nehnar	1976/1977–1984/1985	1.2	3920–4925	Kashmir	N	[80, 81]
4	Tipra Bank	1981/1982–1987/1988	7	3800–5600	Uttarakhand	NW	[82]
7	Kolaho II	1983–1984	11.9	3690–5425	Kashmir	NW	[82]
8	Shishram	1983–1984	9.9	3740–4900	Kashmir	N	[82, 83]
9	Chhota Shigri	2002/2003–2009/2010	15.7	4050–6263	Himachal	N	[26, 84]
11	Dokriani	1992/1995–1997/2000	7	3900–6000	Uttarakhand	NW	[59]
12	Chorabar	2003/2004–2009/2010	6.6	3865–6940	Uttarakhand	S	[85]
13	Gor Garang, Naradu	1976/1977–1984/1985	2	4750–5400	Himachal	SW	[78, 86]
14	Hamtah	2000/2001–2008/2009 2010/2011–2011/2012	3.2	4020–5000	Himachal	NW-N	[87]
16	Patsio Glacier	2010/2011–2016/2017	2.25	4875–5718	Himachal	NE	[88]
18	Chhota Shigri	2010–2017	15.5	4072–5830	Himachal	N	[89]
19	Stok glacier	2014–2017	0.74	5300–5850	Ladakh	NE	[90]
20	Mera glacier	2007–2019	5.1	4910–6390	Nepal	NW	[91]
21	Pokalde glacier	2010–2017	0.1	5430–5690	Nepal	N	[91]
22	Rikha Samba Dhaulagiri	2011–2017	5.7	5416–6515	Nepal	SE	[92]
23	Yala, Langtang	2011–2017	1.61	5168–5661	Nepal	SW	[92]
24	Naradu Glacier	2011–2018	3.8	4700–5249	Himachal	NE-SW	[93]

the hydrological and climatic system [74]. Gara glacier showed a positive balance of mass in the first 2 years and until the past 1 year, but it was negative over the following 5 years. The seasonal study indicates that a positive mass balance in summer consequently leads to a net positive annual mass balance [30, 74]. Only two glaciers underwent mass balance measurements in the 1990s: the Shaune Garang Glacier in 1991 and the Dokriani Glacier in the Garhwal Himalaya [59]. Between 1991 and 2000, the Dokriani glacier recorded a continuous negative mass balance series of 0.83 m w.e. year^{-1}, with a gap between 1996 and 1997. Since 2000/2001, 2002/2003, and 2003/2004, respectively, three glaciers (Chotta Shigri, Hamtah, and Chorabari) have been constantly studied for mass balance [5, 26, 76, 77]. Continuous monitoring by glaciological methods in Chota Shigri, Nardu, Pastio, Rikha, Stok, Mera, Yala, and Pokalde glaciers in the Himalayas from 2007 to 2021 has been carried out by different researchers, some of them are discussed in Table 9.2.

A glaciological method requires repeated field measurements, which must be performed under sometimes difficult conditions, can achieve the highest accuracy and provide the investigator with an insight into the field conditions. Various factors, including the accurate and precise estimation of ice loss at various stake locations, the adjustment of measurements spatially across the entire glacier, accurate measurement of the density of ice/snow, and the representativeness of accumulation pits and stakes affect the accuracy of glaciological mass balance estimates.

9.2.8.2 Mass Balance Through Geodetic Methods

The geodetic methods are an indirect way of calculating the glacier mass balance. The glacier mass balance over the years can be computed by comparing maps of the glacier generated at two distinct times and using the observed variation in glacier thickness. The best method for carrying out remote qualitative measurements is using a gravimetric approach [94]. The geodetic method is simpler and more efficient for the determination of the glacier mass balance of large, glaciated areas over a longer time period. Several research studies have shown that the measurement of long-term (decadal scale) cumulative mass balance by *in situ* mass balance with a geodetic method needs to be checked for prejudices [5].

It is therefore essential to understand glacier development at a large scale in the Himalaya with a remote sensing geodetic approach. In the Western Himalayan basins, glacier mass balance has been assessed with a geodetic method for 11 years from 2000 to 2011. A total of 42 glaciers with an area greater than 1.0 km^2 were chosen out of 89 glaciers in the basin for the research, encompassing about 72%, i.e., 215 km^2 of the entire glaciated area.

The geodetic approach is the ideal way for monitoring glaciers that are huge in extent and challenging to visit, making glaciological mass calculation impossible. The geodetic system is based on topography changes and glacier elevation. In this method, an estimate of the volume shall be based on a difference in glacier surface elevations over two separate periods. Several factors, including the uncertainty in DEM switching, radar signals, uncertainties caused by the void filling of source DEMs, delineation of glacier lines, and mass conversion using a single value for ice density, have led to uncertainty regarding geodetic mass balance estimates. In assessing the uncertainty, all these factors have been taken into consideration under methodology and an additional term has been given for radar penetration errors [95–98].

9.3 Discussion and Conclusion

A short review of the variability and similarity of methodologies has been applied in the Himalayan glaciated region for the glacier mass balance studies. Here, we examine some key strengths and weaknesses of individual approaches, intending to guide an optimal strategy applied from 1974 to 2021 when considering different topographies, feasibility, and some limitations. The heterogeneity in climatic conditions and terrain patterns of the regions is shown by differences in mass balance measurements for each glacier. The Himalayan glacier mass balance data have mostly been characterized by negative and positive years from 1974 to 2021.

In this work, we analyzed the existing methods, and it is applied them to the Himalayan glaciers through *ex situ* and *in situ* techniques. In particular, owing to the presence of supraglacial lakes/ponds, debris, snow fields (continuous and seasonal), dead ice, mountain shadow formation, and identification of ice and clouds, a rapid and reliable delineation of the boundary of glaciers in the Himalaya remains a difficult task [99, 100]. Different techniques for the semiautomated delineation of glacier debris cover have been adopted by several authors [2, 10, 41, 95, 101–105]. Nevertheless, most of these studies do not meet the desired accuracy and rely on some form of manual interpretation [105]. Even with geophysical tools in the field, the precise identification of glacier boundaries can be challenging [105]. In addition, the restricted geographical areas in which these studies are conducted make it difficult to carry out much of them. In most of the studies, a combination of different datasets including remote sensing satellite data, topographic data, temperature, and InSAR coherence was the basic idea with the most promising and accurate results (accuracy of 96% and 91%) [66, 106, 107].

Until now, no method in place can help us make more precise decisions across a wide geographic area and reduce the costs, time, and effort involved. Therefore, there is a need to develop algorithms that will make it possible to map glaciers rapidly and reliably. Visual interpretation rather than high-resolution data, where more precise information is desired, is generally used to determine the glacier limit. The mass balance in the IHR is generally negative except on the Gara, Gor Garang, Shaune Garang, Chhota Shigri, Hamtah, and Chorabari glaciers where there are few positive balances for a year [5, 26, 76, 77].

In comparison to the observed negative value, the positive values obtained from a measured mass balance are not very promising. Thus, it is important to keep the MB measurements continuously to carry out subsequent scenarios of glacier mass gain/loss. To understand the dynamics of the Himalayan glacier, we have an analysis from glaciological and relate it with climatic variability. To gain a deep understanding of glacier dynamics within the Himalayan region, cumulative data collection with high-resolution climatic and remote sensing data will be helpful. Climatic variation directly affects the glacier dynamics of Himalayan glaciers and glacier lakes leading to catastrophic floods and other events [27, 108, 109].

In the 21st century, changing climate has had a significant impact on glacier health, e.g., retreat or expansion of glacier boundaries. Also, it helps to monitor to understand the continuous state of the cryosphere [66]. Furthermore, identifying glacier borders is critical for accurately determining glacier length and predicting glacier-mediated calamities and climate change [65]. Science and society have changed significantly with the introduction of large data techniques such as geospatial and climatic data. Nowadays, growing rapid field data gather huge amounts of data from location-aware instruments. To limit glacier mass balance in a high-elevation accumulation area, the deployment of additional stakes above 5850 m would therefore be advisable. However, it is noted that due to logistical and safety constraints, the ability to obtain such highly reliable observations is inherently difficult. To better understand the regional behavior of Himalayan glaciers in response to climate changes, it is necessary to carry out continuous glacier measurements.

References

1. W. W. Immerzeel, P. Droogers, S. De Jong and M. F. P. Bierkens, "Large-scale monitoring of snow cover and runoff simulation in Himalayan river basins using remote sensing," *Remote Sens. Environ.*, vol. 113, pp. 40–49, Jan. 2009, doi: 10.1016/j.rse.2008.08.010.
2. J. L. Bamber *et al.*, "A review of remote sensing methods for glacier mass balance determination," *Front. Environ. Sci.*, vol. 59, no. 5, pp. 672–704, 2007, doi: 10.1016/j.gloplacha.2006.11.031.
3. D. Scherler, S. Leprince and M. R. Strecker, "Glacier-surface velocities in alpine terrain from optical satellite imagery-Accuracy improvement and quality assessment," *Remote Sens. Environ.*, vol. 112, no. 10, pp. 3806–3819, Oct. 2008, doi: 10.1016/j.rse.2008.05.018.
4. C. A. Scott, F. Zhang, A. Mukherji, W. Immerzeel, D. Mustafa and L. Bharati, "Water in the Hindu Kush Himalaya," in *The Hindu Kush Himalaya Assessment: Mountains, Climate Change, Sustainability and People*, P. Wester, A. Mishra, A. Mukherji, and A. B. Shrestha, Eds., Cham: Springer International Publishing, 2019, pp. 257–299. doi: 10.1007/978-3-319-92288-1_8.
5. B. Pratap, D. P. Dobhal, R. Bhambri, M. Mehta and V. C. Tewari, "Four decades of glacier mass balance observations in the Indian Himalaya," *Reg. Environ. Change*, vol. 16, no. 3. Springer Verlag, pp. 643–658, Mar. 01, 2016. doi: 10.1007/s10113-015-0791-4.
6. S. J. S. Khalsa, M. B. Dyurgerov, T. Khromova, B. H. Raup and R. G. Barry, "Space-based mapping of glacier changes using ASTER and GIS tools," in *IEEE Trans. Geosci. Remote Sens.*, Oct. 2004, pp. 2177–2183. doi: 10.1109/TGRS.2004.834636.
7. A. Aggarwal, S. K. Jain, A. K. Lohani and N. Jain, "Glacial lake outburst flood risk assessment using combined approaches of remote sensing, GIS and dam break modelling," *Geomatics Nat. Hazards Risk*, vol. 7, no. 1, pp. 18–36, Jan. 2016, doi: 10.1080/19475705.2013.862573.
8. S. R. Bajracharya, S. Pradhananga, A. B. Shrestha and R. Thapa, "Future climate and its potential impact on the spatial and temporal hydrological regime in the Koshi Basin, Nepal," *J. Hydrol. Reg. Stud.*, vol. 45, p. 101316, Feb. 2023, doi: 10.1016/J.EJRH.2023.101316.
9. Y. Nie, Q. Liu, J. Wang, Y. Zhang, Y. Sheng and S. Liu, "An inventory of historical glacial lake outburst floods in the Himalayas based on remote sensing observations and geomorphological analysis," *Geomorphology*, vol. 308, Feb. 2018, doi: 10.1016/j.geomorph.2018.02.002.
10. M. Cogley, J. G., Hock, R., Rasmussen, L. A., Arendt, A., Bauder, A., Braithwaite, R., Jansson, P., Kaser, G., Möller, M., Nicholson, L., and Zemp *et al.*, "A review of remote sensing methods for glacier mass balance determination," *Front. Environ. Sci.*, vol. 59, no. 5, pp. 672–704, 2011, doi: 10.1016/j.gloplacha.2006.11.031.

11. A. Linda *et al.*, "Four years of mass balance on Chhota Shigri Glacier , Himachal Pradesh , India , a new benchmark glacier in the western Himalaya Four years of mass balance on Chhota Shigri Glacier , Himachal Pradesh , India , a new benchmark glacier in the western Himala," *J. Glaciol.*, vol. 53, no. 183, p. 603, 2007.
12. A. V. Kulkarni, "Mass balance of Himalayan glaciers using AAR and ELA methods," *J. Glaciol.*, vol. 38, no. 128, pp. 101–104, 1992, doi: 10.3189/s0022143000009631.
13. M. Mehta, D. P. Dobhal and M. P. S. Bisht, "Change of Tipra Glacier in the Garhwal Himalaya, India, between 1962 and 2008," *Prog. Phys. Geogr.: Earth Environ.*, vol. 35, no. 6, pp. 721–738, 2011, doi: 10.1177/0309133311411760.
14. A. Raman, A. V. Kulkarni and V. Prasad, "Glacier mass balance estimation in Garhwal Himalaya using improved accumulation area ratio method," *Environ. Monit. Assess.*, vol. 194, no. 8, p. 583, Jul. 2022, doi: 10.1007/s10661-022-10261-y.
15. J. G. Cogley, "Mass-balance terms revisited," *J. Glaciol.*, vol. 56, no. 200, pp. 997–1001, 2011, doi: 10.3189/002214311796406040.
16. A. Rabatel *et al.*, "Can the snowline be used as an indicator of the equilibrium line and mass balance for glaciers in the outer tropics?," *J. Glaciol.*, vol. 58, no. 212, pp. 1027–1036, 2012, doi: 10.3189/2012JoG12J027.
17. S. Singh and R. Kumar, "A statistical approach to estimate equilibrium line altitude (ELA) and its trend analysis on Naradu Glacier, Himachal Himalaya," *Mater Today Proc.*, vol. 34, pp. 869–874, 2019, doi: 10.1016/j.matpr.2020.08.787.
18. W. W. Immerzeel, P. Droogers, S. M. de Jong and M. F. P. Bierkens, "Large-scale monitoring of snow cover and runoff simulation in Himalayan river basins using remote sensing," *Remote Sens. Environ.*, vol. 113, no. 1, pp. 40–49, Jan. 2009, doi: 10.1016/j.rse.2008.08.010.
19. J. L. Bamber and A. Rivera, "A review of remote sensing methods for glacier mass balance determination," *Glob. Planet. Change*, vol. 59, no. 1–4, pp. 138–148, Oct. 2007, doi: 10.1016/j.gloplacha.2006.11.031.
20. D. SCHERLER, S. LEPRINCE and M. STRECKER, "Glacier-surface velocities in alpine terrain from optical satellite imagery—Accuracy improvement and quality assessment," *Remote Sens. Environ.*, vol. 112, no. 10, pp. 3806–3819, Oct. 2008, doi: 10.1016/j.rse.2008.05.018.
21. D. K. Hall, G. A. Riggs and V. V Salomonson, "Development of Methods for Mapping Global Snow Cover Using Moderate Resolution Imaging Spectroradiometer Data."
22. R. K. Tiwari and M. Arora, "Estimation of surface ice velocity of Chhota-Shigri glacier using sub-pixel ASTER image correlation *Remote Sens. Agroinf.* View project SCATSAT-1 Applications View project." [Online]. Available: https://www.researchgate.net/publication/260993110
23. S. Leprince, S. Barbot, F. Ayoub and J. P. Avouac, "Automatic and precise orthorectification, coregistration, and subpixel correlation of satellite

images, application to ground deformation measurements," *IEEE Trans. Geosci. Remote Sens.*, vol. 45, no. 6, pp. 1529–1558, 2007, doi: 10.1109/TGRS.2006.888937.
24. P. Gantayat, A. V. Kulkarni and J. Srinivasan, "Estimation of ice thickness using surface velocities and slope: Case study at Gangotri Glacier, India," *J. Glaciol.*, vol. 60, no. 220, pp. 277–282, 2014, doi: 10.3189/2014JoG13J078.
25. A. Sattar, A. Goswami, A. V. Kulkarni and P. Das, "Glacier-surface velocity derived ice volume and retreat assessment in the dhauliganga basin, central himalaya – A remote sensing and modeling based approach," *Front. Earth Sci. (Lausanne)*, vol. 7, Apr. 2019, doi: 10.3389/feart.2019.00105.
26. P. Wagnon *et al.*, "Four years of mass balance on Chhota Shigri Glacier, Himachal Pradesh, India, a new benchmark glacier in the western Himalaya," *J. Glaciol.*, vol. 53, no. 183, pp. 603–611, 2007, doi: 10.3189/002214307784409306.
27. S. Singh, R. Kumar and A. P. Dimri, "Mass balance status of Indian Himalayan glaciers: A brief review," *Front. Environ. Sci.*, vol. 6, no. AUG. Frontiers Media S.A., Aug. 21, 2018. doi: 10.3389/fenvs.2018.00030.
28. V. K. Raina, M. K. Kaul and S. Singh, "Mass-Balance Studies of Gara Glacier," *J. Glaciol.*, vol. 18, no. 80, pp. 415–423, 1977, doi: 10.3189/s0022143000021092.
29. K. Marine, W. R. Division and S. A. Centre, "Mass balance of Hilnalayan glaciers using AAR and ELA Dlethods," no. 128, pp. 101–104, 1992.
30. -V K Raina and V. K. Raina, "Himalayan Glaciers A State-of-Art Review of Glacial Studies, Glacial Retreat and Climate Change."
31. D. P. Dobhal, M. Mehta and D. Srivastava, "Influence of debris cover on terminus retreat and mass changes of Chorabari Glacier, Garhwal region, central Himalaya, India," *J. Glaciol.*, vol. 59, no. 217, pp. 961–971, Jul. 2013, doi: 10.3189/2013JoG12J180.
32. M. F. Azam *et al.*, "From balance to imbalance: a shift in the dynamic behaviour of Chhota Shigri glacier, western Himalaya, India," *J. Glaciol.*, vol. 58, no. 208, pp. 315–324, Sep. 2012, doi: 10.3189/2012JoG11J123.
33. W. W. Immerzeel, P. Droogers, S. M. de Jong and M. F. P. Bierkens, "Large-scale monitoring of snow cover and runoff simulation in Himalayan river basins using remote sensing," *Remote Sens. Environ.*, vol. 113, no. 1, pp. 40–49, Jan. 2009, doi: 10.1016/j.rse.2008.08.010.
34. J. Bamber and A. Rivera, "A review of remote sensing methods for glacier mass balance determination," *Glob. Planet. Change*, vol. 59, Oct. 2007, doi: 10.1016/j.gloplacha.2006.11.031.
35. J. L. Bamber and A. Rivera, "A review of remote sensing methods for glacier mass balance determination," *Glob. Planet. Change*, vol. 59, no. 1–4, pp. 138–148, Oct. 2007, doi: 10.1016/j.gloplacha.2006.11.031.
36. S. Kaushik, P. K. Joshi and T. Singh, "Development of glacier mapping in Indian Himalaya: a review of approaches," *Int. J. Remote Sens.*, vol. 40, no. 17, pp. 6607–6634, 2019, doi: 10.1080/01431161.2019.1582114.

37. R. Bhambri and T. Bolch, "Glacier mapping: A review with special reference to the Indian Himalayas," *Prog. Phys. Geogr.*, vol. 33, no. 5, pp. 672–704, 2009, doi: 10.1177/0309133309348112.
38. A. Aggarwal, S. K. Jain, A. K. Lohani and N. Jain, "Glacial lake outburst flood risk assessment using combined approaches of remote sensing, GIS and dam break modelling," *Geomatics Nat. Hazards Risk*, vol. 7, no. 1, pp. 18–36, Jan. 2016, doi: 10.1080/19475705.2013.862573.
39. Y. Nie, Q. Liu, J. Wang, Y. Zhang, Y. Sheng and S. Liu, "An inventory of historical glacial lake outburst floods in the Himalayas based on remote sensing observations and geomorphological analysis," *Geomorphology*, vol. 308, pp. 91–106, May 2018, doi: 10.1016/j.geomorph.2018.02.002.
40. B. Bajracharya, A. B. Shrestha and L. Rajbhandari, "Glacial lake outburst floods in the Sagarmatha region: Hazard assessment using GIS and hydrodynamic modelling," *Mt. Res. Dev.*, vol. 27, no. 4, pp. 336–344, Nov. 2007, doi: 10.1659/mrd.0783.
41. R. Bhambri and T. Bolch, "Glacier mapping: a review with special reference to the Indian Himalayas," *Prog. Phys. Geogr.: Earth Environ.*, vol. 33, no. 5, pp. 672–704, Oct. 2009, doi: 10.1177/0309133309348112.
42. A. V. Kulkarni, "Mass balance of Himalayan glaciers using AAR and ELA methods," *J. Glaciol.*, vol. 38, no. 128, pp. 101–104, 1992, doi: 10.3189/s0022143000009631.
43. A. Raman, A. V Kulkarni and V. Prasad, "Glacier mass balance estimation in Garhwal Himalaya using improved accumulation area ratio method," *Environ. Monit. Assess.*, vol. 194, no. 8, p. 583, 2022, doi: 10.1007/s10661-022-10261-y.
44. R. K. Tiwari, R. P. Gupta and M. K. Arora, "Estimation of surface ice velocity of Chhota-Shigri glacier using sub-pixel ASTER image correlation," *Curr. Sci.*, vol. 106, no. 6, pp. 853–859, 2014, doi: 10.18520/cs/v106/i6/853-859.
45. P. Gantayat, A. V. Kulkarni and J. Srinivasan, "Estimation of ice thickness using surface velocities and slope: Case study at Gangotri Glacier, India," *J. Glaciol.*, vol. 60, no. 220, pp. 277–282, 2014, doi: 10.3189/2014JoG13J078.
46. A. Sattar, A. Goswami, A. V Kulkarni, P. Das and J. J. Fürst, "Glacier-Surface Velocity Derived Ice Volume and Retreat Assessment in the Dhauliganga Basin, Central Himalaya – A Remote Sensing and Modeling Based Approach," vol. 7, no. May, pp. 1–15, 2019, doi: 10.3389/feart.2019.00105.
47. A. Shukla, R. P. Gupta and M. K. Arora, "Delineation of debris-covered glacier boundaries using optical and thermal remote sensing data," *Remote Sens. Lett.*, vol. 1, no. 1, pp. 11–17, Mar. 2010, doi: 10.1080/01431160903159316.
48. P. Pandey, A. V. Kulkarni and G. Venkataraman, "Remote sensing study of snowline altitude at the end of melting season, Chandra-Bhaga basin, Himachal Pradesh, 1980-2007," *Geocarto Int.*, vol. 28, no. 4, pp. 311–322, 2013, doi: 10.1080/10106049.2012.705336.
49. S. N. Remya, A. V. Kulkarni, T. Hassan Syed and H. C. Nainwal, "Glacier mass loss in the Alaknanda basin, Garhwal Himalaya on a decadal scale," *Geocarto Int.*, vol. 37, no. 10, pp. 3014–3032, 2022, doi: 10.1080/10106049.2020.1844309.

50. M. Debnath, M. C. Sharma and H. J. Syiemlieh, "Glacier dynamics in changme khangpu basin, sikkim himalaya, India, between 1975 and 2016," *Geosciences (Switzerland)*, vol. 9, no. 6, Jun. 2019, doi: 10.3390/geosciences9060259.
51. S. Kundu and M. Chakraborty, "Delineation of glacial zones of Gangotri and other glaciers of Central Himalaya using RISAT-1 C-band dual-pol SAR," *Int. J. Remote Sens.*, vol. 36, no. 6, pp. 1529–1550, Mar. 2015, doi: 10.1080/01431161.2015.1014972.
52. V. K. Gaddam *et al.*, "Glacial retreat in Himalaya using Indian Remote Sensing satellite data," *Geocarto Int.*, vol. 35, no. 3, pp. 1179–1198, 2020, doi: 10.1117/12.694004.
53. V. D. Mishra, J. K. Sharma, K. K. Singh, N. K. Thakur and M. Kumar, "Assessment of different topographic corrections in AWiFS satellite imagery of Himalaya terrain," *J. Earth Syst. Sci.*, vol. 118, no. 1, pp. 11–26, 2009, doi: 10.1007/s12040-009-0002-0.
54. A. Shukla, S. Garg, M. Mehta, V. Kumar and U. Kant Shukla, "Temporal inventory of glaciers in the Suru sub-basin, western Himalaya: Impacts of regional climate variability," *Earth Syst. Sci. Data*, vol. 12, no. 2, pp. 1245–1265, 2020, doi: 10.5194/essd-12-1245-2020.
55. E. Berthier *et al.*, "Measuring glacier mass changes from space—a review," *Rep. Prog. Phys.*, vol. 86, no. 3, p. 36801, Feb. 2023, doi: 10.1088/1361-6633/acaf8e.
56. S. Singh, R. Kumar and A. P. Dimri, "Mass balance status of Indian Himalayan glaciers: A brief review," *Front. Environ. Sci.*, vol. 6, no. AUG, 2018, doi: 10.3389/fenvs.2018.00030.
57. R. Pellitero *et al.*, "A GIS tool for automatic calculation of glacier equilibrium-line altitudes," *Comput. Geosci.*, vol. 82, pp. 55–62, 2015, doi: 10.1016/j.cageo.2015.05.005.
58. Y. Mohajerani, S. Jeong, B. Scheuchl, I. Velicogna, E. Rignot and P. Milillo, "Automatic delineation of glacier grounding lines in differential interferometric synthetic-aperture radar data using deep learning," *Sci. Rep.*, vol. 11, no. 1, Dec. 2021, doi: 10.1038/s41598-021-84309-3.
59. D. P. Dobhal, J. T. Gergan and R. J. Thayyen, "Recession and morphogeometrical changes of Dokriani glacier (1962-1995) Garhwal Himalaya, India," *Curr. Sci.*, vol. 86, no. 5, pp. 692–696, 2004.
60. R. K. Tiwari, P. K. Garg, A. Shukla, R. S. Ahluwalia, N. Singh and P. Chauhan, "A geomorphic and morphometric analysis of surface ice velocity variation of different valley type glaciers," *Land Surface and Cryosphere Remote Sensing III*, vol. 9877, p. 98771I, 2016, doi: 10.1117/12.2227108.
61. D. P. Dobhal, B. Pratap, R. Bhambri and M. Mehta, "Mass balance and morphological changes of Dokriani Glacier (1992–2013), Garhwal Himalaya, India," *Quat. Sci. Adv.*, vol. 4, no. May, 2021, doi: 10.1016/j.qsa.2021.100033.
62. Q. X. Man, H. D. Guo, G. Liu and P. L. Dong, "Comparison of different methods for monitoring glacier changes observed by Landsat images," in

IOP Conf. Ser.: Earth Environ. Sci., Institute of Physics Publishing, 2014. doi: 10.1088/1755-1315/17/1/012127.
63. D. Varade, A. K. Maurya, O. Dikshit, G. Singh and S. Manickam, "Snow depth in Dhundi: an estimate based on weighted bias corrected differential phase observations of dual polarimetric bi-temporal Sentinel-1 data," *Int. J. Remote Sens.*, vol. 41, no. 8, pp. 3031–3053, Apr. 2020, doi: 10.1080/01431161.2019.1698076.
64. V. Sood, H. S. Gusain, S. Gupta and S. Singh, "Topographically derived subpixel-based change detection for monitoring changes over rugged terrain Himalayas using AWiFS data," *J. Mt. Sci.*, vol. 18, no. 1, pp. 126–140, 2021, doi: 10.1007/s11629-020-6151-y.
65. V. Sood, S. Singh, A. K. Taloor, S. Prashar and R. Kaur, "Monitoring and mapping of snow cover variability using topographically derived NDSI model over north Indian Himalayas during the period 2008–19," *Appl. Comput. Geosci.*, vol. 8, no. September, p. 100040, 2020, doi: 10.1016/j.acags.2020.100040.
66. V. Sood, R. K. Tiwari, S. Singh, R. Kaur and B. R. Parida, "Glacier Boundary Mapping Using Deep Learning Classification over Bara Shigri Glacier in Western Himalayas," *Sustainability (Switzerland)*, vol. 14, no. 20, 2022, doi: 10.3390/su142013485.
67. Z. Xie, V. K. Asari and U. K. Haritashya, "Evaluating deep-learning models for debris-covered glacier mapping," *Appl. Comput. Geosci.*, vol. 12, p. 100071, Dec. 2021, doi: 10.1016/J.ACAGS.2021.100071.
68. Z. Xie, U. K. Haritashya, V. K. Asari, B. W. Young, M. P. Bishop and J. S. Kargel, "GlacierNet: A Deep-Learning Approach for Debris-Covered Glacier Mapping," *IEEE Access*, vol. 8, pp. 83495–83510, 2020, doi: 10.1109/ACCESS.2020.2991187.
69. Z. Q. Zhao, P. Zheng, S. T. Xu and X. Wu, "Object Detection with Deep Learning: A Review," *IEEE Trans. Neural. Netw. Learn. Syst.*, vol. 30, no. 11, pp. 3212–3232, 2019, doi: 10.1109/TNNLS.2018.2876865.
70. X. X. Zhu *et al.*, "Deep learning in remote sensing: a review," no. december, 2017, doi: 10.1109/MGRS.2017.2762307.
71. J. Bolibar, A. Rabatel, I. Gouttevin, C. Galiez, T. Condom and E. Sauquet, "Glacier evolution modelling with deep learning: challenges and opportunities," no. May, 2020, doi: 10.5194/egusphere-egu2020-9723.
72. A. E. Racoviteanu, M. W. Williams and R. G. Barry, "Optical remote sensing of glacier characteristics: A review with focus on the Himalaya," *Sensors*, vol. 8, no. 5, pp. 3355–3383, May 2008, doi: 10.3390/s8053355.
73. S. Singh, R. Kumar and A. P. Dimri, "Mass balance status of Indian Himalayan glaciers: A brief review," *Front. Environ. Sci.*, vol. 6, no. AUG. Frontiers Media S.A., Aug. 21, 2018. doi: 10.3389/fenvs.2018.00030.
74. V. K. Raina, M. K. Kaul and S. Singh, "Mass-Balance Studies of Gara Glacier," *J. Glaciol.*, vol. 18, no. 80, pp. 415–423, Jan. 1977, doi: 10.3189/S0022143000021092.

75. A. E. Racoviteanu, M. W. Williams and R. G. Barry, "Optical remote sensing of glacier characteristics: A review with focus on the Himalaya," *Sensors*, vol. 8, no. 5, pp. 3355–3383, May 2008, doi: 10.3390/s8053355.
76. M. F. Azam *et al.*, "From balance to imbalance: A shift in the dynamic behaviour of Chhota Shigri glacier, western Himalaya, India," *J. Glaciol.*, vol. 58, no. 208, pp. 315–324, 2012, doi: 10.3189/2012JoG11J123.
77. D. P. Dobhal, M. Mehta and D. Srivastava, "Influence of debris cover on terminus retreat and mass changes of Chorabari Glacier, Garhwal region, central Himalaya, India," *J. Glaciol.*, vol. 59, no. 217, pp. 961–971, 2013, doi: 10.3189/2013JoG12J180.
78. V. K. Raina, M. K. Kaul and S. Singh, "Mass-Balance Studies of Gara Glacier," *J. Glaciol.*, vol. 18, no. 80, pp. 415–423, Jan. 1977, Accessed: Jul. 20, 2023. [Online]. Available: https://www.cambridge.org/core/product/identifier/S0022143000021092/type/journal_article
79. -V K Raina and V. K. Raina, "Himalayan Glaciers A State-of-Art Review of Glacial Studies, Glacial Retreat and Climate Change."
80. M. B. Dyurgerov, "Glacier mass balance and regime: Data of measurements and analysis," *Instaar*, no. 55, p. 267, 2002.
81. M. B. Dyurgerov and M. F. Meier, "Glaciers and the changing Earth system: a 2004 snapshot." 2005.
82. M. Dyurgerov and M. F. Meier, "Glaciers and the Changing Earth System: A 2004 Snapshot. Occasional Paper 58, Institute of Arctic and Alpine Research, University of Colorado, Boulder, CO," 2005.
83. M. B. Dyurgerov and M. F. Meier, "Mass balance of mountain and subpolar glaciers: a new global assessment for 1961–1990," *Arct. Alp. Res.*, vol. 29, no. 4, pp. 379–391, 1997.
84. M. F. Azam *et al.*, "From balance to imbalance: A shift in the dynamic behaviour of Chhota Shigri glacier, western Himalaya, India," *J. Glaciol.*, vol. 58, no. 208, pp. 315–324, Sep. 2012, doi: 10.3189/2012JoG11J123.
85. D. P. P. Dobhal, M. Mehta and D. Srivastava, "Influence of debris cover on terminus retreat and mass changes of Chorabari Glacier, Garhwal region, central Himalaya, India," *J. Glaciol.*, vol. 59, no. 217, pp. 961–971, Jul. 2013, doi: 10.3189/2013JoG12J180.
86. M. B. Dyurgerov, M. F. Meier, B. N. Goswami, J. Shukla, E. K. Schneider and Y. C. Sud, "Study of the dynamics of the intertropical convergence zone with a symmetric version of the GLAS climate model," *Arct. Alp. Res.*, vol. 41, no. 1, pp. 379–391, 2005.
87. H. GUSAIN, V. MISHRA and M. BHUTIYANI, "Winter temperature and snowfall trends in the cryospheric region of north-west Himalaya," *Mausam*, vol. 65, no. 3, pp. 425–432, 2014, doi: 10.54302/mausam.v65i3.1053.
88. T. Angchuk *et al.*, "Annual and seasonal glaciological mass balance of Patsio Glacier, western Himalaya (India) from 2010 to 2017," *J. Glaciol.*, vol. 67, no. 266, pp. 1137–1146, Dec. 2021, doi: 10.1017/jog.2021.60.

89. A. Mandal *et al.*, "13 years (2002-2015) of in-situ mass balance record and meteorological conditions of Chhota Shigri Glacier, Indian Himalaya," in AGU Fall Meeting Abstracts, Dec. 2016, pp. C11C-0776.
90. M. Soheb, A. Ramanathan, T. Angchuk, A. Mandal, N. Kumar and S. Lotus, "Mass-balance observation, reconstruction and sensitivity of Stok glacier, Ladakh region, India, between 1978 and 2019," *J. Glaciol.*, vol. 66, no. 258, pp. 627–642, 2020, doi: 10.1017/jog.2020.34.
91. P. Wagnon *et al.*, "No Title," *J. Glaciol.*, vol. 67, no. 261, pp. 117–125, Feb. 2021, doi: 10.1017/jog.2020.88.
92. D. Stumm, S. P. Joshi, T. R. Gurung and G. Silwal, "Mass balances of Yala and Rikha Samba glaciers, Nepal, from 2000 to 2017," *Earth Syst. Sci. Data*, vol. 13, no. 8, pp. 3791–3818, Aug. 2021, doi: 10.5194/essd-13-3791-2021.
93. R. R. Kumar, S. S. Singh, A. Singh, R. R. Kumar, S. S. Singh and S. S. Randhawa, "Surface mass balance analysis at Naradu Glacier, Western Himalaya, India," *Sci. Rep.*, vol. 11, no. 1, Dec. 2021, doi: 10.1038/s41598-021-91348-3.
94. E. Rignot, A. Rivera and G. Casassa, "Contribution of the Patagonia Icefields of South America to Sea Level Rise," *Science* (1979), vol. 302, no. 5644, pp. 434–437, 2003, doi: 10.1126/science.1087393.
95. H. Nagai *et al.*, "Geodetic and direct mass-balance measurements: Comparison and joint analysis," *Ann. Glaciol.*, vol. 58, no. 3, pp. 96–100, 2014, doi: 10.3189/172756409787769744.
96. M. Huss, "Density assumptions for converting geodetic glacier volume change to mass change," *Cryosphere*, vol. 7, no. 3, pp. 877–887, May 2013, doi: 10.5194/tc-7-877-2013.
97. J. G. Cogley, "Geodetic and direct mass-balance measurements: Comparison and joint analysis," *Ann. Glaciol.*, vol. 50, no. 50, pp. 96–100, 2009, doi: 10.3189/172756409787769744.
98. A. V Kulkarni, "COMPARISON OF MASS BALANCE ESTIMATES BY GEODETIC AND IMPROVED ACCUMULATION AREA RATIO METHOD, WESTERN HIMALAYA," 2019. [Online]. Available: https://www.researchgate.net/publication/334524534
99. B. Raup, A. Racoviteanu, S. J. S. Khalsa, C. Helm, R. Armstrong and Y. Arnaud, "The GLIMS geospatial glacier database: A new tool for studying glacier change," *Glob. Planet. Change*, vol. 56, no. 1–2, pp. 101–110, 2007, doi: 10.1016/j.gloplacha.2006.07.018.
100. A. E. Racoviteanu, M. W. Williams and R. G. Barry, "Optical remote sensing of glacier characteristics: A review with focus on the Himalaya," *Sensors*, vol. 8, no. 5, pp. 3355–3383, 2008, doi: 10.3390/s8053355.
101. A. R. Bajracharya, S. R. Bajracharya, A. B. Shrestha and S. B. Maharjan, "Climate change impact assessment on the hydrological regime of the Kaligandaki Basin, Nepal," *Sci. Total Environ.*, vol. 625, pp. 837–848, Jun. 2018, doi: 10.1016/j.scitotenv.2017.12.332.

102. R. Bhambri, T. Bolch and R. K. Chaujar, "Frontal recession of Gangotri Glacier, Garhwal Himalayas, from 1965 to 2006, measured through highresolution remote sensing data," *Curr. Sci.*, vol. 102, no. 3, pp. 489–494, 2012.
103. B. A. Robson, T. Bolch, S. MacDonell, D. Hölbling, P. Rastner and N. Schaffer, "Automated detection of rock glaciers using deep learning and object-based image analysis," *Remote Sens. Environ.*, vol. 250, no. August, 2020, doi: 10.1016/j.rse.2020.112033.
104. H. Frey *et al.*, "Estimating the volume of glaciers in the Himalayan-Karakoram region using different methods," *Cryosphere*, vol. 8, no. 6, pp. 2313–2333, 2014, doi: 10.5194/tc-8-2313-2014.
105. F. Paul *et al.*, "The glaciers climate change initiative: Methods for creating glacier area, elevation change and velocity products," *Remote Sens. Environ.*, vol. 162, pp. 408–426, 2015, doi: 10.1016/j.rse.2013.07.043.
106. R. Nijhawan, J. Das and B. Raman, "A hybrid of deep learning and hand-crafted features based approach for snow cover mapping," *Int. J. Remote Sens.*, vol. 40, no. 2, pp. 759–773, 2019, doi: 10.1080/01431161.2018.1519277.
107. Y. Shi *et al.*, "A Novel Approach to Subpixel Land-Cover Change Detection Based on a Supervised Back-Propagation Neural Network for Remotely Sensed Images with Different Resolutions," *Remote Sens. (Basel)*, vol. 11, no. 2, pp. 197–212, 2019, doi: 10.18520/cs/v117/i6/1002-1013.
108. M. J. Westoby, N. F. Glasser, M. J. Hambrey, J. Brasington, J. M. Reynolds and M. A. A. M. Hassan, "Reconstructing historic glacial lakeoutburst floods through numerical modelling and geomorphological assessment: Extreme events in the himalaya," *Earth Surf. Processes Landforms*, vol. 39, no. 12, pp. 1675–1692, 2014, doi: 10.1002/esp.3617.
109. A. P. Dimri and U. C. Mohanty, "Simulation of mesoscale features associated with intense western disturbances over western Himalayas," *Meteorol. Appl.*, vol. 16, no. 3, pp. 289–308, 2009, doi: 10.1002/met.117.

10

Geospatial Techniques for Flash Flood Hazard Assessment and Management

Nitesh Godara[1,2], Amit Challana[3], Tarun Bansal[4] and Arun Bawa[5]*

[1]*Norwegian University of Science and Technology, Trondheim, Norway*
[2]*Rambøll, Norway*
[3]*Katholieke Universiteit (KU), Leuven, Belgium*
[4]*Luleå University of Technology, Luleå, Sweden*
[5]*Texas A&M AgriLife Research, Blackland Research & Extension Center, Temple, TX, USA*

Abstract

This chapter explores how geospatial tools such as remote sensing and Geographic Information Systems (GIS) can help handle flash flood risks. Flash floods, known for their sudden and intense nature, require precise monitoring and rapid responses. Remote sensing uses satellite or drone-observed data collected using optical, Radar, and LiDAR sensors to track precipitation patterns, water levels, and dynamic changes in terrain, enabling timely flood detection and assessment. Integration of remote sensing data with GIS techniques enhances spatial analysis, allowing for the creation of flood maps, hydrological models, and real-time monitoring systems which help in emergency response coordination. Recent advancements in flood monitoring using drone mapping have emerged as a game-changing tool in flash flood management. Drones equipped with advanced sensors provide high-resolution imagery and real-time data from hard-to-reach and hazardous areas, facilitate rapid damage assessment, aid in identifying at-risk locations, and support search and rescue operations. Combining drone-derived data with GIS layers results in more accurate 3D flood models, facilitating visualization and decision-making for emergency responders and decision-makers.

The chapter delves into the significance of predictive modeling in flash flood scenarios. Remote sensing data feeds into hydrological models, enabling the simulation of flash flood events and potential outcomes. The generated information helps in preparing for flash floods by identifying vulnerable areas, predicting

**Corresponding author*: arun.bawa@ag.tamu.edu

Disha Thakur, Sanjay Kumar, Har Amrit Singh Sandhu and Chander Prakash (eds.) Sustainable Development Using Geospatial Techniques, (241–262) © 2024 Scrivener Publishing LLC

flood extents, and evaluating potential impacts on infrastructure and communities. The resulting data can be integrated into decision support systems, offering actionable insights for issuing timely alerts, managing evacuations, and allocating resources efficiently. Additionally, the chapter explores the socioeconomic aspects of integrating geospatial techniques in flash flood management, emphasizing the role of collaboration among disaster management agencies, meteorologists, supply chain analysts, and local communities. The power of geospatial visualizations in communicating flood risks to the public is underscored promoting awareness, preparedness, and resilience. The authorities can better reduce the effects of flash floods on vulnerable regions and communities by integrating geospatial tools such as remote sensing, GIS, and drone mapping.

Keywords: Geospatial tools, remote sensing, Geographic information systems, drone mapping, UAV, flood monitoring

10.1 Understanding Flash Flood Hazards

10.1.1 Introduction to Flash Floods

One of the most formidable and life-threatening natural disasters that can strike with little warning, often catching communities and individuals off guard, is flash floods. These events, characterized by their sudden and intense nature, occur when an extreme amount of rainfall occurs over a specific area in a short period [1], or when a rapid release of water, typically from sources like dam breaks or sudden snowmelt, inundates a region [2]. Unlike the more predictable riverine floods, flash floods are known for their quickness and unpredictability, leaving minimal time for preparation or response. In addition, the severity of flash flood hazards also lies in their devastating consequences. Flash floods can rapidly transform quiet streams or dry riverbeds into raging torrents that tear through landscapes, carrying debris and mud, and causing an immediate danger to life and property.

These hazards are not confined to any specific geographical region; flash floods can occur in various terrains, from urban areas with impermeable surfaces that accelerate runoff [3] to mountainous regions where steep slopes carry the rainfall water rapidly into river valleys [4]. They can emerge from a multitude of meteorological triggers, including intense thunderstorms, tropical cyclones, or prolonged, extreme rainfall events.

As per the August 2021 World Meteorological Organization report [5], weather, climate, and water hazards were responsible for 50% of all disasters, 45% of all recorded deaths, and 74% of all recorded economic losses between 1970 and 2019. In addition to 58,700 fatalities, the floods caused

US $115 billion in economic losses. Globally, floods have been linked to 44% of all recorded disasters, with riverine floods accounting for 24% and general floods for 14%. In terms of financial impact, 31% of economic losses are attributable to various flood types, including riverine floods (20%), general floods (8%), and flash floods (3%).

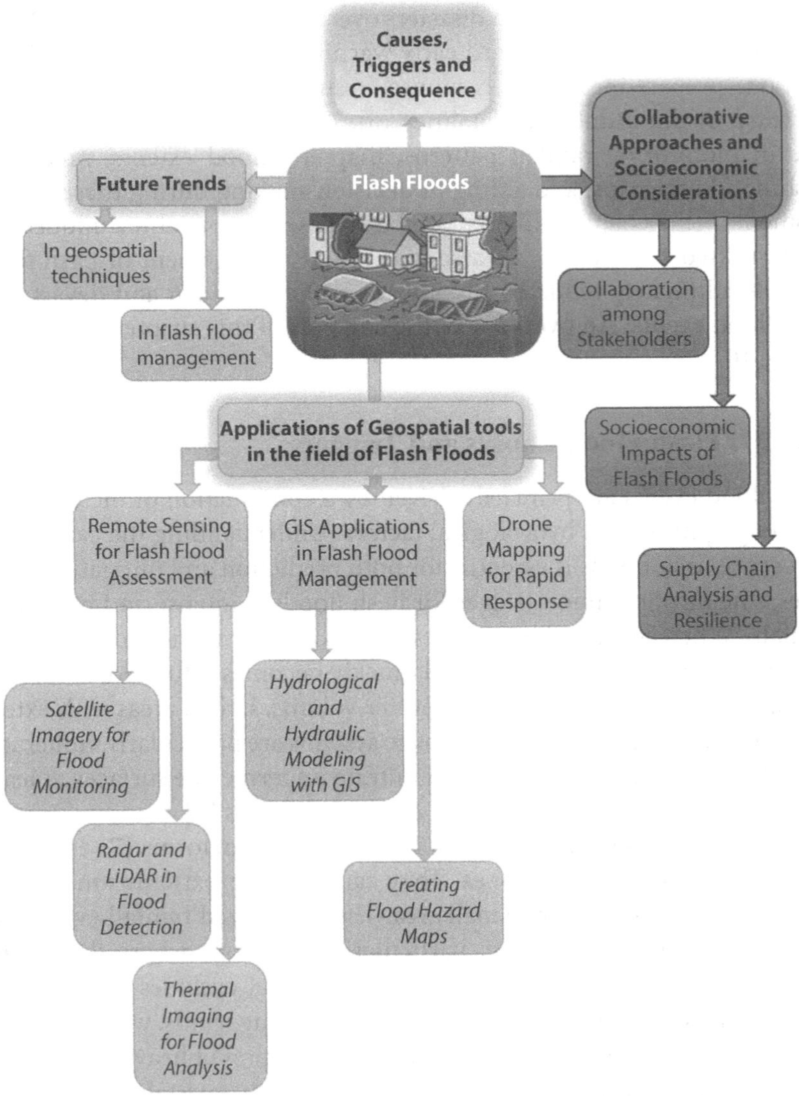

Figure 10.1 Applications of remote sensing and geospatial tools for flash flood monitoring and assessment.

In essence, understanding flash flood hazards is crucial for disaster preparedness and management. These events emphasize the need for accurate and timely monitoring, early warning systems, and robust emergency response mechanisms. Furthermore, the emergence of geospatial tools, such as Geographic Information Systems (GIS), remote sensing (RS), and drone (also known as unmanned aerial vehicles; UAV) mapping, has significantly improved the ability to assess and manage flash flood hazards. Despite a five-fold increase in disasters over the past 50 years (1970–2019), the betterment of multi-hazard early warning systems has improved prevention, preparedness, and response, resulting in an almost threefold decrease in the number of fatalities [5]. Geospatial tools play a pivotal role in monitoring rainfall patterns, mapping flood extents, predicting flash flood scenarios, and aiding in decision-making during these critical moments. This chapter will explore how these geospatial techniques can be harnessed to enhance the assessment and management of flash flood hazards, ultimately reducing the impact of these sudden and devastating flash flood events. An overview of the contents of this chapter is shown in Figure 10.1.

10.1.2 Flash Flood Causes and Triggers

Flash flood events are primarily driven by a combination of meteorological, geographical, and hydrological factors. Understanding the causes and triggers of flash floods is essential for both prediction and mitigation measures. The most common trigger for flash floods is intense and prolonged rainfall. When an area experiences a rapid accumulation of rainwater, especially in a short period, the soil quickly becomes saturated, and rivers and streams get a rapid increase in water volume. Urban areas with extensive pavement and inadequate drainage systems are particularly vulnerable to flash floods. Since water cannot infiltrate impermeable surfaces, it leads to rapid runoff that can overwhelm drainage systems [6].

Another common cause of flash floods is thunderstorms [7] especially slow-moving or stationary ones. They can result in extreme amounts of rainfall concentrated over a small area. These localized rainfall events can quickly lead to flash flooding, particularly in regions where the terrain cannot absorb the excess water. Likewise, tropical cyclones are powerful storms that not only bring heavy rain but also strong winds, which can topple trees and disrupt infrastructure. The combination of heavy rainfall and wind damage can lead to flash flooding, especially in coastal and low-lying areas.

In mountainous regions, flash floods can be triggered by rapid snowmelt [4]. A sudden increase in temperature or extreme rainfall on the already available snow (also called rain-on-snow events) can cause the snow to melt rapidly, leading to surges of water rushing downstream, often leading to catastrophic damage. In colder climates, ice jams can form in rivers during winter. When these jams break free due to warming temperatures or other factors, they can release a significant volume of water, potentially causing flash flooding downstream. In cold regions, rapid ice or glacier melt due to elevated temperatures or volcanic activity can also lead to flash floods in downstream areas. Glacial lake outburst floods (GLOFs) represent another factor behind flash floods [8]. GLOFs are a type of flash flood where a lake, primarily supplied by the meltwater from glaciers, breaches its dam and releases massive amounts of water and sediment downstream.

The sudden release of water from a normal dam breach or failure can unleash a wall of water downstream, leading to a flash flood [9]. This is particularly dangerous as it can occur without any preceding weather event. Flash floods can carry large amounts of debris, including rocks, trees, and sediment, which can worsen the damage. Debris flow is especially common in steep terrains and can lead to landslides and blocked watercourses.

All these causes show that flash floods can be triggered by a range of natural and human-induced factors, all of which share the common characteristic of rapid overwhelming of natural or man-made drainage capacity of an area. These events are challenging to predict due to their sudden nature, emphasizing the importance of monitoring, early warning systems, and the use of geospatial techniques to assess and manage flash flood hazards effectively.

10.2 Geospatial Tools Overview

10.2.1 Remote Sensing Fundamentals

The science of gathering information about the Earth's surface without direct physical contact is known as RS [10]. RS data collection involves the use of various techniques such as optical, hyperspectral, thermal infrared, microwave, LiDAR, and Radar sensors installed in different platforms such as satellites, drones, airplanes, or ground-based systems [11]. These sensors interact with the electromagnetic radiation from the Earth's surface and atmosphere to capture its spatial, temporal, and spectral characteristics. The electromagnetic spectrum contains a broad range of wavelengths, from gamma rays and X-rays to radio waves [11]. Different sensors, mentioned above, are designed to capture specific portions of this spectrum. Usually, in flash flood monitoring and predictions, LiDAR-, Radar-, optical-, and infrared-based RS sensors are utilized to provide valuable data for mapping and forecasting. A few of these sensors and their applications in flash flood monitoring are discussed in section 10.3.

10.2.2 Geographic Information Systems (GIS) Basics

GIS is another powerful tool for flash flood monitoring that allows the collection, management, analysis, and visualization of geospatial data [12–14]. GIS operates with spatial data that represents the geographic location and characteristics of features on the Earth's surface. GIS techniques, along with RS data, enable the integration of various data layers and types, later to be utilized to provide improved insights on flash flood impact monitoring. For example, integrating geospatial information on elevation data, floodplain boundaries, rainfall measurements, river networks, and other relevant information related to flash floods in a common geographic framework and facilitating a comprehensive understanding of flash flood dynamics [15]. Additionally, a range of GIS-enabled geospatial analysis techniques can be employed to derive meaningful information and relationships from integrated spatial. Some examples are proximity analysis to identify areas at risk, overlay analysis to assess the intersection of flood vulnerability and critical infrastructure, and terrain analysis to identify flow paths and potential flood zones.

10.3 Remote Sensing for Flash Flood Assessment

10.3.1 Satellite Imagery for Flood Monitoring

Satellite imagery has revolutionized the flood monitoring and management field, offering a comprehensive and global perspective on these natural disasters. This powerful tool utilizes a network of Earth-observing satellites, tailored with various sensors and instruments, to capture high-resolution pictures and data of our planet's surface [16]. When it comes to monitoring floods, satellite imagery serves as a critical tool for several reasons. One of the most significant advantages of satellite imagery is its ability to provide coverage over vast geographical areas [17]. Satellites orbiting Earth can capture images of entire regions, countries, or even continents in a single pass. This extensive coverage is invaluable for monitoring floods that can affect large and remote areas, making it possible to assess the scale and impact of the event quickly.

Satellite imagery can be acquired and analyzed in near real time. This rapid response capability is crucial during the early stages of a flood event when timely information is needed for decision-making, evacuations, and disaster response. Agencies and organizations can access these images within a few hours of the capture, significantly enhancing their ability to react swiftly to unfolding flood events. Unlike ground-based monitoring systems, satellites are not hindered by adverse weather conditions such as heavy rainfall or cloud cover. This means that they can capture data during and after a flood event, providing crucial information even in challenging conditions. Synthetic aperture radar (SAR) sensors can penetrate clouds and rain to collect crucial data.

Satellites have been in use for decades, which means that historical satellite imagery is available for trend analysis. By comparing current imagery with past data, scientists and disaster management agencies can identify long-term patterns and assess changes in flood frequency, extent, and impact. This information is invaluable for long-term flood risk assessment and mitigation planning. Furthermore, satellite imagery enables the creation of flood extent maps, which outline the areas affected by floods. By comparing pre-flood and post-flood images, GIS and RS experts can produce accurate flood maps that help authorities understand the spatial distribution of flooding. These maps are crucial for estimating damage, identifying at-risk areas, and planning relief measures. Beyond flood monitoring, satellite imagery aids in assessing the environmental impact of floods. It can help identify changes in land cover, erosion, sediment transport, and water quality. This information guides post-flood rehabilitation and restoration measures.

Many countries and international organizations share satellite data and imagery, promoting collaboration in disaster response and management. In the event of a transboundary flood, such cooperation is vital for ensuring that affected regions receive the necessary support. In summary, satellite imagery plays a pivotal role in flood monitoring by providing extensive coverage, rapid response capabilities, and historical data for trend analysis. It aids in flood extent mapping, damage assessment, and environmental impact evaluation, enhancing our understanding of these complex natural events and ultimately contributing to more effective flood management and disaster resilience.

10.3.2 Radar and LiDAR in Flood Detection

Radar (Radio Detection and Ranging) and LiDAR (Light Detection and Ranging) are advanced RS technologies that have greatly improved our ability to detect and monitor floods. These technologies operate on the principle of emitting signals and measuring their return times to create detailed topographic and hydrological data. When it comes to flood detection, both Radar and LiDAR offer unique advantages and complement each other effectively.

10.3.2.1 Radar Technology

Radar systems use radio waves to penetrate through cloud cover and heavy rainfall, allowing them to collect data even during adverse weather conditions. This capability is crucial for continuous monitoring when traditional optical sensors might be obstructed. Radar can provide highly accurate elevation data, which is vital for understanding the topography of flood-prone areas. By comparing pre-flood and post-flood radar data, researchers can determine the extent of flooding and water depths. A specialized radar technology called SAR processes multiple radar returns to produce high-resolution images [18]. SAR is particularly useful in flood detection because it can create detailed flood extent maps, even in densely vegetated areas where optical imagery may be less effective. Many radar systems operate continuously, allowing for real-time flood monitoring. These real-time data are crucial for issuing flood alerts, managing evacuations, and coordinating disaster response efforts [19].

10.3.2.2 LiDAR Technology

LiDAR technology uses laser pulses to create precise three-dimensional high-resolution maps of the Earth's surface. These maps include elevation data [20] which is essential for flood risk assessment and floodplain modeling. LiDAR-derived elevation data can be integrated into hydrological models to simulate flood scenarios. By analyzing how floodwaters would flow across the landscape based on LiDAR data, researchers can predict potential flood extents and impacts. LiDAR is particularly effective in urban areas [21] where building heights and complex terrain can impact flooding patterns. This technology helps urban planners and disaster managers identify vulnerable areas and optimize flood mitigation strategies. In addition, LiDAR can penetrate vegetation cover to some extent, allowing for flood detection in forested or vegetated regions where other RS methods might encounter challenges. After a flood event, LiDAR can be used to rapidly assess damage to infrastructure and natural landscapes. This information guides recovery and rehabilitation efforts.

In conclusion, both Radar and LiDAR technologies significantly enhance flood detection and monitoring capabilities. They provide valuable data on flood extent, elevation, and topography, aiding in early warning systems, floodplain modeling, and post-flood assessments. Combining these technologies with other geospatial tools like GIS and satellite imagery provides a comprehensive approach to flood detection and management, ultimately contributing to increased resilience in flood-prone regions.

10.3.3 Thermal Imaging for Flood Analysis

Thermal imaging, a technology that detects and captures the infrared radiation (heat) emitted by objects and surfaces, has emerged as a valuable tool for flood analysis. While it may not be the primary tool for flood detection, it plays an important role in comprehending the dynamics and consequences of floods [22], particularly in situations where traditional visual imagery and data may fall short. Furthermore, thermal imaging cameras are sensitive to even small temperature variations on the Earth's surface. During a flood event, the thermal properties of surfaces change. For instance, water bodies have different thermal characteristics compared to dry land. By capturing these variations, thermal imaging can help identify flood-affected areas by highlighting the differences in temperature between the water and surrounding land. Some other applications of thermal imaging in the field of flood hazards are as follows:

Monitoring Water Flow and Speed: Thermal imaging is particularly useful in rivers and streams where it can monitor the speed and flow of water. As floodwaters move, they can cause cooling in their wake due to evaporation and convection. Thermal cameras can detect these differences in temperature, helping researchers and emergency responders estimate the speed and direction of floodwaters in real time.

Flood Impact Assessment: Thermal imaging offers vital information to evaluate the effects of floods on communities and infrastructure. After a flood event, thermal cameras can detect residual moisture and dampness in buildings, which can assist in locating potential regions that can develop mold growth or structural damage. This is especially valuable in post-flood damage assessments.

Search and Rescue: During and after a flood event, thermal imaging can assist in search and rescue operations. Thermal cameras can detect the

heat signatures of individuals or animals in need of assistance, even in low-visibility conditions or when they are partially submerged in floodwaters.
Vegetation and Wildlife Analysis: Floods can impact vegetation and wildlife habitats significantly. Thermal imaging can be used to assess the health of vegetation after flooding, helping to determine the extent of damage. Additionally, it can locate and monitor displaced wildlife, aiding in conservation efforts.
Night-Time Monitoring: One of the significant advantages of thermal imaging is its capability to capture data at night. Since floods can occur at any time, this technology ensures continuous monitoring and assessment, even in darkness.
Integration with GIS: Thermal imagery can be integrated with GIS and other geospatial data to create comprehensive flood analysis models. Combining thermal data with topographical and hydrological information can enhance flood modeling and risk assessment.

In summary, thermal imaging plays a crucial role in flood analysis by detecting temperature variations and providing valuable information on water flow, flood impact, search and rescue, vegetation, and wildlife assessment. Its ability to operate in low-visibility conditions and at night makes it a versatile tool for comprehensive flood management and analysis, complementing other geospatial techniques in understanding and mitigating flood hazards.

10.4 GIS Applications in Flash Flood Management

10.4.1 Hydrological and Hydraulic Modeling with GIS

Hydrological and hydraulic/hydrodynamic modeling with GIS is a powerful approach that integrates spatial data, simulation techniques, and analytical tools to understand and manage water-related processes in a geographic context [23, 24]. This interdisciplinary methodology combines the strengths of hydrology, which focuses on the movement and distribution of water, and hydraulics, which studies the behavior of liquids, particularly in confined environments [25]. The application of GIS in this context enhances the modeling process by providing a spatial dimension to the data, allowing for a more accurate and dynamic representation of water systems.

GIS facilitates the integration of rainfall data, land cover information, soil characteristics, and terrain data to simulate and analyze the complex relationship between rainfall and runoff. This aids in understanding how precipitation translates into surface water flow, enabling better predictions of potential flooding events. GIS also allows for the delineation of watersheds and sub-basins, providing insights into the flow of water within a particular geographic area. This is fundamental for determining how human activities, climatic variations, and changes in land use affect the hydrological cycle. Furthermore, hydrological models integrated with GIS can simulate and map floodplains accurately. This is vital for identifying areas prone to flooding, estimating flood extents, and developing effective floodplain management strategies.

Hydraulic models combined with GIS can simulate the water flow through river channels, considering factors such as channel geometry, roughness, and flow velocity [4]. This helps in predicting how water will behave in rivers and streams under different scenarios. GIS-based hydraulic models are valuable for designing stormwater management systems. By incorporating spatial data on land use, slope, and infrastructure, these models can simulate the movement of stormwater through urban environments, aiding in the design of effective drainage systems. GIS also aids in the planning and evaluation of hydraulic infrastructure, including dams, culverts, and bridges. By integrating spatial information on terrain, water flow, and structural characteristics, engineers can optimize the design and assess the resilience of these structures under various hydrological conditions.

GIS serves as a platform for organizing and visualizing spatial data related to hydrology and hydraulics. This includes digital elevation models, land cover maps, river networks, and other critical datasets. GIS tools enable the analysis of spatial relationships, allowing for the identification of patterns and trends in hydrological and hydraulic data. This spatial analysis enhances decision-making by providing a comprehensive understanding of water-related processes. In addition, GIS can be integrated with real-time data from sensors and RS technologies, allowing for dynamic monitoring of water systems. This is particularly valuable for emergency response during flood events. In other words, hydrological and hydraulic modeling with GIS provides a comprehensive and spatially informed approach to understanding and managing water resources. This integration enhances our ability to model, simulate, and visualize complex water-related processes, ultimately contributing to more effective water resource management, flood prediction, and infrastructure design.

10.4.2 Creating Flood Hazard Maps

GIS has become an indispensable tool in the field of flash flood management, offering precise and dynamic solutions for assessing, mapping, and mitigating flood hazards. Making maps of potential flood hazards is one of the main applications of GIS in this context, which is essential for understanding flood-prone areas, risk assessment, and informed decision-making. GIS applications are explained below briefly:

Data Integration and Visualization: GIS allows for the integration of various spatial datasets, such as topography, hydrology, rainfall patterns, and land use, into a unified system. These data layers can be visualized and analyzed collectively, enabling flood managers to identify the critical locations for flash flooding. By combining data from various sources, GIS can provide a holistic view of the flood hazard landscape.

Floodplain Mapping: GIS tools are instrumental in delineating floodplains accurately. Floodplain maps generated through GIS show the extent to which an area is likely to flood during various flood scenarios. By considering factors like river discharge, rainfall intensity, and terrain characteristics, GIS-based floodplain maps provide a reliable basis for land-use planning and zoning regulations, helping to minimize flood risks.

Hydrological Modeling: GIS software incorporates hydrological modeling capabilities, allowing for the simulation of rainfall-runoff processes and flood scenarios. By using rainfall, soil type, vegetation cover, and land use data as input, GIS can predict the water flow across the landscape during heavy rain events. This information aids in understanding the potential pathways of flash floods and assists in flood prediction and early warning systems.

Risk Assessment and Vulnerability Analysis: GIS-based flood hazard maps are crucial for risk assessment and vulnerability analysis. These maps help identify critical infrastructure, densely populated areas, and environmentally sensitive regions within flood-prone zones. By quantifying potential losses and estimating the exposure of assets and populations to flash floods, GIS supports risk reduction and disaster preparedness planning.

Emergency Response and Evacuation Planning: In the event of a flash flood, GIS can be used to facilitate emergency response measures. Real-time data integration allows emergency responders to monitor flood progression, assess the affected areas, and coordinate rescue operations efficiently. GIS-based evacuation models can also help identify safe evacuation routes and locations for displaced individuals.

Public Awareness and Decision Support: Flood hazard maps created using GIS are powerful tools for public awareness and decision support. They provide clear visualizations of flood-prone areas, fostering understanding and engagement among local communities, policymakers, and stakeholders. GIS-based decision support systems assist authorities in making informed choices during flash flood events, such as issuing timely alerts, allocating resources, and directing evacuations.
Adaptation and Resilience Planning: GIS-based flood hazard maps play a critical role in long-term adaptation and resilience planning. By identifying areas susceptible to flash floods and evaluating potential future changes in flood risk due to factors such as climate change and urbanization, GIS supports the development of strategies and policies aimed at reducing vulnerability and enhancing community resilience.

In summary, GIS applications in flash flood management, particularly in creating flood hazard maps, provide a comprehensive and data-driven approach to understanding, mitigating, and responding to flash flood hazards. These maps serve as foundational tools for risk assessment, emergency response, and long-term planning, ultimately contributing to the safety and resilience of flood-prone communities.

10.5 Drone Mapping for Rapid Response

In recent years, on-demand data collection using drone technology has revolutionized various RS-based sectors and one of its most significant applications is rapid responses. Drones with mapping capabilities have proven essential for accelerating response time, expanding situational awareness, and reducing hazards in a range of emergency scenarios, such as disaster management, search and rescue operations, and emergency services [26, 27].

Drones in Disaster Management: Drones can quickly reach and survey remote or difficult-to-reach locations [27]. They can traverse extensive areas of land considerably faster than the ground-based teams, allowing for an accurate and quick evaluation of the situation [28]. The on-demand and high-speed data collection by drones [27] is especially important during natural disasters, allowing for quick decision-making and the ability to adapt response strategies as the situation evolves [26]. For example, drones equipped with thermal imaging and LiDAR sensors can locate survivors, assess infrastructure damage, track the evolution of natural disasters, and assist in the planning and execution of rescue and relief measures.

High-Resolution Imagery for Damage Assessment: Drone mapping provides a comprehensive view of the affected area at a much higher spatial resolution than satellite mapping. Drone-captured high-resolution data provide an unprecedented level of information, allowing responders and authorities to focus on specific areas or structures to better assess the scale and severity of the situation [29]. With precise information on the extent and severity of the damage, authorities can allocate resources more effectively, including deploying search and rescue teams to specific locations, prioritizing aid to areas with the greatest need, and mobilizing equipment for infrastructure repair. High-resolution drone imagery is valuable not only for immediate response but also for long-term planning as it can serve as a baseline for monitoring the progress of recovery efforts and evaluating the success of post-flood rehabilitation initiatives.

10.6 Collaborative Approaches and Socioeconomic Considerations

Since flash floods are characterized by their sudden onset and intense impact, they demand a multifaceted approach that extends beyond the application of geospatial tools. It is crucial for various stakeholders to collaborate and work together for effective flash flood management. Additionally, understanding the socioeconomic aspects of flash floods and their impact on communities is paramount for resilience and preparedness.

10.6.1 Collaboration Among Stakeholders

Collaboration is a key aspect of flash flood management. It demands involvement from various disaster management agencies such as meteorologists, supply chain analysts, and other local municipal communities. These stakeholders play distinct yet interconnected roles in mitigating flash flood risks.

Disaster management agencies are at the forefront of flash flood response. They coordinate emergency services, evacuation plans, and relief efforts. Collaborating with geospatial experts and meteorologists, they can use real-time data from RS and GIS to make informed decisions during a flood event, ensuring a quick and coordinated response. For instance, during the Leh flash floods in India, in August 2010, the Indian government responded immediately. The Indian Army was leading the rescue operation with the help of the Central Reserve Police Force (CRPF), the Indo-Tibetan Border Police (ITBP), and the State Government of Jammu and Kashmir [30].

UAVs (unmanned aerial vehicles) can play a pivotal role in disaster management. They can significantly reduce response times and enhance situational awareness. They act as an efficient means of assessing the disaster-affected areas through aerial surveillance, especially for the areas where conventional ground-based assessment is unavailable. UAVs are often granted special authorizations by the regulatory authorities to expedite the response efforts [31].

Hazards such as flash floods do not respect borders. Therefore, it is important for various authorities to cooperate and work together to save lives and protect livelihoods. Issuing early warnings for hazardous weather phenomena such as flash floods and cyclones is critical. It requires real-time data sharing and coordination across all government levels, along with regional and international partners in meteorology. In the United States, National Meteorological and Hydrological Services (NHMSs) typically issue warnings, but weak connections with other agencies are common. Initiatives by organizations like the United Nations World Meteorological Organization (WMO), the U.S. Agency for International Development/Office of U.S. Foreign Disaster Assistance, and others aim to enhance coordination, improve warning systems, and reduce loss of life and property, particularly in Southern Africa [32].

From a managerial perspective, sustainability presents significant challenges for businesses, with climate change and extreme weather events being a major concern. These events disrupt business operations, particularly affecting small- and medium-sized enterprises (SMEs), which are often unprepared for disasters like flash floods [33]. Collaboration with supply chain analysts helps identify vulnerable points in the supply chain, enabling proactive measures to ensure continuity during flood events. Geospatial data can aid in identifying alternative routes and distribution centers.

The negative impacts also extend to local communities. Local communities are on the front lines during flash flood events. Raising awareness and fostering community engagement are crucial. Geospatial visualizations of flood risks can help communities understand the potential dangers and make informed decisions. Collaborative efforts include the advancement of community-based early warning systems and evacuation plans. Identifying areas vulnerable to heavy rainfall requires precise data, for example, precipitation measurements, land cover, and soil type. In less developed regions, these data are often scarce. A simpler and quicker approach is to use GIS for spatial analysis, adapting to available data [34].

10.6.2 Socioeconomic Impacts of Flash Floods

Understanding the socioeconomic impacts of flash floods is essential for effective mitigation and recovery strategies. Flash floods can lead to a range of social and economic consequences, including loss of life and property, economic losses, and environmental damage. Environmental damage from flash floods includes habitat destruction, water pollution, and soil erosion. Assessing these impacts is essential for environmental remediation and habitat restoration efforts.

Flash floods can result in casualties and injuries, affecting the health and well-being of individuals. The long-term health impacts of exposure to floodwaters, such as waterborne diseases, must be considered in disaster response and recovery. In addition, flash floods have the potential to seriously harm public buildings, commercial buildings, and residential buildings. Economic losses can be substantial, affecting local economies and livelihoods. Assessing the economic impact is vital for recovery planning. Moreover, displaced populations require temporary shelter and support. Understanding the number of displaced individuals and their needs is essential for resource allocation and ensuring their safety and well-being. Flash floods can also have lasting psychosocial effects on affected individuals and communities. Trauma, stress, and mental health issues may emerge in the aftermath of a flood. Providing psychological support and counseling services is integral to recovery.

Flooding, as recent events have vividly demonstrated in Pakistan, stands as one of the most devastating natural calamities. To mitigate the threats, it poses to both economic stability and human lives, as well as to preserve the environment, flood control, and regulation policies become imperative. The escalating severity and prolonged duration of floods are pervasive global concerns, exerting adverse socioeconomic consequences on the affected nations [35]. While it is impossible to eliminate the occurrence of floods entirely, prudent planning and effective training can significantly curtail their detrimental impacts.

10.6.3 Supply Chain Analysis and Resilience

Supply chains are particularly vulnerable to disruptions caused by flash floods. Collaborative efforts between supply chain analysts, businesses, and government agencies are necessary to build resilience and ensure continuity. Key considerations include the following:

Supply Chain Vulnerability Assessment: Identifying critical points in the supply chain that are susceptible to flash flood impacts, such as transportation routes, warehouses, and manufacturing facilities.
Risk Mitigation Strategies: Developing and implementing strategies to mitigate supply chain disruptions, including diversifying transportation routes, enhancing warehouse infrastructure, and stockpiling critical goods.
Coordination and Communication: Effective coordination between supply chain stakeholders, government agencies, and disaster management organizations is vital for timely response and resource allocation.
Public-Private Partnerships: Collaboration between businesses and government entities can lead to public-private partnerships that enhance supply chain resilience. Sharing geospatial data and early warnings can be a part of this collaboration.
Post-Event Evaluation: After a flash flood event, assessing the supply chain's response and recovery efforts is crucial for identifying areas of improvement and enhancing future resilience.

In conclusion, collaborative approaches and socioeconomic considerations are integral components of flash flood management. By fostering cooperation among stakeholders and understanding the broader impacts of flash floods on communities and supply chains, authorities can better prepare for, respond to, and recover from these sudden and devastating natural disasters. Geospatial tools play a pivotal role in supporting these efforts by providing accurate data for decision-making and enhancing community awareness and resilience.

10.7 Future Trends and Conclusion

As the field of managing flash floods keeps developing, several emerging trends and advancements in geospatial techniques are shaping the way we understand, predict, and respond to these sudden and devastating natural disasters. These trends hold the potential to further enhance our ability to reduce the impact of flash floods and build resilience in vulnerable regions.

10.7.1 Emerging Trends in Geospatial Techniques

Machine learning algorithms and artificial intelligence are being integrated into geospatial tools to improve the accuracy and speed of flood prediction. These technologies can analyze vast datasets, from historical weather patterns and RS data to enhancing early warning systems and providing more precise flood forecasts. Crowdsourcing platforms and social media are also increasingly used to gather real-time information about flash floods. The public's reports, images, and videos can enhance official data sources and avail newer important information to the emergency responders.

Furthermore, advances in drone technology and LiDAR sensors are enabling hyper-resolution mapping of terrain and floodplains. These high-resolution datasets are invaluable for detailed flood modeling, risk assessment, and damage analysis. Additionally, geospatial tools are becoming more accessible and user-friendly. The public's understanding of flood risks is being enhanced by improved data visualization and user interfaces, enabling communities to take proactive measures and make well-informed decisions.

Floods are a widespread and devastating natural disaster, resulting in many fatalities as well as economic losses. Recent scholarly efforts have explored the potential of using camera images and wireless sensors in IoT networks to enhance flood management. Water level estimation, modeling, mapping, and real-time flood monitoring are all possible with computer vision and IoT sensor technologies. For future research, the under-explored potential of these technologies in monitoring and managing coastal lagoons can be particularly emphasized [36]. This data can be integrated with geospatial systems for more accurate flood monitoring and early warning.

10.7.2 Advancements in Flash Flood Management

Flash flood management is increasingly being integrated into a broader multi-hazard approach, considering the interplay between various natural disasters that can cause flash floods, such as landslides, hurricanes, and wildfires. Furthermore, there is a growing emphasis on building community resilience through education, preparedness, and community-based early warning systems. Geospatial tools are essential for increasing awareness and motivating local communities to act. As climate change leads to more extreme weather events, such as increased rainfall and changing precipitation patterns, flash flood management strategies are evolving to adapt to these new challenges. Geospatial data is instrumental in understanding climate impacts.

Transboundary flash floods require international collaboration and data sharing. Regional and international organizations are working together to ensure timely responses and support for affected regions. Moreover, Geospatial data is increasingly integrated into policy and decision-making processes. Governments and agencies are recognizing the value of geospatial tools in planning and resource allocation for flash flood management.

10.7.3 Concluding Remarks and Future Directions

Geospatial techniques, including RS, GIS, and drone mapping, have significantly improved our ability to assess and manage flash flood hazards. These tools provide timely and accurate data for flood monitoring, prediction, and response. Collaborative approaches among stakeholders are essential for effective flash flood management, and an understanding of the socioeconomic impacts of flash floods is crucial for building resilience.

As we look to the future, emerging trends in geospatial techniques and advancements in flash flood management will continue to shape the field. With a multi-hazard approach, community resilience, and a focus on climate change adaptation, we can better prepare for the challenges posed by flash floods. International collaboration and policy integration will play key roles in improving our ability to respond to these natural disasters.

In conclusion, by seamlessly integrating geospatial tools, collaborative efforts, and forward-thinking strategies, authorities can enhance their capacity to lessen the effect of flash floods on vulnerable regions and communities. As technology and knowledge continue to advance, our ability to manage and adapt to flash floods will only improve, ultimately saving lives and protecting livelihoods.

References

1. Sweeney, T. L. (1992). Modernized areal flash flood guidance (NOAA Technical Memorandum NWS HYDRO). Office of Hydrology. https://repository.library.noaa.gov/view/noaa/13498/noaa_13498_DS1.pdf
2. Al-Fugara, A., Mabe, A. N., Alayyash, S., & Khasawneh, A. (2023). Hydrological and Hydrodynamic Modeling for Flash Flood and Embankment Dam Break Scenario: Hazard Mapping of Extreme Storm Events. *Sustainability*, 15(3), 1758. https://doi.org/10.3390/su15031758
3. Cutter, S. L., Emrich, C. T., Gall, M., & Reeves, R. (2018). Flash Flood Risk and the Paradox of Urban Development. *Nat. Hazard. Rev.*, 19(1), 05017005. https://doi.org/10.1061/(ASCE)NH.1527-6996.0000268
4. Godara, N., Bruland, O., & Alfredsen, K. (2023). Simulation of flash flood peaks in a small and steep catchment using rain-on-grid technique. *J. Flood Risk Manage.*, 16(3), e12898. https://doi.org/10.1111/jfr3.12898

5. WMO. (2023, July 23). Water-related hazards dominate disasters in the past 50 years [World Meteorological Organization]. https://public.wmo.int/en/media/press-release/water-related-hazards-dominate-disasters-past-50-years
6. Abd-Elaty, I., Kuriqi, A., Pugliese, L., Zelenakova, M., & El Shinawi, A. (2023). Mitigation of urban waterlogging from flash floods hazards in vulnerable watersheds. *J. Hydrol.: Reg. Stud.*, 47, 101429. https://doi.org/10.1016/j.ejrh.2023.101429
7. Gupta, H., Sorooshian, S., Gao, X., Imam, B., Hus, K.-L., Bastidas, L., Li, J., & Mahani, S. (2002). The challenge of predicting flash floods from thunderstorm rainfall. *Philos. Trans. R. Soc. London, Ser. A* 360(1796), 1363–1371. https://doi.org/10.1098/rsta.2002.1015
8. Jha, L. K., & Khare, D. (2016). Glacial lake outburst flood (GLOF) study of Dhauliganga basin in the Himalaya. *Cogent Environ. Sci.*, 2(1), 1249107. https://doi.org/10.1080/23311843.2016.1249107
9. Psomiadis, E., Tomanis, L., Kavvadias, A., Soulis, K. X., Charizopoulos, N., & Michas, S. (2021). Potential Dam Breach Analysis and Flood Wave Risk Assessment Using HEC-RAS and Remote Sensing Data: A Multicriteria Approach. *Water*, 13(3), 364. https://doi.org/10.3390/w13030364
10. Joseph, G. (2005). Fundamentals of remote sensing. Universities Press.
11. Sabins Jr., F. F., & Ellis, J. M. (2020). *Remote sensing: Principles, interpretation, and applications*. Waveland Press.
12. Sowmya, K., John, C. M., & Shrivasthava, N. K. (2015). Urban flood vulnerability zoning of Cochin City, southwest coast of India, using remote sensing and GIS. *Nat. Hazards*, 75, 1271-1286.
13. Schwarz, B., Pestre, G., Tellman, B., Sullivan, J., Kuhn, C., Mahtta, R., ... & Hammett, L. (2018). Mapping floods and assessing flood vulnerability for disaster decision-making: A case study remote sensing application in Senegal. *Earth observation open science and innovation*, 293-300.
14. Harley, P., & Samanta, S. (2018). Modeling of inland flood vulnerability zones through remote sensing and GIS techniques in the highland region of Papua New Guinea. *Appl. Geomatics*, 10(2), 159-171.
15. Pradhan, B. (2010). Flood susceptible mapping and risk area delineation using logistic regression, GIS and remote sensing. *J. Spat. Hydrol.*, 9(2).
16. Sandhu, H. A. S., Gusain, H. S., Arora, M. K., & Bawa, A. (2018). Mass balance estimation of dokriani glacier in central Indian himalaya using remote sensing data. *J. Indian Soc. Remote Sens.*, 46, 1835-1840.
17. Wang, Y., Hong, H., Chen, W., Li, S., Pamučar, D., Gigović, L., ... & Duan, H. (2018). A hybrid GIS multi-criteria decision-making method for flood susceptibility mapping at Shangyou, China. *Remote Sens.*, 11(1), 62.
18. Kussul, N., Shelestov, A., & Skakun, S. (2011). Flood Monitoring from SAR Data. In F. Kogan, A. Powell, & O. Fedorov (Eds.), *Use of Satellite and In-Situ Data to Improve Sustainability* (pp. 19–29). Springer Netherlands. https://doi.org/10.1007/978-90-481-9618-0_3

19. Novák, P., Kyznarová, H., Pecha, M., Šercl, P., Svoboda, V., & Ledvinka, O. (2021). Utilization of Weather Radar Data for the Flash Flood Indicator Application in the Czech Republic. *Remote Sens.*, 13(16), 3184. https://doi.org/10.3390/rs13163184
20. Muhadi, N. A., Abdullah, A. F., Bejo, S. K., Mahadi, M. R., & Mijic, A. (2020). The Use of LiDAR-Derived DEM in Flood Applications: A Review. *Remote Sens.*, 12(14), 2308. https://doi.org/10.3390/rs12142308
21. Abdullah, A. F. B. (2020). *A Methodology for Processing Raw Lidar Data to Support Urban Flood Modelling Framework* (1st ed.). CRC Press. https://doi.org/10.1201/9781003059295
22. Bartsch, J., & Schuetz, T. (2023). Mapping surface flow pathways in urban areas using UAV-based thermal imaging in combination with flooding experiments [Other]. oral. https://doi.org/10.5194/egusphere-egu23-13431
23. Popescu, C., & Bărbulescu, A. (2023). Floods Simulation on the Vedea River (Romania) Using Hydraulic Modeling and GIS Software: A Case Study. *Water*, 15(3), 483. https://doi.org/10.3390/w15030483
24. Feng, B., Wang, J., Zhang, Y., Hall, B., & Zeng, C. (2020). Urban flood hazard mapping using a hydraulic–GIS combined model. *Nat. Hazards*, 100(3), 1089–1104. https://doi.org/10.1007/s11069-019-03850-7
25. Kiba, L. G., Nengzouzam, G., & Ranjan, P. (2023). Flood Hazard Mapping Using Hydraulic Models and GIS: A Review. In M. Pandey, A. K. Gupta, & G. Oliveto (Eds.), *River, Sediment and Hydrological Extremes: Causes, Impacts and Management* (pp. 65–72). Springer Nature Singapore. https://doi.org/10.1007/978-981-99-4811-6_4
26. Karamuz, E., Romanowicz, R. J., & Doroszkiewicz, J. (2020). The use of unmanned aerial vehicles in flood hazard assessment. *J. Flood Risk Manage.*, 13(4), e12622.
27. Munawar, H. S., Ullah, F., Qayyum, S., Khan, S. I., & Mojtahedi, M. (2021). Uavs in disaster management: Application of integrated aerial imagery and convolutional neural network for flood detection. *Sustainability*, 13(14), 7547.
28. Loli, M., Manousakis, J., Mitoulis, S. A., & Zekkos, D. (2022, June). UAV-enabled flood damage assessment and recovery monitoring of bridges following Medicane Ianos. In *Proceedings of the 11th International Conference on Bridge Maintainance, Safety and Management.*
29. Akay, S. S., Özcan, O., & Şanlı, F. B. (2022). Quantification and visualization of flood-induced morphological changes in meander structures by UAV-based monitoring. *Eng. Sci. Technol. Int. J.*, 27, 101016.
30. Gupta, P., Khanna, A., & Majumdar, S. (2012). Disaster management in flash floods in leh (ladakh): A case study. *Indian. J. Community Med.*, 37(3), 185–190. https://doi.org/10.4103/0970-0218.99928
31. Erdelj, M., Natalizio, E., Chowdhury, K. R., & Akyildiz, I. F. (2017). Help from the Sky: Leveraging UAVs for Disaster Management. *IEEE Pervasive Comput.*, 16(1), 24–32. https://doi.org/10.1109/MPRV.2017.11

32. Jubach, R., & Tokar, A. (2016). International Severe Weather and Flash Flood Hazard Early Warning Systems—Leveraging Coordination, Cooperation, and Partnerships through a Hydrometeorological Project in Southern Africa. *Water*, 8(6), 258. https://doi.org/10.3390/w8060258
33. Skouloudis, A., Tsalis, T., Nikolaou, I., Evangelinos, K., & Leal Filho, W. (2020). Small & Medium-Sized Enterprises, Organizational Resilience Capacity and Flash Floods: Insights from a Literature Review. *Sustainability*, 12(18), 7437. https://doi.org/10.3390/su12187437
34. Wicht, M., & Osinska-Skotak, K. (2016). Identifying urban areas prone to flash floods using GIS – preliminary results [Preprint]. Urban Hydrology/ Remote Sensing and GIS. https://doi.org/10.5194/hess-2016-518
35. Manzoor, Z., Ehsan, M., Khan, M. B., Manzoor, A., Akhter, M. M., Sohail, M. T., Hussain, A., Shafi, A., Abu-Alam, T., & Abioui, M. (2022). Floods and flood management and its socio-economic impact on Pakistan: A review of the empirical literature. *Front. Environ. Sci.*, 10, 1021862. https://doi.org/10.3389/fenvs.2022.1021862
36. Arshad, B., Ogie, R., Barthelemy, J., Pradhan, B., Verstaevel, N., & Perez, P. (2019). Computer Vision and IoT-Based Sensors in Flood Monitoring and Mapping: A Systematic Review. *Sensors*, 19(22), 5012. https://doi.org/10.3390/s19225012

11

Prediction of Water Hardness Using Machine Learning and Model Interpretation

Aarti Kochhar*, Harpinder Singh, Shashikant Patel, P.K. Litoria and Brijendra Pateriya

Punjab Remote Sensing Centre, Ludhiana, Punjab, India

Abstract

Traditional methods to evaluate and access water quality take much effort and are usually impossible for consumers and farmers in developing countries. These arduous methods are based on lab-centered analysis and are expensive. This defies the notion of the requirement of speedy action for deteriorating and dire conditions of the quality of water. The machine learning (ML)–based application resolves this issue through prediction models. For efficient governance of groundwater, policymakers should be empowered by tools based on ML. With the same motivation, this paper aims to predict water hardness based on twelve input parameters, i.e., turbidity, electrical conductivity (EC), pH, total dissolved solids (TDS), total alkalinity, chloride, sulfate, nitrate, fluoride, calcium, magnesium, and iron. A total of 5842 samples spread over the Punjab state of India have been used for the development and validation of six ML models. Performance of random forest (RF), k-nearest neighbors (kNN), support vector regression (SVR), gradient boosting regression (GBR), partial least square regression (PLSR), and multilayer perceptron (MLP) has been evaluated and compared in terms of mean square error and coefficient of determination. The developed models have a promising performance to confirm the prospect of their use for the development of a low-cost prediction system for water hardness. Interpretations on the basis of explainable artificial intelligence (XAI) further add to the reliability, understanding, and management of ML results of this research.

**Corresponding author*: aarti.kochhar92@gmail.com

Disha Thakur, Sanjay Kumar, Har Amrit Singh Sandhu and Chander Prakash (eds.) Sustainable Development Using Geospatial Techniques, (263–284) © 2024 Scrivener Publishing LLC

***Keywords*:** Machine learning, explainable artificial intelligence, water hardness, prediction

11.1 Introduction

The main supply of fresh water for drinking and irrigation comes from groundwater for around 50% of the world's population [1]. With the availability of clean water being the 6th goal of sustainable development established by the United Nations (UN), the role of groundwater is essential in achieving it [2]. Growing industries, large-scale application of pesticides, unregulated disposal of waste, and other geological metamorphic activities contaminate groundwater resources [3, 4]. Intake of contaminated water increases the risk of cancer in the population [4], and Punjab already has its cancer woes, with one of the highest cancer cases in India [5, 6]. As said by the Secretary-General of the UN in an Environment Day message [7], drinking contaminated water accounts for 80% of all ailments and mortality in developing countries. Contaminated water from growing industries and unmanaged chemical fertilizers for high yields of crops contribute to the high rates [8]. Crop health also deteriorates if contaminated water is used. Hardness is a crucial factor in evaluating the quality of water [9]. Hard water, a significant factor worldwide, is causing serious diseases like cardiovascular disease, cancer, cerebrovascular mortality, central nervous system malformations, childhood atopic dermatitis, Alzheimer's disease, stones in the kidney, diabetes, reproductive and digestive health issues, and constipation [10]. Hard water forms scales or layers of salt on plant roots and prevents further absorption of minerals and water [11, 12]. The hardness of water leads to clogging of water pipes because of scale build-up. It reduces the efficiency and life of domestic and industrial equipment and increases the cost of boiling and cleaning [13, 14].

So, groundwater quality control is crucial for the preservation of the health of the ecosystem and human population. Water quality estimation and monitoring are essential for developing preventative plans and formulating policy. It requires designated staff to collect samples and process them in a laboratory. The sample is usually taken from dug wells, hand pumps, or tube wells. These wells are either production wells or particularly monitoring wells. The process is time-consuming and expensive. To incorporate more radical preventive measures, there is a requirement for more advanced technologies to integrate decision-making and control actions. Supervised ML has opened up more innovative means for the efficient prediction of water quality. ML has empowered effective and

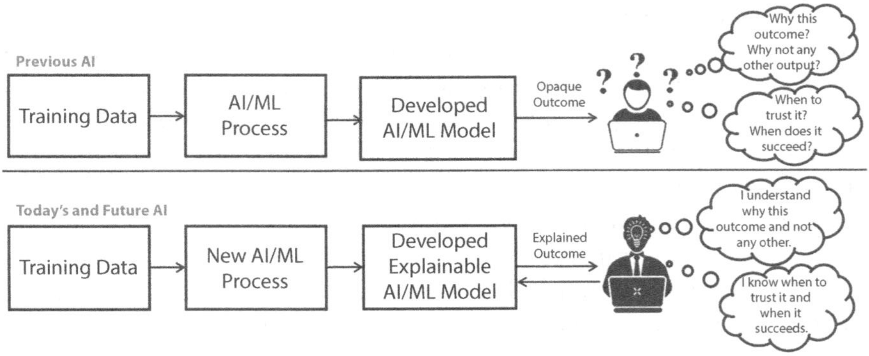

Figure 11.1 Need for XAI.

informed policymaking for governance. Prediction-based approaches are more cost-effective than conventional approaches in terms of human resources and expenses. Although prediction models for groundwater have been proliferating, there are still many unexplored aspects regarding the models and parameters. There is still a fallback in declarative awareness and context understanding by humans for these models. The "third wave of AI", XAI, has picked up pace in the past 2 to 3 years to add transparency to artificial intelligence (AI) models [15]. XAI explains the reason for every decision and answers the why, when, and how of the outcomes. Figure 11.1 explains the need for XAI.

Physical, biological, and chemical aspects of groundwater quality can be modeled independently or with other parameters. This work explains the hardness of groundwater concerning twelve other parameters, *viz.*, turbidity, EC, pH, TDS, total alkalinity, chloride, nitrate, magnesium, sulfate, fluoride, iron, and calcium. The results are further interpreted using the SHAP (Shapley Additive exPlanations) framework of XAI for analyzing water hardness prediction models.

11.2 Literature Survey

Evaluation of water suitability using index-based approaches and statistical methods is an essential part of groundwater studies. In the past few years, the application of AI/ML has proved to be more robust for predictions and forecasts related to groundwater. Tiyasha *et al.* [16] studied the role of ML in river water quality prediction. While few research papers focused on implementing ML algorithms for irrigation water quality assessment [17,

18], others focused on water quality in the rivers [19, 20]. Meyers *et al.* [21] explored ML algorithms to predict turbidity in the main trunk network of the water distribution system in the United Kingdom (UK). Fijani *et al.* [22] and Chou *et al.* [23] observed reservoir water quality using ML techniques. Shah *et al.* [24] predicted the TDS and EC of surface water from the Indus River basin using eight independent variables.

Jafari *et al.* [25] analyzed the parameters of the groundwater aquifer from Tabriz Plain, Iran, for the prediction of TDS. Sodium, bicarbonate, calcium, magnesium, and sulfate were input variables. The performance of four prediction models, MLP, support vector machine (SVM), gene expression programming (GEP), and adaptive neuro-fuzzy inference system (ANFIS), were compared. Out of 1742 samples, 1220 (70%) were used for training, and 522 (30%) samples were used to test the algorithms. El Bilali *et al.* [26] worked on the prediction of irrigation water quality indicators like TDS, potential salinity (PS), exchangeable sodium percentage (ESP), sodium adsorption ratio (SAR), residual sodium carbonate (RSC), and magnesium adsorption ratio (MAR) using three physical parameters temperature, EC, and pH as input. Five hundred twenty samples from the Berrechid aquifer, Morocco, were used to compare the prediction performance of the artificial neural network (ANN), SVR, RF, and adaptive boosting (AdaBoost).

Rather than simulating the prediction from samples of one location, a few authors diversified the study by sampling data from multiple locations. Sunayana *et al.* [27] assessed water quality at solid waste disposal sanitary landfill sites by sampling 10 diverse locations near the site. The potential of gradient descent with momentum backpropagation (GDM), Levenberg-Marquardt (LM), and BFGS (Broyden, Fletcher, Goldfarb, and Shanno) quasi-Newton backpropagation was explored for the prediction of groundwater hardness. Along with physiochemical parameters, the effect of time and location of the sample was also quantified.

Along with predicting a parameter, formulation of the singular index and assigning a class for general quality assessment provide a different perspective. Ahmed *et al.* [28] employed temperature, pH, turbidity, TDS, fecal coliform, and nitrate to calculate the singular water quality index (WQI). Each parameter was assigned an appropriate weight, and samples were classified into excellent to poor classes. Polynomial regression outperformed for the prediction of WQI, and MLP outperformed for the classification of the sample. Mosavi *et al.* [29] predicted groundwater hardness susceptibility with land use, distance from the sea, elevation, depth to groundwater, curvature, distance from the river, pH, evaporation, precipitation, groundwater level, and lithology as input predictor variables. Sixteen years of data from 135 ground monitoring wells were investigated to analyze the performance of RF and boosted regression trees (BRT) algorithms.

However, most of these results were either non-interpretable or were interpretable by domain experts. XAI was proposed for the justification of predictions by ML systems. Adadi and Berrada [30] reviewed XAI to understand its importance in gaining insights into the opacity of ML. The authors cover the definition, need, application domains, challenges, methods, and all aspects of XAI. Ever since its dawn, many others have tried XAI over various applications like hydro-climatology [31], medical [32, 33], cybersecurity [34], human behavior analysis [35], speech signal analysis, and mortality risk analysis [36]. This paper illustrates the potential and novel implementation of the XAI technique for water quality prediction and analysis. XAI approaches like feature importance, and feature effects can provide a new paradigm for policymaking and industrial processes related to clean water. To explore XAI in this paper, 12 parameters have been included in predicting the hardness of the water.

11.3 Materials and Methods

11.3.1 Parameters

A total of 12 variables have been used as predictor variables for the prediction of hardness. Turbidity, EC, pH, TDS, total alkalinity, nitrate, chloride, fluoride, sulfate, magnesium, calcium, and iron are input variables. Turbidity expresses optical water clarity, i.e., how well light passes through water. Represented by TDS, the level of dissolved solids affects clarity as well as the taste of water. These dissolved solids may be organic or inorganic. So, TDS and turbidity both measure the clarity of the water but in different ways. pH represents the intensity of the acidity of water. The ability to neutralize deviations in pH is alkalinity. Nitrate, chloride, fluoride, sulfate, magnesium, calcium, and iron represent the concentration of the corresponding component in water. Table 11.1 presents the statistical analysis of all parameters, including the output variable, i.e., hardness. As shown in Figure 11.2, 5842 samples spread across the Punjab state of India have been used to develop machine learning models. Figure 11.3 shows the correlation matrix of input and output parameters. Output parameter hardness seems to be correlated with dissolved minerals, and physical and chemical parameters of water.

Table 11.1 Statistical analysis of all parameters.

S. no.	Parameter	Units	Average	Acceptable limit	Permissible limit	Standard deviation
1	Turbidity	NTU	0.24976	1	5	2.038278807
2	pH	No unit	8.0062	6.5–8.5	No relaxation	0.517927275
3	EC	µS/cm	636.548			416.2491069
4	TDS	mg/L	372.9372	500	2000	241.7989794
5	Total alkalinity	mg/L	279.0315	200	600	108.7811936
6	Chloride	mg/L	31.32283	250	1000	38.68755775
7	Sulfate	mg/L	29.87316	200	400	141.6900201
8	Nitrate	mg/L	5.150976	45	No relaxation	7.93733848
9	Fluoride	mg/L	0.630077	1	1.5	0.813429742
10	Calcium	mg/L	37.72321	75	200	20.3535664
11	Magnesium	mg/L	23.43889	30	100	13.88572411
12	Iron	mg/L	0.233208	0.3	No relaxation	1.012215892
13	Hardness	mg/L	191.57	200	600	79.47310423

11.3.2 Model Description

Six models have been deployed for the prediction of hardness using input predictors. Table 11.2 represents the hyperparameters of the six ML models. This section describes the six models and related hyperparameters.

kNN is a basic supervised machine learning model. In order to forecast the values of the output parameter, it captures feature similarity. For prediction, the distance of a new point from each point in the training data is calculated. The distance used for this work is Manhattan distance, i.e., the summation of the absolute difference between all point coordinates. Equation 11.1 shows the Manhattan distance formula where X and Y are points with coordinates as $X=[x_1, x_{2,} x_{3......} x_n]$ and $Y=[y_1, y_{2,} y_{3......} y_n]$.

$$d(X,Y) = \sum_{i=1}^{n} |x_i - y_i| \tag{11.1}$$

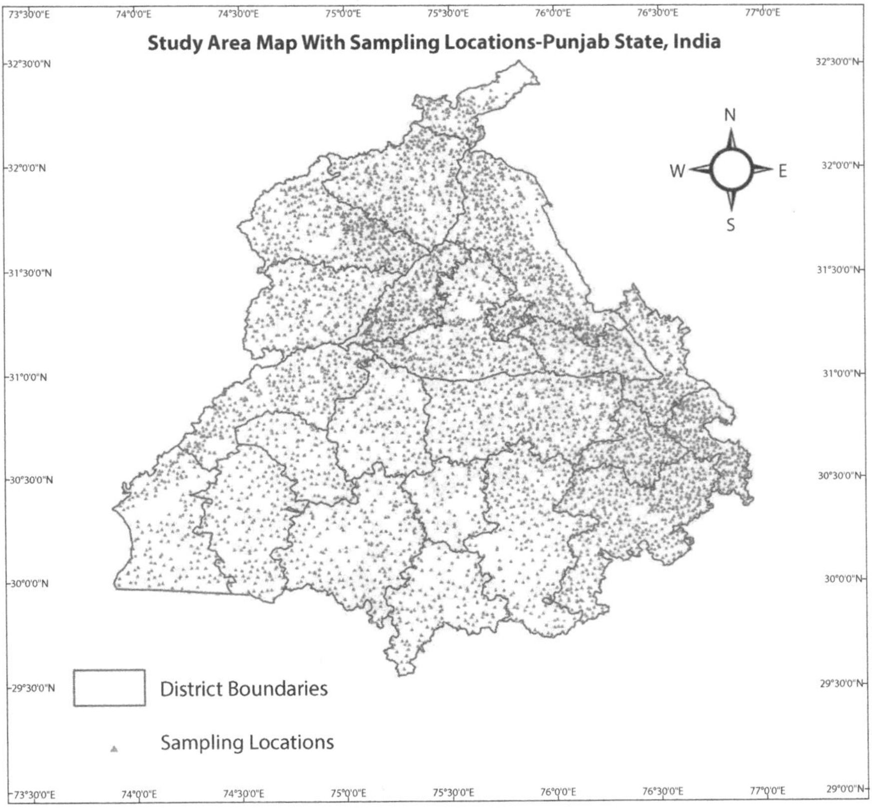

Figure 11.2 Sampling locations in Punjab.

After this calculation, the *k* numbers of nearest neighbors are selected based on minimum distance. In this work, based on hyperparameter tuning, the number of nearest neighbors selected is 5. The average of these five data points is the prediction of a new point. Algorithm 'auto' ensures the appropriate choice of algorithm automatically based on data points. Leaf size represents the number of nodes of a leaf node in an algorithm tree and affects the algorithm's speed.

RF is a collection of subsets of decision trees to provide an accurate and stable solution to complex problems. Each subset of the decision tree predicts a data point. The average of all the decision trees' predictions is the final forecast. Figure 11.4 represents the algorithm of random forest. Two hundred trees are there in the forest. A total of 20 levels are there until the end leaf node, from the root node. A minimum of two samples are

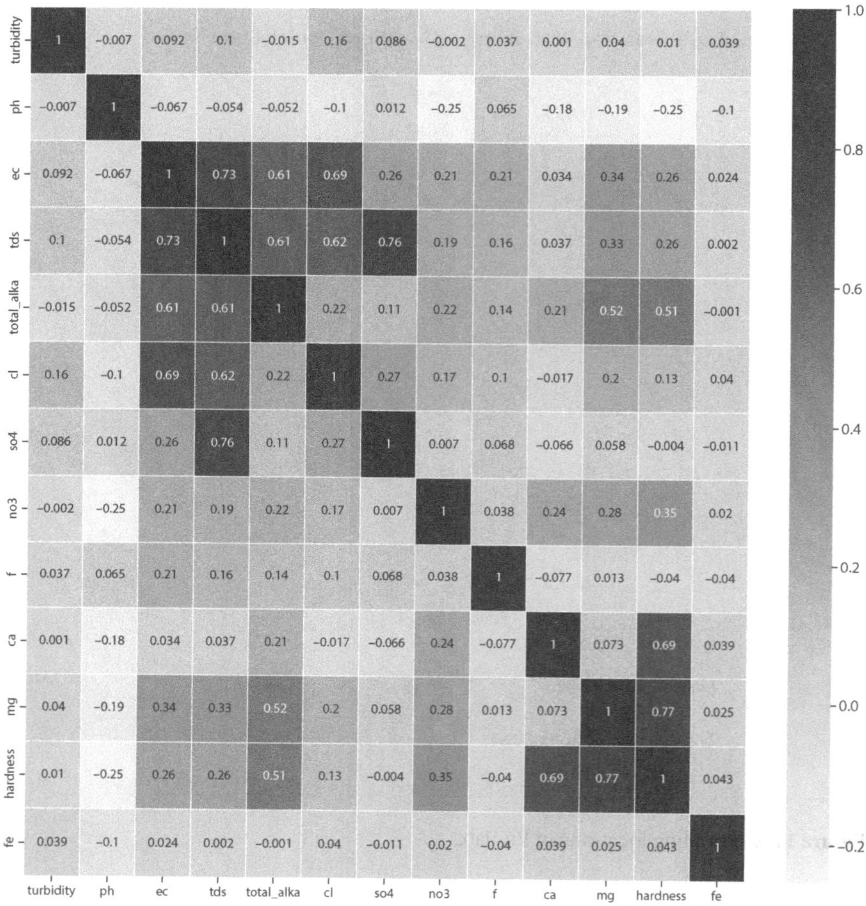

Figure 11.3 Correlation matrix (total_alka: total alkalinity, so4: sulfate, cl: chloride, no3: nitrate, f: fluoride, mg: magnesium, ca: calcium, and fe: Iron).

required for the splitting of a node. There must be at least two samples at each leaf node.

PLSR develops a relationship between the X variable and Y variable based on covariance. The tolerance for the PLSR algorithm for this research work is $1e^{-6}$.

SVR is a supervised ML model that analyzes the training data to form decision boundaries called hyperplanes. Data points near these hyperplanes are called support vectors and help in the prediction of the output parameter. The necessary format for input data is created using the kernel. Eventually, the kernel helps in finding effective hyperplanes.

Table 11.2 Machine learning models and selected hyperparameters.

S. no.	Machine learning model	Hyperparameters
1	k-Nearest neighbors (kNN)	algorithm= 'auto', metric='manhattan', leaf_size=30, weights='distance', n_neighbors=5
2	Gradient boosting regression (GBR)	learning_rate= 0.1, n_estimators=1000, max_features=0.5, min_samples_leaf=3, max_depth=4
3	Random forest (RF)	max_depth=20, min_samples_leaf=2, max_features='auto', n_estimators=200, min_samples_split=2
4	Partial least square regression (PLSR)	copy=True, scale=True, max_iter=500, n_components=4, tol=1e-06
5	Support vector regression (SVR)	kernel='linear', epsilon=.0001, C=5
6	Multilayer perceptron (MLP)	activation: 'relu', alpha: 0.0001, hidden_layer_sizes: (1000,), solver: 'lbfgs', 'learning_rate': 'constant'

Boundary lines are at a distance of ε (epsilon) from hyperplanes and affect the accuracy of the prediction model. These boundary lines decide the acceptable error margin of the model. C, the regularization or optimization parameter, defines the tolerance, i.e., acceptance of data points outside the error margin, and controls the model's generalization capability [37].

MLP forms a deep neural network (DNN) with multiple hidden layers containing an interconnection of multiple neurons. Figure 11.5 shows an example of an input, output, and hidden layer of the MLP neural network.

Figure 11.6 shows the internal structure and output of a single neuron. Interconnections among different layers have adaptive weights (w_1, w_2 ...... w_n). An activation function tunes these weights. Equation 11.2 represents the ReLU (rectified linear unit) activation function implemented for MLP for this research work.

$$f(y) = \max(0, y) \tag{11.2}$$

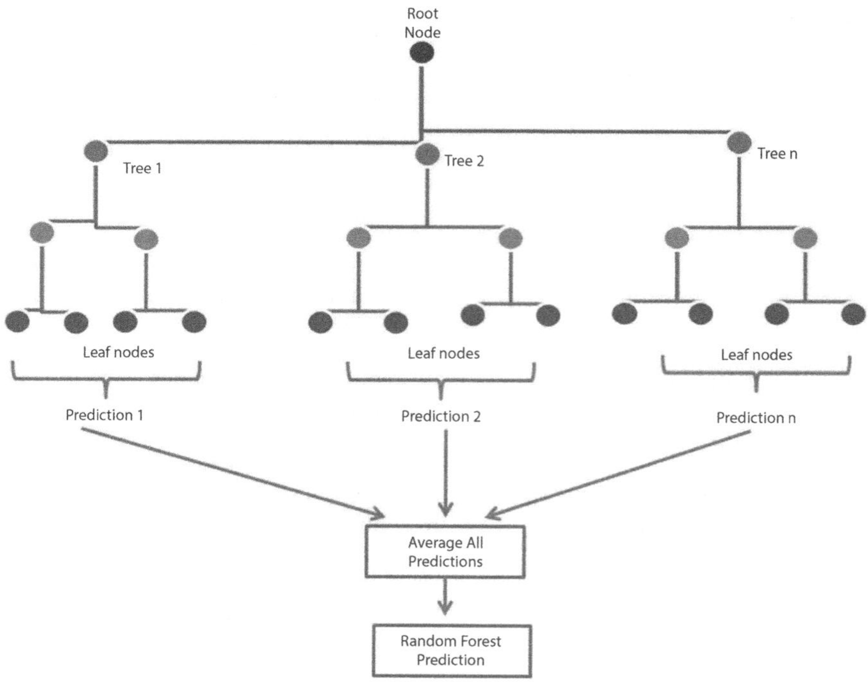

Figure 11.4 Random forest algorithm.

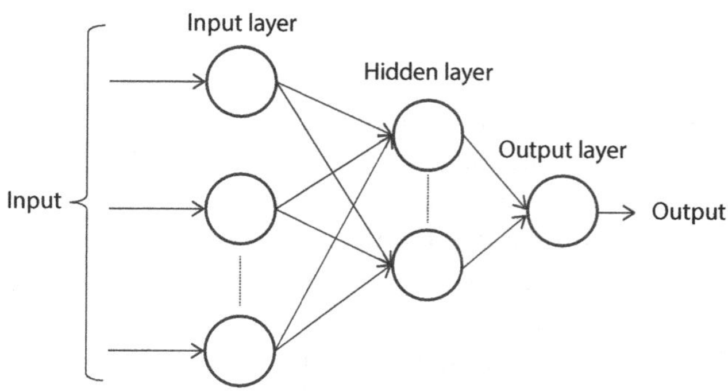

Figure 11.5 Typical MLP neural network.

For linear representations, single-layer perceptrons perform admirably, but MLP is necessary for nonlinear problems. For the hidden layer, the number of neurons in a hidden layer is 1000 for this research work. Solver 'lbfgs' is used for weight optimization for faster convergence. Weights are

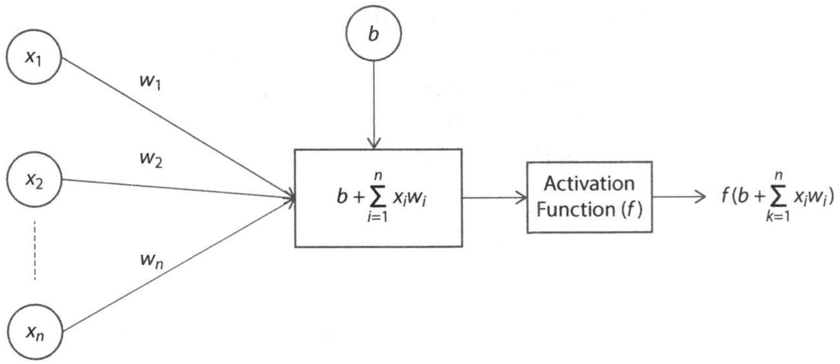

Figure 11.6 Internal structure and output of a single neuron.

updated as per the constant learning rate. The alpha parameters regularize over-fitting by controlling weights.

11.3.3 Modeling Procedure

The complete modeling procedure followed for all the models has been illustrated in Figure 11.7. The input dataset was checked and pre-processed for null values and appropriate data types. For normalization, the distribution of all the parameters is rescaled. All parameters are normalized by deducting the parameter mean and dividing the result by the standard deviation. After pre-processing, the training and testing dataset was generated in a ratio of 80:20, respectively. ML models are implemented using the Scikit-Learn library in Python. Models "learn" over the training dataset and are validated over the testing dataset. Best input parameters are identified for each model with all possible parameter combinations. Using a grid search hyper-tuning algorithm, the models used in this study are

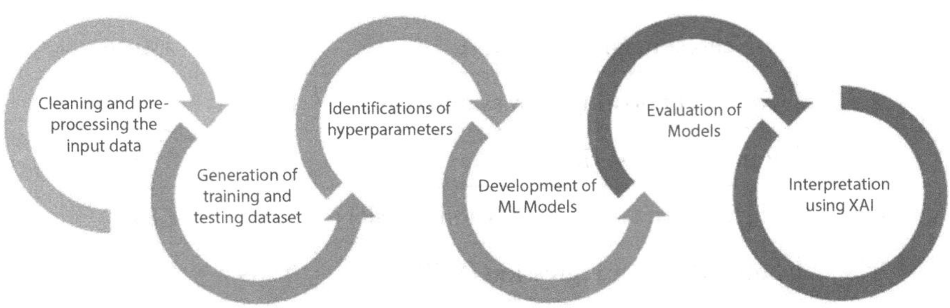

Figure 11.7 Model flow diagram.

optimized for architecture, functions, and parameters. Following the training and testing of the created model on the provided dataset, the models' accuracy is evaluated and contrasted using prediction error plots, mean square error (MSE), and coefficient of determination (R^2). At last, as a key contribution of this research, models are characterized using XAI.

11.4 Results and Discussions

In this section, the research results are first introduced, and subsequently, the detailed discussions on results derived from various models have been illustrated. Results of the various machine learning models along with model explanations and internal relations between the water quality components are presented.

11.4.1 Results: Machine Learning Models

The performance of the six machine learning regression techniques discussed above has been compared to estimate the hardness value of water. It was found that the MLP gave the best performance ($R^2 = 0.99$). The MLP was followed by SVR, RF, PLSR, and GBR in terms of R^2. These four models gave almost similar results (R^2) of 0.98. The last performance was of the kNN technique ($R^2 = 0.93$). The R^2 and MSE for all the models are shown in Table 11.3.

Furthermore, the error can be depicted using prediction error plots. Prediction error plots visualize the target values (y) against predicted values ($\hat{y}$).

Table 11.3 Results (R^2 and MSE) of the machine learning models.

		Model results	
S. no.	**Machine learning model**	**R^2**	**MSE**
1	k-Nearest neighbors (kNN)	0.93	450.26
2	Gradient boosting regression (GBR)	0.985	92.23
3	Random forest (RF)	0.987	82.57
4	Partial least square regression (PLSR)	0.987	2.89
5	Support vector regression (SVR)	0.989	67.45
6	Multilayer perceptron (MLP)	0.99	5.43

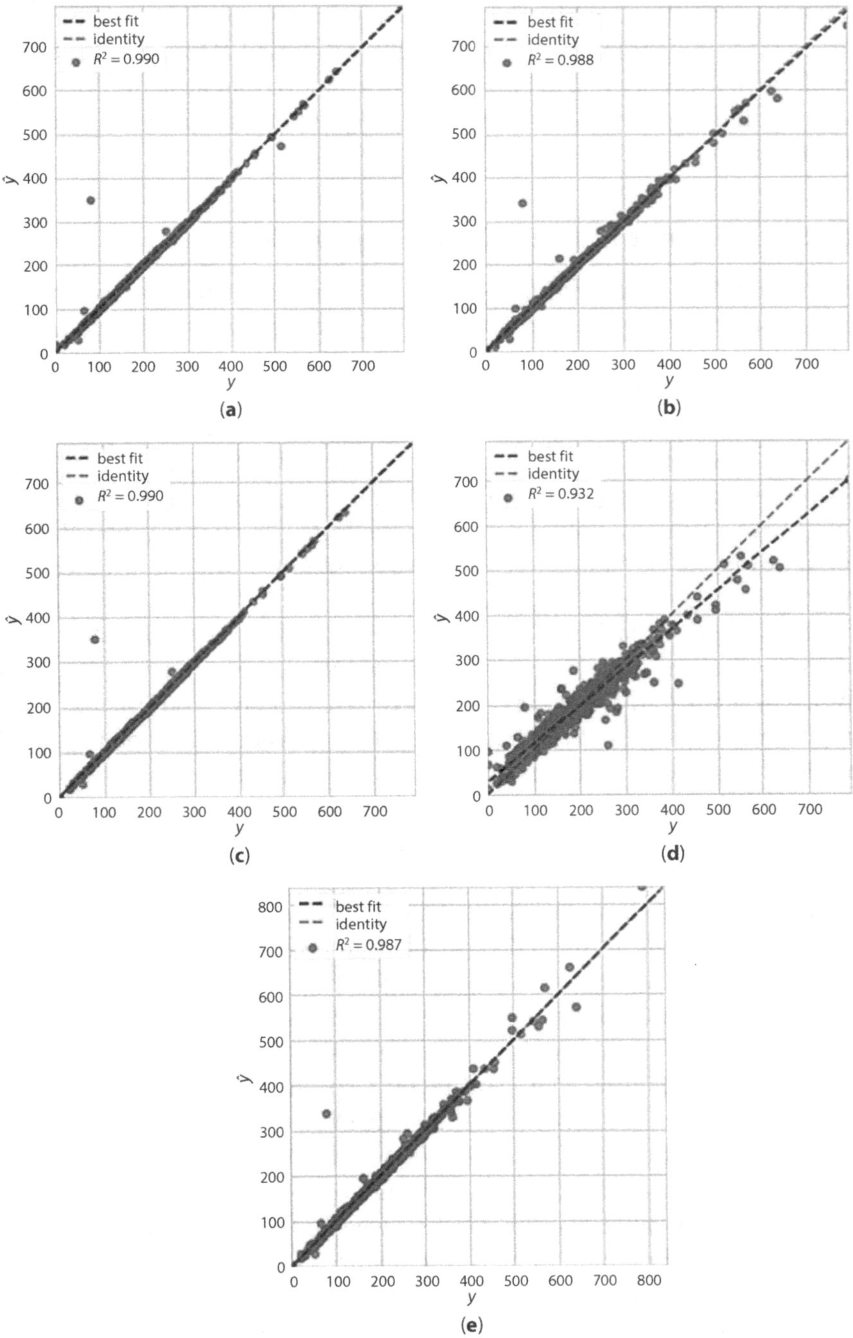

Figure 11.8 Prediction error plot of (a) MLP, (b) RF, (c) SVR, (d) kNN, and (e) GBR.

It helps in visualizing variance in the output of the model. Figure 11.8 represents the prediction error plot of all the ML-based algorithms. The best fit of the model is plotted against an identity line to diagnose the error. The identity line represents the state where the prediction exactly matches the actual values.

11.4.2 Machine Learning Model Explanation

XAI demystifies the black-box behavior of ML models and helps in explaining and interpreting a model's behavior. XAI enhances trust in models by added interpretations and hence provides an improvement in management and applicability of results.

SHAP is a popular framework of XAI [38]. At first feature importance of parameters is calculated using SHAP called Shapley values. More Shapley value means more important feature, i.e., more impact on predictions. SHAP value is plotted with feature effects. SHAP works by exhausting a power set of parameters for the assessment of the importance of a parameter.

The top three machine learning models (MLP, SVR, and PLSR) with higher R^2 values were explained using SHAP values. The SHAP values for every feature in every sample are displayed to give a summary of the important features in the SHAP variable importance plot of the MLP model (Figure 11.9). The y-axis of the plot shows the feature and the x-axis shows the Shapley value. The important features are ranked in descending order (vertically). Since a wider range of SHAP values denotes more differentiation in the model output, the feature is more significant. Additionally, the plot illustrates how each feature inclines to affect model predictions. Overlapping points are spread vertically in the y-axis at the same point to represent per feature distribution of importance, i.e., Shapley values. The scale's red and blue colors correspond to the feature's high and low values, respectively. The SHAP variable importance plots for SVR and PLSR models are shown in Figures 11.10 and 11.11, respectively. The SHAP function "Kernel Explainer" has explained the MLP, SVR, and PLSR models.

The important parameters magnesium and calcium are related to water hardness for all three models. Lower values of hardness are indicated by a negative Shapley value. Hardness is displayed on the axis of Shapley values. Blue illustrates the lower value of the feature. So, blue values of magnesium and calcium are associated with negative Shapley values, i.e., lower values of magnesium and calcium represent lower values of hardness. The higher (red) the value of magnesium and calcium, the higher (positive) the

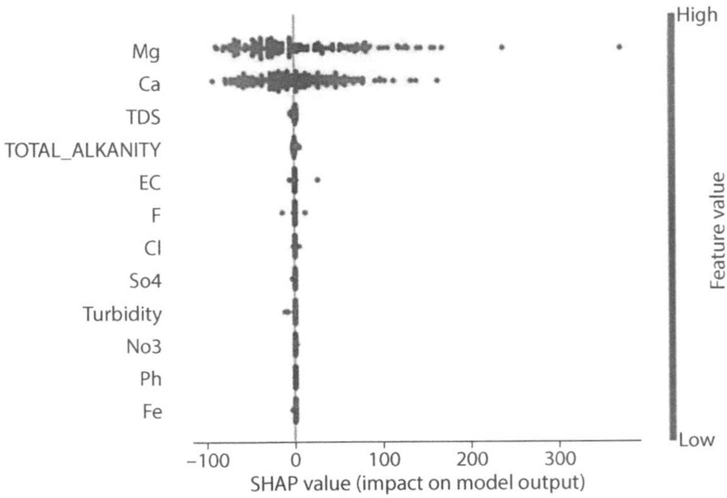

Figure 11.9 SHAP variable importance plot (MLP).

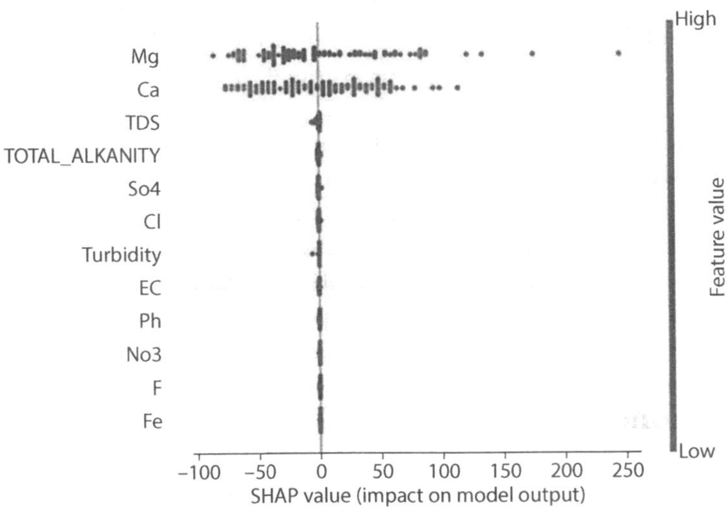

Figure 11.10 SHAP variable importance plot (SVR).

hardness value. The other important parameters are TDS, total alkalinity, EC, and Cl.

Several authors have widely established the influence of magnesium and calcium in the prediction of hardness [39–41] and the World Health Organization [42]. Along with magnesium and calcium, Shapely values

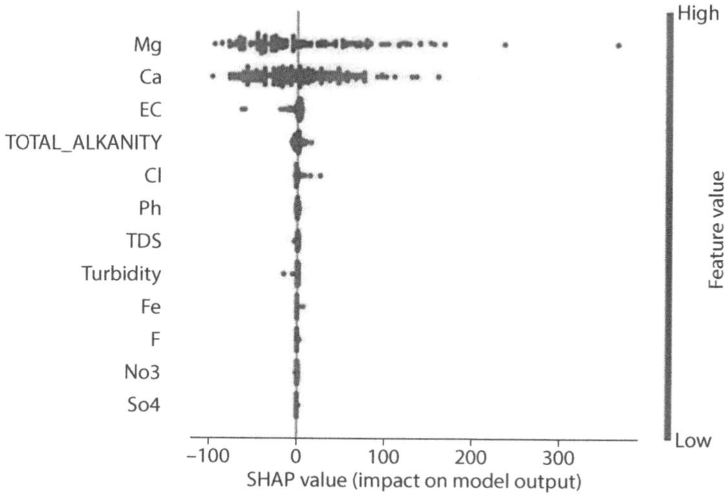

Figure 11.11 SHAP variable importance plot (PLSR).

from this research work signify that the prominence of other parameters like TDS, total alkalinity, EC, and Cl, and turbidity is also substantial. So, to emphasize patterns, a visual guide is added by plotting hardness against these prominent features in Figure 11.12.

The study's findings indicate that using the other parameters of water samples, ML can be utilized to predict the hardness condition of water. This study shows the potential reliability of results for predicting hardness using machine learning techniques rather than the chemical analysis of the samples. The sophisticated black box ML models used to determine the hardness content of water can be well explained using the XAI tools.

11.5 Conclusion

The study's major goal is to assess the competence of ML models for predicting water hardness and the potential of XAI for the interpretation of results. Prediction is made using available input parameters like turbidity, EC, pH, TDS, total alkalinity, nitrate, chloride, fluoride, sulfate, magnesium, calcium, and iron. ML algorithms like kNN, GBR, RF, PLSR, SVR, and MLP are developed to check the applicability of water hardness prediction in Punjab. Results based on more than 5000 diverse data points indicate the robustness of this research work. The paper discusses the basics of XAI, followed by statistics and correlations of parameters. The key point

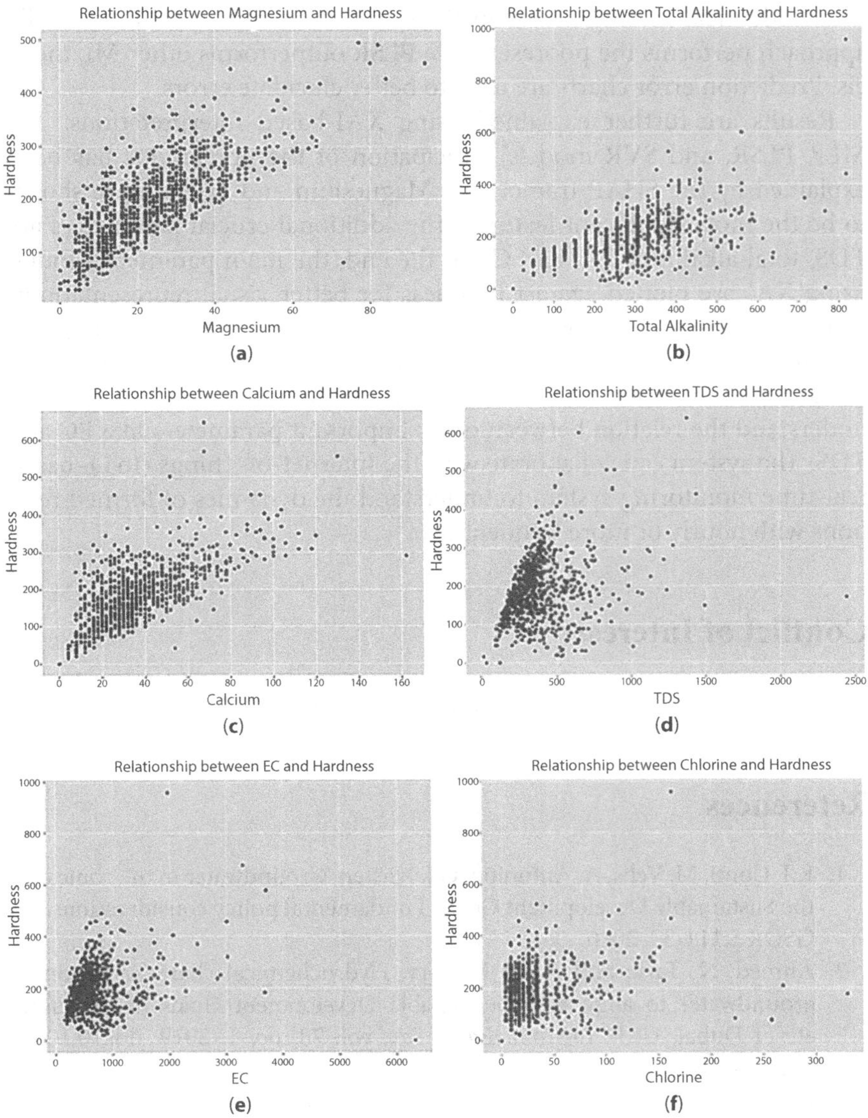

Figure 11.12 Relationship between hardness and prominent features.

of this research is that the hyperparameters used to implement ML models and their values are widely elaborated and discussed. Results show the reliability of MLP, RF, PLSR, and SVR for predicting hardness in terms of R^2 and MSE. It was discovered that the MLP performed the best in terms of R^2, followed by SVR, RF, PLSR, and GBR. These four models had nearly identical findings, with an R^2 of 0.98. In terms of MSE, the kNN approach performs the poorest, while PLSR outperforms other ML models. Prediction error charts are used to better elucidate errors.

Results are further explained using XAI-based interpretations. The MLP, PLSR, and SVR models' anticipation of the predictions has been explained by the SHAP framework. Magnesium and calcium are shown to be the most significant features. The additional crucial parameters are TDS, total alkalinity, EC, and Cl. In the end, the main parameters found using XAI are plotted against hardness for better visual representations. Hence, the paper does not merely focus on predicting water hardness but also supplements hyperparameter elaborations, XAI interpretations, and visual representations. The same flow can be followed in future work to understand the relation between other important parameters like EC and TDS. The system can collaborate with the Internet-of-Things (IoT)–based real-time monitoring system to understand the dynamics of formed relations with hourly or more frequent data.

Conflict of Interest

The authors have no conflict of interest to declare.

References

1. K.I. Conti, M. Velis, A. Antoniou, G.J. Nijsten, Groundwater in the context of the Sustainable Development Goals: Fundamental policy considerations, Br. GSDR 5:111-33, 2016
2. Ahmed, N. Tariq and A. Al Muhery, Hydrochemical characterization of groundwater to align with Sustainable Development Goals in the emirate of Dubai, UAE, *Environ. Earth Sci.*, vol. 78, no. 1, 2019. doi:10.1007/s12665-018-8030-1
3. D. Thakur D., R. Ganguly R., A. K. Gupta. Characterization and waste to energy techniques for improving Municipal Solid Waste Management in Una Town, Himachal Pradesh, India – a Case Study", *J. Solid Waste Technol. Manage.*, 46(4):547-562, 2020.
4. G. Kaur, R. Kumar, S. Mittal, P. K. Sahoo and U. Vaid, Ground/drinking water contaminants and cancer incidence: A case study of rural areas of South West Punjab, India, *Hum. Ecol. Risk Assess.: Int. J.*, vol. 27, no. 1, pp. 205–226, Dec. 2019, doi: 10.1080/10807039.2019.1705145.
5. J. Singh, https://www.downtoearth.org.in/news/punjab-cancer-capital-of-india-40255, 2013

6. D. Singla, A. Singla, M. Kaur, D.S.Walia, Incidence of Malignant Diseases in Punjab: A Retrospective Study in a Tertiary Level Government Hospital in Punjab, *J. Clin. Diagn. Res.*, Vol. 15, No. 3, 2021.
7. United Nations Press Release SG/SM/8707-OBV/348, https://www.un.org/press/en/2003/sgsm8707.doc.htm, 2003
8. N. S. Singh, R. Sharma, T. Parween and P. K. Patanjali, "Pesticide Contamination and Human Health Risk Factor," in M. Oves, M.Z. Khan, I.M.I. Ismail eds., *Modern Age Environmental Problems and their Remediation*, pp. 49–68, Oct. 2017, doi: 10.1007/978-3-319-64501-8_3.
9. A. G. D. Amarasooriya and T. Kawakami, "Removal of fluoride, hardness and alkalinity from groundwater by electrolysis," *Groundwater Sustain. Dev.*, vol. 9, p. 100231, Oct. 2019, doi: https://doi.org/10.1016/j.gsd.2019.100231.
10. P. Sengupta, Potential health impacts of hard water. *Int. J. Prev. Med.*, Vol. 4, No. 8, p. 866, 2013.
11. R.S. Ayers, D.W. Westcot, Water quality for agriculture. Rome: Food and Agriculture Organization of the United Nations, 1985
12. Water Quality for Crop Production, Center for Agriculture, Food, and the Environment. https://ag.umass.edu/greenhouse-floriculture/greenhouse-best-management-practices-bmp-manual/water-quality-for-crop.
13. M. Haddad, B. Barbeau, Hybrid hollow fiber nanofiltration–calcite contactor: A novel point-of-entry treatment for removal of dissolved Mn, Fe, NOM and hardness from domestic groundwater supplies. *Membranes*, Vol. 9, No. 7, p.90, 2019.
14. W. Liu, R. P. Singh, S. Jothivel and D. Fu, Evaluation of groundwater hardness removal using activated clinoptilolite, *Environ. Sci. Pollut. Res.*, vol. 27, no. 15, pp. 17541–17549, Sep. 2019, doi: 10.1007/s11356-019-06193-9.
15. D. Gunning and D. Aha, DARPA's Explainable Artificial Intelligence (XAI) Program, *AI Mag.*, vol. 40, no. 2, pp. 44–58, Jun. 2019, doi: 10.1609/aimag.v40i2.2850.
16. Tiyasha, T. M. Tung and Z. M. Yaseen, A survey on river water quality modelling using artificial intelligence models: 2000–2020, *J. Hydrol.*, vol. 585, p. 124670, Jun. 2020, doi: 10.1016/j.jhydrol.2020.124670.
17. S. H. Ewaid, S. A. Kadhum, S. A. Abed and R. M. Salih, Development and evaluation of irrigation water quality guide using IWQG V.1 software: A case study of Al-Gharraf Canal, Southern Iraq, *Environ. Technol. Innovation*, vol. 13, pp. 224–232, Feb. 2019, doi: 10.1016/j.eti.2018.12.001.
18. H. S. Jahin, A. S. Abuzaid and A. D. Abdellatif, Using multivariate analysis to develop irrigation water quality index for surface water in Kafr El-Sheikh Governorate, Egypt, *Environ. Technol. Innovation*, vol. 17, p. 100532, Feb. 2020, doi: 10.1016/j.eti.2019.100532.
19. H. Lu and X. Ma, Hybrid decision tree-based machine learning models for short-term water quality prediction, *Chemosphere*, vol. 249, p. 126169, Jun. 2020, doi: 10.1016/j.chemosphere.2020.126169.

20. M. Castrillo and Á. L. García, Estimation of high frequency nutrient concentrations from water quality surrogates using machine learning methods, *Water Res.*, vol. 172, p. 115490, Apr. 2020, doi: 10.1016/j.watres.2020.115490.
21. G. Meyers, Z. Kapelan and E. Keedwell, Short-term forecasting of turbidity in trunk main networks, *Water Res.*, vol. 124, pp. 67–76, Nov. 2017, doi: 10.1016/j.watres.2017.07.035.
22. E. Fijani, R. Barzegar, B. Liu, E. Tziritis and K. Skordas, Design and implementation of a hybrid model based on two-layer decomposition method coupled with extreme learning machines to support real-time environmental monitoring of water quality parameters, *Sci. Total Environ.*, vol. 648, pp. 839–853, Jan. 2019, doi: 10.1016/j.scitotenv.2018.08.221.
23. J.S. Chou, C.C. Ho and H.S. Hoang, Determining quality of water in reservoir using machine learning, *Ecol. Inf.*, vol. 44, pp. 57–75, Mar. 2018, doi: 10.1016/j.ecoinf.2018.01.005.
24. M. I. Shah, M. F. Javed and T. Abunama, Proposed formulation of surface water quality and modelling using gene expression, machine learning, and regression techniques, *Environ. Sci. Pollut. Res.*, vol. 28, no. 11, pp. 13202–13220, Nov. 2020, doi: 10.1007/s11356-020-11490-9.
25. R. Jafari, A. Torabian, M. Ghorbani, Seyed Ahmad Mirbagheri, and Amir, Prediction of groundwater quality parameter in the Tabriz plain, Iran using soft computing methods, *J. Water Supply: Res. Technol. AQUA*, vol. 68, no. 7, pp. 573–584, Aug. 2019, doi: 10.2166/aqua.2019.062
26. El Bilali, A. Taleb and Y. Brouziyne, Groundwater quality forecasting using machine learning algorithms for irrigation purposes, *Agric. Water Manage.*, vol. 245, p. 106625, Feb. 2021, doi: /10.1016/j.agwat.2020.106625.
27. Sunayana, K. Kalawapudi, O. Dube and R. Sharma, Use of neural networks and spatial interpolation to predict groundwater quality, *Environ. Dev. Sustain.*, vol. 22, no. 4, pp. 2801–2816, Feb. 2019, doi: 10.1007/s10668-019-00319-2.
28. U. Ahmed, R. Mumtaz, H. Anwar, A. A. Shah, R. Irfan and J. García-Nieto, Efficient Water Quality Prediction Using Supervised Machine Learning, *Water*, vol. 11, no. 11, p. 2210, Nov. 2019, doi: 10.3390/w11112210.
29. Mosavi *et al.*, Susceptibility Prediction of Groundwater Hardness Using Ensemble Machine Learning Models, *Water*, vol. 12, no. 10, p. 2770, Oct. 2020, doi: 10.3390/w12102770.
30. Adadi and M. Berrada, Peeking Inside the Black-Box: A Survey on Explainable Artificial Intelligence (XAI), *IEEE Access*, vol. 6, pp. 52138–52160, 2018, doi: 10.1109/access.2018.2870052.
31. D. Chakraborty, H. Başağaoğlu and J. Winterle, Interpretable vs. noninterpretable machine learning models for data-driven hydro-climatological process modeling, *Expert Syst. Appl.*, vol. 170, p. 114498, May 2021, doi: 10.1016/j.eswa.2020.114498.
32. E. Tjoa and C. Guan, A Survey on Explainable Artificial Intelligence (XAI): Toward Medical XAI, *IEEE Trans. Neural Networks Learn. Syst.*, vol. 32, no. 11, pp. 1–21, 2020, doi: 10.1109/tnnls.2020.3027314.

33. J.A. Duell, A Comparative Approach to Explainable Artificial Intelligence Methods in Application to High-Dimensional Electronic Health Records: Examining the Usability of XAI, arXiv:2103.04951, 2021
34. Kuppa and N. Le-Khac, "Black Box Attacks on Explainable Artificial Intelligence(XAI) methods in Cyber Security", in *International Joint Conference on Neural Networks (IJCNN)*, IEEE, Glasgow, United Kingdom, pp 1-8, 2019 doi: 10.1109/IJCNN48605.2020.9206780
35. Heimerl, K. Weitz, T. Baur and E. Andre, Unraveling ML Models of Emotion with NOVA: Multi-Level Explainable AI for Non-Experts, *IEEE Trans. Affective Comput.*, pp. 1–1, 2020, doi: 10.1109/taffc.2020.3043603.
36. W. Samek, G. Montavon, S. Lapuschkin, C. J. Anders and K.-R. Muller, "Explaining Deep Neural Networks and Beyond: A Review of Methods and Applications," *Proc. IEEE*, vol. 109, no. 3, pp. 247–278, Mar. 2021, doi: 10.1109/jproc.2021.3060483.
37. Kumar, S., and Tarlochan Kaur. "Efficient solar radiation estimation using cohesive artificial neural network technique with optimal synaptic weights." *Proc. Inst. Mech. Eng., Part A: J. Power Energy* 234, no. 6 pp.862-873 2020.
38. S. M. Lundberg and S.I. Lee, "A unified approach to interpreting model predictions," in *Proceedings of the 31st International Conference on Neural Information Processing Systems*, Long Beach, CA, USA, vol. 30, pp. 4768–4777, Dec. 2017.
39. Nouraki, M. Alavi, M. Golabi and M. Albaji, Prediction of water quality parameters using machine learning models: a case study of the Karun River, Iran, *Environ. Sci. Pollut. Res.*, vol. 28, no. 40, pp. 57060–57072, Jun. 2021, doi: 10.1007/s11356-021-14560-8.
40. K. Ravindra, P. S. Thind, S. Mor, T. Singh and S. Mor, Evaluation of groundwater contamination in Chandigarh: Source identification and health risk assessment, *Environ. Pollut.*, vol. 255, p. 113062, Dec. 2019, doi: 10.1016/j.envpol.2019.113062.
41. S. Rapant, V. Cvečková, K. Fajčíková, D. Sedláková and B. Stehlíková, Impact of Calcium and Magnesium in Groundwater and Drinking Water on the Health of Inhabitants of the Slovak Republic, *Int. J. Environ. Res. Public Health*, vol. 14, no. 3, Mar. 2017, doi: 10.3390/ijerph14030278.
42. World Health Organization, No. WHO/HSE/WSH/10.01/10/Rev/1, https://www.who.int/water_sanitation_health/dwq/chemicals/hardness.pdf., 2010

12

Air Quality Mapping Using GIS for Kanpur City, India

Varun Yadav and Rajiv Ganguly*

Department of Civil Engineering, Harcourt Butler Technical University, Kanpur, Uttar Pradesh, India

Abstract
Geospatial techniques are widely used for representing the concentration of air pollutants in a city. A detailed spatial distribution of particulate matter (PM) in Kanpur city has been presented using two interpolation techniques, namely, kriging and inverse distance weighting (IDW). Geospatial maps representing air pollutant concentrations in the winter seasons of 2016 have been reported. It was inferred from the study that the concentration of PM was significantly higher in areas with heavy traffic and industrial activities. The spatial distribution maps generated using the two interpolation techniques showed significant variations in PM concentrations across different parts of the city. The study utilizes two distinct spatial interpolation methods for creating detailed maps of PM concentration in Kanpur city, used for monitoring and management of air pollution in the area. The study also highlights the importance of incorporating spatial distribution maps into policymaking processes to alleviate the effects of air pollution on public health and the environment.

***Keywords*:** Air pollution, spatial interpolation, GIS, ordinary kriging (OK), inverse distance weighted (IDW)

12.1 Introduction

Air pollution is a pressing environmental issue that has significant implications for public health and the well-being of ecosystems. It is a complex

**Corresponding author*: rganguly@hbtu.ac.in

Disha Thakur, Sanjay Kumar, Har Amrit Singh Sandhu and Chander Prakash (eds.) Sustainable Development Using Geospatial Techniques, (285–298) © 2024 Scrivener Publishing LLC

problem that arises from various sources, including industrial emissions, vehicular exhaust, and the burning of fossil fuels. The consequences of air pollution are far-reaching and can manifest in respiratory diseases, cardiovascular disorders, and environmental degradation [1]. In India, rapid urbanization and industrial growth have led to alarming levels of air pollution, particularly in densely populated cities. Kanpur, located in the state of Uttar Pradesh, India, is one such city grappling with severe air pollution. With a population of over 3 million, Kanpur is known for its high industrial activity and heavy traffic congestion [2–4]. These factors contribute significantly to the emission of air pollutants, including particulate matter (PM), nitrogen dioxide (NO_2), sulfur dioxide (SO_2), and volatile organic compounds (VOCs). The detrimental effects of air pollution on public health and the environment necessitate comprehensive assessments and targeted interventions to mitigate its impact.

The objectives of the study were to assess air pollution in Kanpur city using geospatial techniques, specifically inverse distance weighting (IDW) and kriging interpolation methods. By utilizing these techniques, this research aims to provide a detailed understanding of the spatial distribution of air pollutant concentrations across the city during the winter season for the year 2016–2017. The findings will contribute to the development of high-resolution air pollution maps, which can serve as valuable tools for policymakers, urban planners, and environmental agencies in formulating effective strategies for pollution control and management.

By exploring the effectiveness of these methods in capturing the spatial variability of air pollutant concentrations, this research will enhance the understanding of their applicability in similar urban contexts. It will also provide valuable insights into the strengths and limitations of these techniques, facilitating future research and improving air pollution modeling methodologies.

12.2 Methodology

12.2.1 Site Selection and Data Collection

The first step in the methodology was the selection of a site for the assessment. In Kanpur, there are a total of eight manual monitoring stations that are in operation as per the National Ambient Monitoring Program (NAMP) located throughout the city as shown in Figure 12.1. Air pollutant concentration data were obtained from existing monitoring stations These monitoring stations were selected based on their representativeness

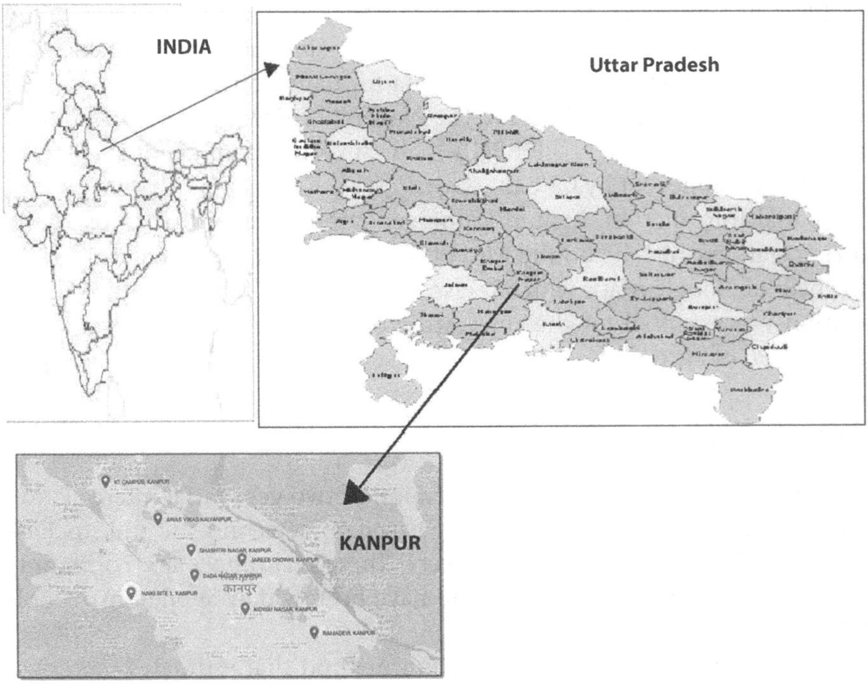

Figure 12.1 Locations of air monitoring stations in Kanpur city, Uttar Pradesh, India.

Table 12.1 Monitoring sites operated by UPPCB (Uttar Pradesh Pollution Control Board).

Location	Site code	Location (longitude, latitude)	Type of site
Kidwai Nagar	KDN	80.337537E, 26.424953N	Commercial
Jareeb Chowki	JCH	80.323798E, 26.462878N	Housing
Panki Site 1	PNK	80.260878E, 26.447848N	Industrial
Shashtri Nagar	SSN	80.298239E, 26.466816N	Housing
Awas Vikas Kalyanpur	AVK	80.243256E, 26.480815N	Housing
Dada Nagar	DDN	80.293425E, 26.453306N	Housing
IIT CAMPUS	IIT	80.233243E, 26.512857N	Housing
Ramadevi	RM	80.386794E, 26.411065N	Commercial

of different pollution sources such as industrial commercial and residential areas and major roadways. These monitoring sites are listed in Table 12.1 along with their classifications. The data for daily concentrations of PM_{10} were obtained from UPPCB for the November, December, and January months of 2016–2017 [5]. UPPCB maintains the monitoring stations as per the guidelines set by CPCB.

Spatial information, including the geographical coordinates of monitoring stations and the boundaries of Kanpur city, was also collected with the help of Google Maps which is also mentioned in Table 12.1. This information was essential for the application of geospatial techniques and the generation of air pollution maps.

12.2.2 Geospatial Techniques

The present study includes the application of two geospatial interpolation techniques, namely, IDW and kriging based on deterministic and stochastic interpolation methods, respectively. These techniques are commonly used for geospatial interpolation and have been widely applied in air pollution assessment.

IDW is a deterministic interpolation method that estimates unknown values based on the weighted average of neighboring known values. The weights are determined based on the inverse distance between the known and unknown points. The IDW algorithm assumes that closer points have a stronger influence on the estimation. In this study, IDW was applied to generate air pollution concentration maps for Kanpur city.

Kriging, on the other hand, is a geostatistical interpolation method that takes into account both the spatial autocorrelation and the variability of the data. Kriging estimates unknown values based on a linear combination of neighboring data points, with the weights determined by a variogram model. Kriging provides more accurate estimations compared to IDW, particularly in areas with complex spatial patterns and unevenly distributed monitoring stations. In this study, kriging was employed to generate air pollution concentration maps as well.

Using both kriging and IDW allows for intercomparison of both the chosen spatial interpolation techniques. While both techniques have their unique advantages and disadvantages, the IDW and kriging spatial interpolation capture various kinds of spatial patterns due to their use of different variogram models and assumptions, respectively. By using both, it is possible to determine which of these approaches are more accurate and better represent the existing conditions in the selected study location. Kriging is based on the spatial autocorrelation of the data, while IDW evaluates the inverse distance relationship. Using both methods presents a detailed picture of the data and highlights distinct facets of the spatial variation in cases when the dataset displays complicated geographical patterns and the underlying spatial process is not well known. By using these two approaches, the underlying techniques for the evaluation of these spatial interpolation methods in the context of the dataset can be compared by the maps generated with the help of ArcGIS software.

12.2.3 Software and Tools

We employed the "geostatistical analyst" tool in ArcGIS to analyze data inventory that varies spatially. By utilizing advanced statistical techniques and measured data, we generated a two-dimensional surface. Spatial interpolation was utilized to obtain the most accurate representation of the surface and predict values for unmonitored sampling points, resulting in a comprehensive surface [6]. These techniques were helpful for data visualization, spatial analysis, and map generation. These software platforms provided the necessary functionalities for conducting IDW and kriging interpolations, as well as overlaying the results on spatial maps [7]. Analyzing the pollutant in terms of space and time aids in making well-informed policy adjustments and focusing mitigation efforts on crucial locations that can be identified [8].

12.3 Data Analysis and Results

12.3.1 Descriptive Analysis

The data collected for air pollution assessment in Kanpur city underwent descriptive analysis to gain an understanding of the air pollutant concentrations and spatial information. Descriptive statistics such as mean and standard deviation were calculated for the considered study period is mentioned in Table 12.2. The highest monthly average concentration of PM_{10} was observed as 547.7 $\mu g/m^3$ at the RM (commercial) site in November, also the lowest concentration of PM_{10} was recorded as 147.1 $\mu g/m^3$ at the IIT (housing) site in January 2017. Meanwhile, the maximum standard deviation can be observed at the RM (commercial) site from Table 12.2.

The analysis revealed the average concentrations of particulate matter (PM_{10}) in Kanpur city during the study period exceeded the prescribed standards [9]. The descriptive analysis provided an overview of the

Table 12.2 PM_{10} ($\mu g/m^3$) descriptive statistics (mean and standard deviation).

Stations	Nov	Dec	Jan	Mean	SD
KDN	235.1	238.5	237.1	236.9	1.39523
JCH	243.4	237	247.7	242.7	4.39621
PNK	230.6	252.2	265.7	249.5	14.45614
SSN	234.6	247.4	220	234	11.19405
AVK	220.1	225.6	260.3	235.3333	17.79632
DDN	334	330.3	343.3	335.8667	5.468902
IIT	238	200.5	147.1	195.2	37.29853
RM	547.7	300.3	374.1	407.3667	103.7037

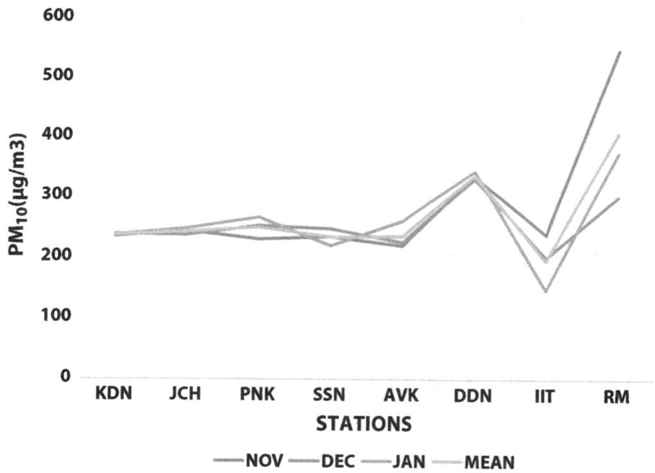

Figure 12.2 Winter months (Nov 16–Jan 17) concentrations of PM_{10} from eight monitoring stations.

pollution levels and their variability across different monitoring stations in Figure 12.2 during the study period.

12.3.2 Interpolation Results

The IDW and kriging interpolation techniques were applied to estimate air pollutant concentrations at unmonitored locations across Kanpur city.

The interpolation results provided detailed air pollution concentration maps, allowing for the identification of spatial patterns and pollution hotspots are shown in Figure 12.3. Figure 12.3 shows the predicted surfaces for the study period, i.e., November 2016 to January 2017. The trend observed in the predicted value is like the trend observed in the monitored values, apart from the variations in magnitude. From the results, it can be observed that the commercial sites (such as RM and KDN) and industrial areas (such as PNK) were highly polluted.

The spatial interpolation technique IDW is more adept in identifying AQ hotspots from the maps generated through the application of software [10]. Alternatively, kriging applies a 'smoothing' technique removing any outliers in the data; however, its effectiveness increases when a higher number of input points are used [11]. The IDW interpolation method estimated pollutant concentrations based on the inverse distance-weighted average of neighboring monitoring stations. It assumed that closer stations have a greater influence on the estimation [12]. The IDW maps depicted the spatial distribution of air pollutants in Kanpur city, highlighting areas with high pollution levels and areas that were relatively cleaner.

The kriging interpolation method, on the other hand, considered the variability of the air pollutant data. It provided more accurate estimations compared to IDW, particularly in areas with complex spatial patterns and unevenly distributed monitoring stations [13]. The kriging maps captured

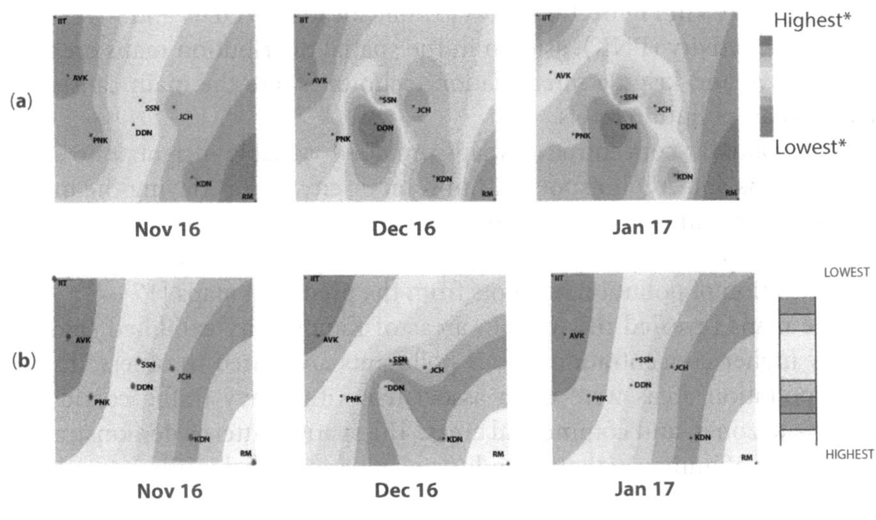

Figure 12.3 Predicted surfaces using (a) IDW and (b) kriging for November 2016 to January 2017.

the fine-grained variations in air pollutant concentrations across different parts of Kanpur city.

12.3.3 Comparison of Interpolation Techniques

One important aspect in the comparison of the interpolation technique is that both methods are valuable in estimating air pollutant concentrations at unmonitored locations. However, there were notable differences in their performance and applicability. IDW, being a simple technique, provided reasonable estimates of pollution levels. It relied on the proximity of monitoring stations, giving more weight to the closest stations [14, 15]. IDW was particularly effective in areas with relatively homogeneous pollution patterns and evenly distributed monitoring stations. On the other hand, kriging demonstrated superior accuracy, especially in areas with complex spatial patterns and unevenly distributed monitoring stations. By considering the spatial autocorrelation and variability of the data, kriging captured fine-grained variations in pollution concentrations [16]. This method proved to be more robust and reliable, particularly in urban environments where pollution sources are diverse and non-uniformly distributed.

12.3.4 Hotspots of Air Pollution

With greater PM_{10} concentrations detected in various regions of Kanpur City, the research identified major hotspots of air pollution which have been presented in Figure 12.4. Heavy traffic locations (DDN and JCH) and industrial activity (PNK), as seen in the spatial distribution maps created using IDW and kriging interpolation techniques, are the main causes of elevated air pollution levels.

The pollutant concentration maps provided valuable insights into spatial patterns and AQ hotspots within Kanpur city. By overlaying the maps on the geographical boundaries of the city, areas with high pollution levels were identified. IDW is more efficient in comparison to kriging for the identification of pollution hotspots from the predicted maps [17].

The maps revealed that certain areas of Kanpur city exhibited significantly higher concentrations of air pollutants compared to others. These high-pollution areas were often associated with heavy traffic corridors, industrial zones, and commercial areas. The spatial patterns demonstrated the impact of human activities, industrial emissions, and traffic congestion on air quality within the city.

The identification of pollution hotspots allowed for targeted interventions and mitigation strategies. By focusing resources and implementing

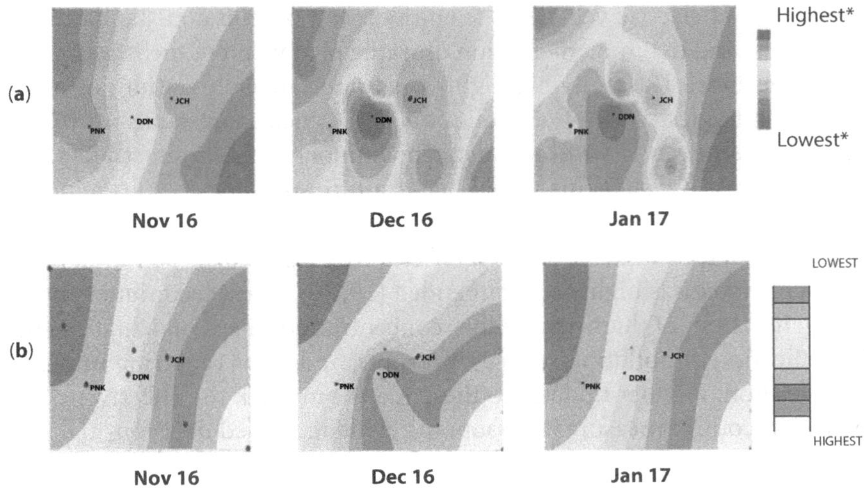

Figure 12.4 Hotspots of air pollution.

pollution control measures in these hotspots, it was possible to have a more significant impact on reducing air pollution levels in Kanpur city.

12.3.5 Policy Implications and Recommendations

The data analysis and results obtained from the assessment of air pollution using IDW and kriging interpolation techniques in Kanpur city have significant policy implications. The spatial distribution maps and hotspot identification can serve as valuable tools for policymakers and urban planners in formulating effective air pollution control strategies. By incorporating spatial information and pollution concentration maps into policymaking processes, it is possible to prioritize pollution control measures in areas with high pollution levels. This targeted approach can lead to a more efficient allocation of resources and the implementation of measures such as emission controls, traffic management, and industrial regulations. Furthermore, the findings of the study underscore the importance of continuous monitoring and assessment of air pollution in Kanpur city. The use of geospatial techniques provides a high-resolution understanding of pollution patterns, enabling policymakers to track progress, evaluate the effectiveness of interventions, and make informed decisions regarding air quality management [18]. To address the high pollution levels in areas with heavy traffic and industrial activities, comprehensive strategies need to be implemented. These may include the improvement of public transportation systems, the promotion of electric vehicles, the enforcement of emission standards for industries, and the implementation of traffic management measures.

Furthermore, the study highlights the importance of continuous monitoring and assessment of air pollution. It is crucial to establish a robust monitoring network with an adequate number of stations strategically located across the city. This would help capture spatial variations in pollution levels and provide real-time data for decision-making. The integration of geospatial techniques, such as IDW and Kriging, into air pollution management systems is highly recommended [19, 20]. These techniques enable the generation of high-resolution concentration maps, facilitating the identification of pollution sources, hotspots, and areas requiring immediate attention. Regular updates to these maps can track the effectiveness of pollution control measures and support evidence-based decision-making.

12.4 Discussions and Conclusion

12.4.1 Discussion

The assessment of air pollution using IDW and kriging interpolation techniques in Kanpur city, Uttar Pradesh, India, has yielded significant insights into the spatial patterns of air pollutant concentrations. The study revealed that air pollution in Kanpur city is characterized by significant spatial variations, with areas experiencing heavy traffic and industrial activities exhibiting higher pollution levels. The spatial distribution maps generated through IDW and kriging interpolation techniques help in identifying the pollution hotspots and areas requiring immediate attention.

The results obtained from the study may have important policy implications for managing air pollution within Kanpur city. The identification of pollution hotspots and spatial variations in pollution levels provides valuable information for policymakers and urban planners to prioritize pollution control measures. By focusing resources and implementing targeted interventions in areas with high pollution levels, the impact of pollution control efforts can be maximized.

12.4.2 Conclusion

In conclusion, this study provides a comprehensive assessment of air pollution in Kanpur city, Uttar Pradesh, India, using IDW and kriging interpolation techniques. The findings highlight the spatial variations in air pollutant concentrations and the identification of pollution hotspots.

The commercial sites such as RM and KDN and industrial sites PNK were observed as the hotspots in the city. The policy implications emphasize the importance of targeted pollution control measures and the integration of geospatial techniques into air pollution management strategies. It is important to recognize that air pollution is a complex issue influenced by various factors, including emissions from industries, vehicular traffic, construction activities, and residential sources. To effectively address air pollution in Kanpur city, a multidimensional approach is required, involving stakeholders from government agencies, industries, communities, and academia. Continuous monitoring, data analysis, and research are essential to track progress, evaluate the effectiveness of interventions, and adapt strategies to changing pollution patterns. By implementing the recommended measures and collaborating with relevant stakeholders, Kanpur city can take significant strides toward improving air quality and protecting public health. The concerted efforts of all stakeholders are crucial in creating a sustainable and healthy environment for the residents of Kanpur city.

Furthermore, the findings of this study can serve as a valuable reference for other cities facing similar air pollution challenges. The methodologies employed, such as IDW and kriging interpolation techniques, can be applied in different geographical contexts to assess air pollution and develop targeted mitigation strategies.

References

1. Mannucci, P. M., & Franchini, M. (2017). Health effects of ambient air pollution in developing countries. *Int. J. Environ. Res. Public Health*, *14*(9), 1048. https://doi.org/10.3390/ijerph14091048
2. Sharma, M., Pandey, R., Maheshwari, M., Sengupta, B., Shukla, B. P., Gupta, N. K., & Johri, S. (2003). Interpretation of air quality data using an air quality index for the city of Kanpur, India. *J. Environ. Eng. Sci.*, *2*(6), 453-462. https://doi.org/10.1139/s03-047.
3. Gautam, A. S., Kumar, S., Gautam, S., Anand, A., Kumar, R., Joshi, A., ... & Singh, K. (2021). Pandemic induced lockdown as a boon to the Environment: trends in air pollution concentration across India. *Asia-Pac. J. Atmos. Sci.*, 1-16. https://doi.org/10.1007/s13143-021-00232-7.
4. Sarfraz, M., Mohsin, M., Naseem, S., & Kumar, A. (2021). Modeling the relationship between carbon emissions and environmental sustainability during COVID-19: new evidence from asymmetric ARDL cointegration approach. *Environ. Dev. Sustain.*, 1-19. https://doi.org/10.1007/s10668-021-01324-0.
5. Uttar Pradesh Pollution Control Board. (http://www.uppcb.com/ambient_quality.htm).

6. Johnston, K., Ver Hoef, J. M., Krivoruchko, K., & Lucas, N. (2001). Using ArcGIS geostatistical analyst (Vol. 380). Redlands: Esri.
7. Janowski, D., & Leski, T. (2023). Landscape-scale mapping of soil fungal distribution: proposing a new NGS-based approach. *Sci. Rep.*, *13*(1), 10280. https://doi.org/10.1038/s41598-023-37538-7.
8. Boiné, K., Demers, C. M., & Potvin, A. (2018). Spatio-temporal promenades as representations of urban atmospheres. *Sustain. Cities Soc.*, *42*, 674-687. https://doi.org/10.1016/j.scs.2018.04.028
9. National Ambient Air Quality Standards (NAAQS), 2019. https://cpcb.nic.in/upload/NAAQS_2019.pdf
10. Shukla, K., Kumar, P., Mann, G. S., & Khare, M. (2020). Mapping spatial distribution of particulate matter using Kriging and Inverse Distance Weighting at supersites of megacity Delhi. *Sustain. Cities Soc.*, *54*, 101997. https://doi.org/10.1016/j.scs.2019.101997.
11. Bezyk, Y., Sówka, I., Górka, M., & Blachowski, J. (2021). Gis-based approach to spatio-temporal interpolation of atmospheric CO2 concentrations in limited monitoring dataset. *Atmosphere*, *12*(3), 384. https://doi.org/10.3390/atmos12030384.
12. Zhao, R., Zhan, L., Yao, M., & Yang, L. (2020). A geographically weighted regression model augmented by Geodetector analysis and principal component analysis for the spatial distribution of PM2. 5. *Sustain. Cities Soc.*, *56*, 102106. https://doi.org/10.1016/j.scs.2020.102106.
13. Hu, Y., Chen, X., & Xia, H. (2023). A hybrid prediction model of air quality for sparse stations based on spatio-temporal feature extraction. *Atmos. Pollut. Res.*, *14*(6), 101765. https://doi.org/10.1016/j.apr.2023.101765.
14. Glenk, K., Johnston, R. J., Meyerhoff, J., & Sagebiel, J. (2020). Spatial dimensions of stated preference valuation in environmental and resource economics: methods, trends and challenges. *Environ. Resour. Econ.*, *75*, 215-242. https://doi.org/10.1007/s10640-018-00311-w.
15. Han, Y., Zhang, F., Huang, L., Peng, K., & Wang, X. (2021). Does industrial upgrading promote eco-efficiency?— A panel space estimation based on Chinese evidence. *Energy Policy*, *154*, 112286. https://doi.org/10.1016/j.enpol.2021.112286.
16. Zimmerman, D. L., & Homer, K. E. (1991). A network design criterion for estimating selected attributes of the semivariogram. *Environmetrics*, *2*(4), 425-441. https://doi.org/10.1002/env.3770020403.
17. Kamboj, K., Sisodiya, S., Mathur, A. K., Zare, A., & Verma, P. (2022). Assessment and spatial distribution mapping of criteria pollutants. *Water Air Soil Pollut.*, *233*(3), 82. https://doi.org/10.1007/s11270-022-05522-y.
18. Kaginalkar, A., Kumar, S., Gargava, P., & Niyogi, D. (2021). Review of urban computing in air quality management as smart city service: An integrated IoT, AI, and cloud technology perspective. *Urban Clim.*, *39*, 100972. https://doi.org/10.1016/j.uclim.2021.100972.

19. Hou, D., O'Connor, D., Igalavithana, A. D., Alessi, D. S., Luo, J., Tsang, D. C., & Ok, Y. S. (2020). Metal contamination and bioremediation of agricultural soils for food safety and sustainability. *Nat. Rev. Earth Environ.*, *1*(7), 366-381. https://doi.org/10.1038/s43017-020-0061-y.
20. Jin, Y., O'Connor, D., Ok, Y. S., Tsang, D. C., Liu, A., & Hou, D. (2019). Assessment of sources of heavy metals in soil and dust at children's playgrounds in Beijing using GIS and multivariate statistical analysis. *Environ. Int.*, *124*, 320-328. https://doi.org/10.1016/j.envint.2019.01.024.

13

Geospatial Modeling Approach and Characteristics Study of Graphene-Anchored Cu-Nanoferrites and Their Potential in Arsenic Containing Wastewater Treatment

G. Murtaza* and Ahmad Ayyaz

Centre for Advanced Studies in Physics, GC University, Lahore, Punjab, Pakistan

Abstract

Industrial waste and growth contribute to water pollution in emerging nations like Pakistan, particularly due to a lack of rules awareness, and environmental standards. This research aimed to assess the environmental risk of arsenic in the industrial region of different areas across Pakistan. The synthesis of $CuFe_2O_4$ nanoparticles was conducted via the sol-gel route, resulting in the formation of a hybrid nanocomposite containing graphene with a layered structure. XRD determined that $CuFe_2O_4$ nanoparticles kept crystallite sizes up to 53 nm, which vary with ferrite and graphene concentration. The FTIR analysis reveals the three well-defined bands at wavenumbers of 1649, 1579, and 589 cm^{-1}. These bands can be attributed to specific molecular vibrations, namely, the stretching of C=O associated with carboxyl functionality, the stretching of C=C characteristic of aromatic compounds, and the stretching of Fe-O, respectively. The SEM pattern depicts the evidence of the surface morphology of both $CuFe_2O_4$ nanoparticles as well as their nanocomposite. To quantify the elemental composition and to find the respective content of ferrites and graphene in prepared nanocomposites, EDX spectra were also analyzed. The magnetic order of the studied nanocomposites was investigated using VSM. Results demonstrate that the magnetic $CuFe_2O_4$ nanoparticles possess a magnetic saturation (Ms) value of approximately 16.69 emu/g, accompanied by a significant remanence of approximately 29.22 emu/g. The saturation magnetization (Ms) and remanence exhibit a decreasing trend as the graphene content

**Corresponding author*: gmrai@gcu.edu.pk

Disha Thakur, Sanjay Kumar, Har Amrit Singh Sandhu and Chander Prakash (eds.) Sustainable Development Using Geospatial Techniques, (299–326) © 2024 Scrivener Publishing LLC

within the graphene/$CuFe_2O_4$ composite increases. Furthermore, the research focused on evaluating the composite material's capability to adsorb As(III) as well as As(V) present in aqueous solutions. This is very important to assess the studied nanocomposite's potential for practical usage. The findings indicate that the combination of graphene/$CuFe_2O_4$ composite has a synergistic effect in achieving remarkable adsorption characteristics. In summary, the present investigation will be helpful in the utilization of resources regarding environmental studies and their potential in As-containing wastewater treatment.

***Keywords*:** Geospatial modeling, wastewater, ferrites, nanocomposites, adsorption properties, removal of arsenic

13.1 Introduction

In light of the growing unsuitability of surface waters, groundwater has been recognized as the most reliable global drinking water resource. Many human endeavors, including urbanization, industry, and agriculture, are putting pressure on the world's surface water supplies. The aquatic environment suffers as a result of the increased pollution in water resources brought about by the immediate release of waste entering the water system [1–4]. Especially in areas with few water resources, it is crucial to preserve the purity of the groundwater that is already present. Ensuring long-term exploitation of water resources requires more than just formulating appropriate usage strategies based on water quantity. In addition, the stoichiometry and purity of water reserves must align with desired circumstances and exhibit favorable characteristics.

A comprehensive awareness of the hydrochemistry and purity of groundwater is necessary for the effective administration and preservation of this valuable resource. The significance of water chemistry lies in its capacity to assess the water's quality and its appropriateness for various applications such as drinking, agriculture, and industry [5, 6]. Evaluation of groundwater quality, significant metal pollution, and health risks can be achieved by the application of geospatial techniques and index methods [7]. The developing worldwide focus is on the ecological impacts associated with industrial wastewater polluted with heavy metals such as lead, chromium, arsenic, and mercury. Arsenic (As) is considered among the dangerous contaminants due to its frequent discharge into wastewater as well as soil as a consequence of activities such as pesticide spraying, semiconductor manufacturing operations, and mineral dissolution [8].

Arsenic is considered to be one of the elements with high chemical volatility and carcinogenic properties. The presence of arsenic in the atmosphere presents substantial obstacles to global public health since it makes a considerable contribution to the contamination of natural water sources [9, 10]. The occurrence of arsenic in drinking water corresponds to many biological mechanisms, including the dissolution and precipitation of minerals, and weathering mechanisms [11, 12]. There is a prevailing belief that inorganic arsenic compounds, specifically arsenite As(V) and arsenate As(V), exhibit greater toxicity compared to the organic variant [13]. Both As(III) as well as As(V) can be found in water sources, with the latter being characterized by higher volatility and lower solubility compared to the former [14]. Exposure to high levels of arsenic and prolonged interaction with small amounts of arsenic might potentially lead to a range of chronic health effects, including dermatological, cardiac, and respiratory illnesses [15]. Consequently, several regions and institutions have established more stringent standards for arsenic levels in water to be consumed. Notably, China, the United States, and the World Health Organization (WHO) have introduced regulations that limit the arsenic concentration to 10 $\mu g\ L^{-1}$ [16, 17]. For the sake of reducing the environmental impact and minimizing the adverse health effects associated with arsenic contamination in water, some technologically advanced approaches were examined for arsenic removal. These methods include the nanofiltration process [18], membrane separation [19], the exchange of ions [20, 21], the reverse osmosis process [22], electrochemical approaches [23], and adsorption [24, 25]. The absorption process holds significant significance for both commercial and scientific purposes, primarily due to its convenient availability for adsorption, user-friendly operation, and cost-effectiveness [26, 27]. The adsorption approach relies mostly on the exceptional adsorptive capacity of materials, which can be related to their elevated surface energy and extensive particular areas of surface [28–30]. A wide range of sorbents, synthetic as well as natural, are utilized to separate arsenic from water, as indicated by numerous studies [31–33].

Recent studies explored the performance of metal oxide absorbents, with a particular focus on metals such as aluminum (Al), manganese (Mn), zirconium (Zr), and iron (Fe) [34–36]. Iron oxide, namely, Fe_3O_4, has been extensively researched because of its strong attraction to arsenic, economical nature, and compatibility with ecologically friendly species [37, 38]. In contrast, it has been observed that magnetic Fe_3O_4 nanoparticles and their various nanostructures often exhibit reduced adsorption efficiency as a result of their low particular region of surface [39]. Research into the creation of iron-based binary sorbents has surged in popularity in the past few years, including Fe-Mn, Fe-Ti, Fe-Zr, and Fe-Ce. These binary compounds have been recognized for their synergistic capacity to efficiently remove arsenic from drinking water that is polluted [40, 41]. In the present context, Zhang *et al.* [42] published a study on a composite material consisting of bimetallic oxide Fe-Ce. The study highlights that this composite has a notably higher potential for adsorption for As(V) in comparison to the individual cerium oxide and iron oxide. This finding suggests that the composite material possesses a synergistic and promising characteristic. In a separate investigation, Martinson and colleagues [43] conducted the synthesis of CuO nanoparticles to investigate the adsorption effectiveness of As(III) as well as As(V) in samples of groundwater. The researchers documented a highly effective removal of arsenic efficiency of less than 3 μg/L, indicating their significant potential. Based on the findings presented in these papers, it appears that the combination of Cu and Fe-based oxides could be a viable option for the development of an adsorbent capable of removing arsenic. This is due to the favorable affinity exhibited by both Fe^{3+} and Cu^{2+} toward inorganic arsenic species, as well as the observed insensitivity of this combination to changes in pH levels. Zhang *et al.* [44] have synthesized a nanostructured Fe-Cu using the co-precipitation technique. An outstanding arsenic adsorption efficiency has been observed in As(III) with a value of 122.3 mg/g and for As(V) 82.7 mg/g with a pH level of 7.0, respectively. Furthermore, the investigation of iron oxide-derived adsorbents involves the incorporation of iron oxides onto various substrates. These adsorbents have shown promise in efficiently adsorbing arsenic. Commonly employed substrates achieving this objective encompass reactive carbons [45], graphene [46], cellulose [47], carbon nanotubes [48], macroporous silica [49], and nano-porous carbons [50], among others. Graphene emerges as a highly appealing contender because of its extensive range of potential uses across diverse disciplines. Graphene is a two-dimensional honeycomb lattice composed of closely arranged sp^2-bonded carbon atoms, exhibiting distinctive mechanical and electrical characteristics, chemical stability, a substantial surface area, and exceptional absorptive capabilities [51–53]. Graphene, functioning as an elastic material for electron reception and transfer, exhibits the capability to facilitate the transmission of photo-generated or light-induced electrons. Consequently, it not only enhances the adsorption of pollutants but also augments the transportation and distinction between charges. Furthermore, it possesses a notable capacity to separate various ions, solvents made from organic materials, and lubricants. Additionally, it exhibits a remarkable capability for repeated use and showcases a cost-efficient potential for expansion on a larger scale [54, 55].

The primary objective of the investigation was to ascertain the origins and connections of heavy metals, i.e., arsenic in soil, as well as to investigate their geographical variability via an examination of Geographical Information

Systems (GIS) approaches reported in the literature. Additionally, the adsorptive characteristics and durability of arsenic removal from water for consumption were assessed by BET and time-dependent studies. In this study, we have utilized the sol-gel route to synthesize magnetic $CuFe_2O_4$ nanoparticles/graphene nanocomposites, taking into consideration the favorable properties of both graphene and iron-oxides. The weight ratio of $CuFe_2O_4$ and graphene components was varied to investigate the efficacy of removing arsenic from the water solution. A range of characterizations, including XRD analysis, FE-SEM micrographs, FTIR analysis, BET study, and VSM analysis, have been used to evaluate the texture, structure, morphology, and magnetization of the nanocomposites. This study demonstrates the strong potential of $CuFe_2O_4$ and even graphene nanocomposites in effectively mitigating arsenic contamination in wastewater.

13.2 Experimentation

Overview of GIS Modeling in Pakistan

To comprehend the present situation of the groundwater across the evaluated area, real-time groundwater observation and testing give insight into the quality and amount of accessible water sources [56]. The process of groundwater observation and sampling is characterized by its complexity, limited applicability to large geographical areas, and high costs in comparison to groundwater modeling. Groundwater modeling offers a more simplified and cost-effective approach, enabling the assessment of larger geographical regions [57].

In the context of in-ground modeling, random sampling of groundwater is conducted by obtaining samples from multiple locations. These samples are then analyzed in a laboratory setting to determine their characteristics, including temperature, color, the degree of hardness pH, chloride, sulfate, arsenic, and alkalinity. The purpose of this analysis is to assess the spatial distribution of these elements and identify any potential pollution in the aquifer across the entire area. The findings from this study can provide valuable insights for effective groundwater management strategies [58, 59]. Numerous investigations about the quality of groundwater are being carried out in India, Bangladesh, China, Iran, and Pakistan, using diverse methodologies [60–63]. The methodologies involve regression modeling [62, 64] as well as modeling techniques that geospatial facts, especially GIS as well as remote sensing (RS) [61, 65–68].

Pakistan is now facing a critical situation characterized by a paucity of water resources [69]. The water purity level falls below the criteria established by the World Health Organization (WHO) [70]. The yearly average groundwater capacity of Pakistan is recorded as 67,841.4 m^3, with an abstraction rate of 51,189.42 m^3. The Punjab province exhibits the largest level of abstraction, accounting for 81% of the total. Within this region, around 80% of the water sources are characterized as fresh and clean, while the remaining 20% include salty water. Additionally, the province of Sindh has a groundwater abstraction rate of 12%, with 23% of the extracted

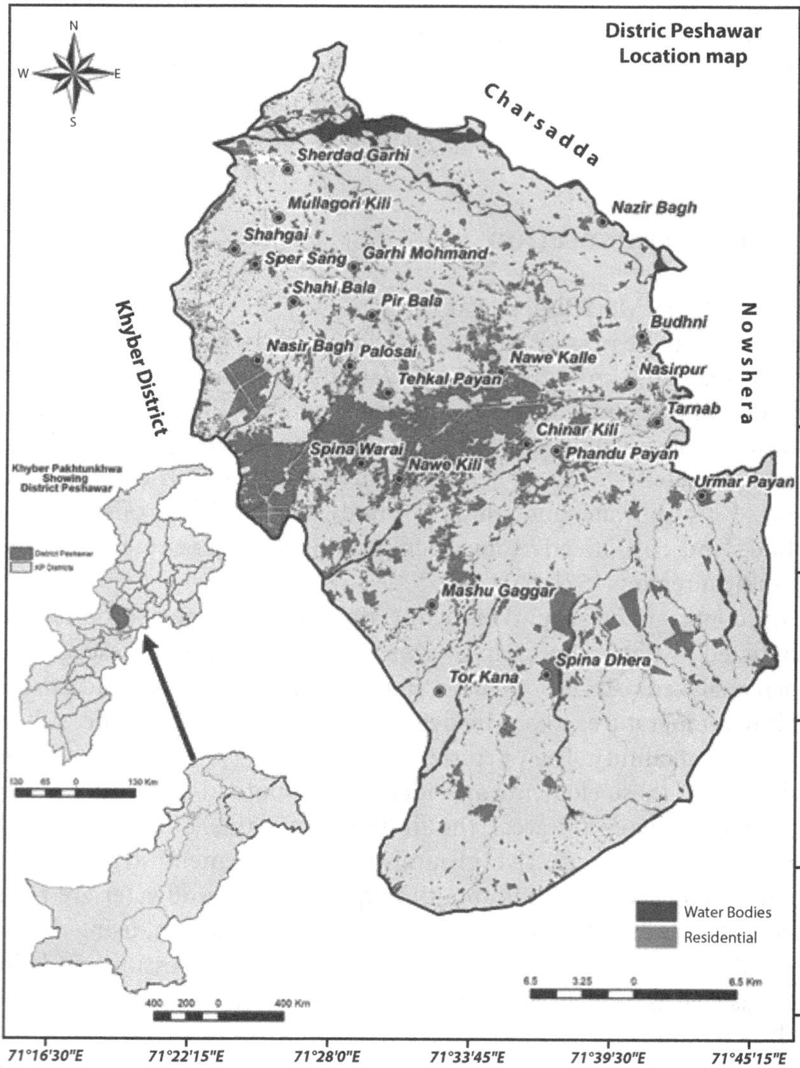

Figure 13.1 Sampling location for GIS modeling in Peshawar. Source: Ahmad *et al.* [72].

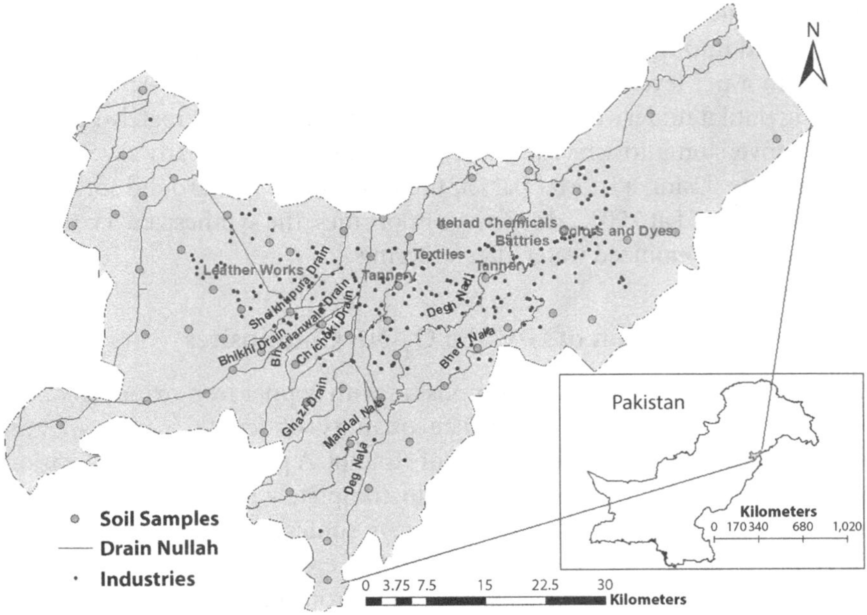

Figure 13.2 Study area and sampling location in Sheikhupura. Source: Shaheen *et al.* [73].

water being freshwater and the remaining 77% being salty. The province of Khyber Pakhtunkhwa has a 5% rate of groundwater abstraction. Pakistan, as a growing nation, must establish comprehensive strategies to address aspects such as the distribution of water, quality, consumption, and safeguarding against waterborne illnesses arising from pollution [71]. Therefore, a lot of areas have been explored by researchers including the Peshawar and Seikhupura districts of Pakistan to analyze the water quality by GIS modeling, as mentioned in Figures 13.1 and 13.2. We have come to know that arsenic is one of the major water pollutants that should be removed. Hence, in this study, we have discussed the arsenic removal from waste by graphene-anchored Cu nano ferrites.

13.2.1 Preparation of $CuFe_2O_4$

In the experiment, three chemical compounds were used: iron nitrate nonahydrate [$Fe(NO_3)_2 \cdot 9H_2O$, 99.9% purity], copper nitrate-3-hydrate [$Cu(NO_3)_2 \cdot 2H_2O$, 99.9% purity], and citric acid anhydrous ($C_6H_8O_7$, 99.9% purity). These compounds were mixed individually in 100 ml of deionized water, with molarities of 0.2, 0.1, and 0.45, respectively. The various precursors were homogeneously blended through a magnetic stirrer operating at 60°C with a stirring rate of 250 revolutions per minute. Ammonia was incrementally introduced into the solution under constant stirring conditions in order to maintain a pH level of 7. The solution underwent continual agitation and heating until it underwent a transition into a viscous gel, ultimately resulting in its conversion into ash. The entire procedure required nearly seven hours to complete. Using a mortal passel, the ash sample was ground into powder, and sintered at 750°C. To remove impurities, the synthesized ferrite was washed using deionized water after sintering.

13.2.2 Preparation of G/$CuFe_2O_4$-NP Composites

In order to produce graphene composites with copper iron oxide nanoparticles, a mixture of the appropriate quantity of ferrite ($CuFe_2O_4$) and graphene was combined with 100 ml of ethanol. A powerful ultrasonication procedure was used for half an hour to completely incorporate graphene with copper ferrite. The mixture underwent sonication before being heated to 100°C for a day in a microwave oven. $CuFe_2O_4$ nanoparticles anchored graphene composite materials were milled from the dried mixture upon heating. The schematic representation in Figure 13.3 illustrates the synthesis mechanism.

13.2.3 Characterization

The structural features have been examined by employing an X-ray diffractometer operating at a voltage of 40 kV and 40 mA, utilizing Cu-K_α

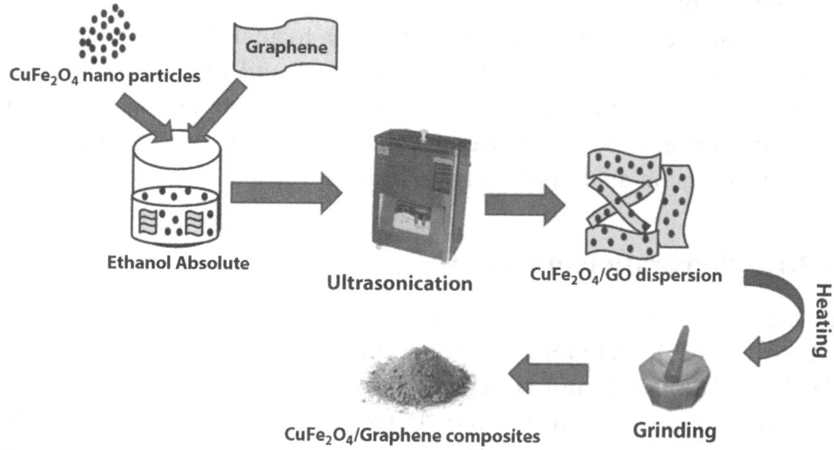

Figure 13.3 Synthesis procedure of G/$CuFe_2O_4$-NP composites.

radiations having a wavelength of 1.5406 Å. The Nova NanoSEM 450 field-emission scanning electron microscopy (FE-SEM) and Oxford energy dispersive x-ray (EDX) spectroscopy techniques were utilized to determine the surface morphology, microstructure, and composition of elements. FTIR spectra were obtained at ambient temperature utilizing a spectrometer manufactured by Shimadzu, a company based in Kyoto, Japan. The magnetic features, such as coercivity, saturation, and remanence, were determined through measurements conducted with the vibrating sample magnetometer (VSM) model PPMS-9, manufactured by Quantum Design. The estimation of the surface area of the as-synthesized specimens was conducted at a temperature of 273 K. This estimation was based on the nitrogen adsorption data obtained by the Brunauer-Emmett-Teller (BET) method. The Quantachrome NovaWin software, specifically version 11.03, was utilized for the analysis. Furthermore, the levels of arsenic were determined using an atomic fluorescence spectrometer known as the LUMINA 3400 AFS.

13.2.4 Arsenic Removal

The stock solutions of As(III) and As(V) were generated with $NaAsO_2$ as well as $Na_3AsO_4 \cdot 12H_2O$ in dilute water to examine the possible making use of the processed materials as adsorbents of arsenic (III and V). The investigation on adsorption and removal was carried out at 30°C, with a pH of 8.0. The initial concentration of arsenic (III and V) was chosen at 20 mg/L. The determination of the efficiency of removal as well as saturation time involved measuring the quantity of material adsorbed at 5, 20, 40, 60, 80, 120, 180, 320, and 380 min.

13.3 Results and Discussion

13.3.1 X-Ray Diffraction (XRD)

Figure 13.4 displays X-ray diffraction patterns of both pure copper ferrite produced in its original form and composites of $CuFe_2O_4$ and graphene. The relative quantity of $CuFe_2O_4$ and graphene was varied in weight ratios varying from 20% to 100%. The diffraction patterns obtained from the (220), (311), (222), (440), and (511) planes of $CuFe_2O_4$ spinel cubic crystal were measured at 2θ angles of 29.9°, 34.5°, 37.1°, 57.1°, and 62°, respectively. These outcomes align with the observations documented in the existing literature [74]. Furthermore, the XRD data exhibit a match with

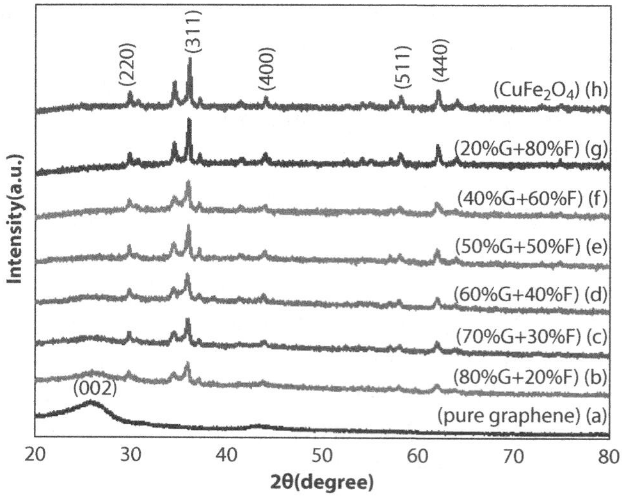

Figure 13.4 Pattern of XRD for a series of G/$CuFe_2O_4$-NP composites with varied concentrations from 20% to 100%.

$CuFe_2O_4$ nanoparticles JCPCDS 00-025-0283. All samples exhibit a prominent peak in the vicinity of the (311) plane, indicating the presence of the spinel phase in the specimens as they were manufactured. In the case of pristine graphene samples, an observable diffraction peak was detected at an angle of 25.9°, for which the (002) diffraction plane is responsible.

In the context of ferrite-Gr composites, it is seen that all X-ray diffraction (XRD) peaks correspond to the cubic crystal structure of $CuFe_2O_4$, with less notable graphene's diffraction peak at around 25.9°. The exfoliation of graphene is ultimately caused by the formation of CuFe2O4 crystals within the interlayers of graphene sheets [75]. The anchoring of CuFe2O4 nanoparticles onto the surface of the graphene sheet is postulated, hence providing an additional barrier against the reaggregation of exfoliated nanosheets of graphene. By utilizing XRD data, it is possible to estimate a range of structural parameters. The calculation of the cell volume and crystallite size was performed using the following mathematical relationship:

$$V_{cell} = a \times a \times a \tag{13.1}$$

$$D = \frac{k\lambda}{\beta cos\theta} \tag{13.2}$$

In this context, the variable D represents the crystallite size, β denotes the full width at half maximum, θ represents the diffraction angle that is measured in radians, λ symbolizes the wavelength of CuKα radiation, whereas K represents the shape factor with a value of around 0.9 for the cubic structure. Table 13.1 lists calculated structural parameters.

The study reveals that the sizes of the crystallites varied between 34.57 and 53.01 nm. Notably, the smallest size was observed in the $CuFe_2O_4$-NP (20 wt.%)/G (80 wt.%) sample, while the largest size was discovered in the pure $CuFe_2O_4$ sample, as indicated in Table 13.1. These findings highlight the nanostructural characteristics of the compounds that were created. Moreover, the computed lattice parameters also demonstrate a substantial alteration in the lattice constant associated with an increased $CuFe_2O_4$ percentage in the composite of G/$CuFe_2O_4$-NPs. Furthermore, the increase in $CuFe_2O_4$ content is associated with changes in crystallite size and volume of unit cells, as indicated in Table 13.1. The surge in $CuFe_2O_4$ content at graphene locations has resulted in an upward trend in cell volume, from 553.38 to 607.64 Å^3.

Table 13.1 Cell volume, lattice constant, and crystallite size of $CuFe_2O_4$ and G/$CuFe_2O_4$-NPs composites.

Samples	**Lattice parameter a (Å)**	**Volume V (Å^3)**	**Crystallite size D (nm)**
$CuFe_2O_4$-NPs (20 wt.%)/G (80 wt.%)	8.21	553.38	34.57
$CuFe_2O_4$-NPs (30 wt.%)/G (70 wt.%)	8.38	588.48	39.25
$CuFe_2O_4$-NPs (40 wt.%)/G (60 wt.%)	8.42	596.94	42.61
$CuFe_2O_4$-NPs (50 wt.%)/G (50 wt.%)	8.43	599.07	45.82
$CuFe_2O_4$-NPs (60 wt.%)/G (40 wt.%)	8.45	603.35	46.60
$CuFe_2O_4$-NPs (80 wt.%)/G (20 wt.%)	8.46	605.49	48.81
Pure Ferrite	8.47	607.64	53.01

13.3.2 Fourier Transform Infrared Spectroscopy (FTIR)

Figure 13.5 displays the Fourier transform infrared spectra of both pure $CuFe_2O_4$ and G/$CuFe_2O_4$-NP composites. The most prominent peak is observed at roughly 589 cm^{-1}, resulting in a link to the Fe-O bond. On the other hand, there are notable peaks observed at around 1157.25 and 1350 cm^{-1}, representing the bending of C-O single bonds in alkoxy groups and the bending of O=C-O bonds in carboxyl groups, respectively [76]. The presence of C=C aromatic vibrational modes and C=O carboxyl bending can be observed at peaks approximately at 1579 and 1649 cm^{-1}, respectively [77, 78]. The observed bands in the FTIR spectrum observations provide support for the existence of metals such as iron, copper, and graphene within both the pure $CuFe_2O_4$ and G/$CuFe_2O_4$-NP composites.

13.3.3 Surface Morphological Analysis

Figure 13.6 displays the SEM micrographs of several mixtures of G/$CuFe_2O_4$-NPs and pure $CuFe_2O_4$-NPs. The scanning electron microscopy (SEM) image depicted in Figure 13.4(a) showcases a composite material consisting of 80 wt.% graphene and 20 wt.% copper ferrite nanoparticles ($CuFe_2O_4$-NPs). The image reveals the presence of predominantly thin graphitic sheets-like structures, with a minor agglomeration of $CuFe_2O_4$-NPs. SEM images [Figures 13.6(a–d)] reveal the presence of a significant quantity of graphene flakes exhibiting a crumpled and slightly wrinkled surface morphology, with diameters in the range of tens of microns. Nevertheless, as the concentration of ferrite increases, the surface morphology undergoes

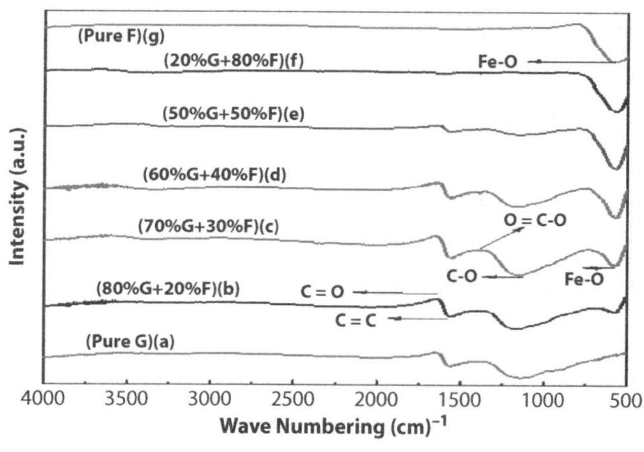

Figure 13.5 FTIR spectra obtained for pure and composite samples.

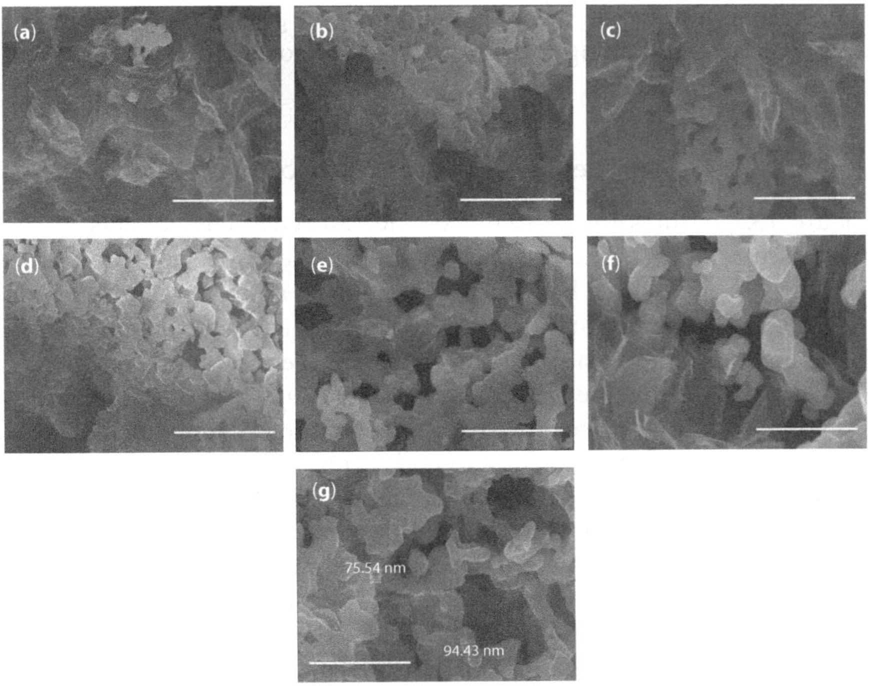

Figure 13.6 SEM images of graphene to $CuFe_2O_4$: (a) 80% graphene 20% $CuFe_2O_4$, (b) 70% graphene 30% $CuFe_2O_4$, (c) 60% graphene 40% $CuFe_2O_4$, (d) 50% graphene 50% $CuFe_2O_4$, (e) 40% graphene 60% $CuFe_2O_4$, (f) 20% graphene 80% $CuFe_2O_4$, and (g) pure ferrite (scale bar – 500 nm).

a gradual transformation, resulting in the emergence of a more uniform agglomerated globular morphological structure [as depicted in Figures 13.6(e–g)] [79]. The determination of the size of $CuFe_2O_4$-NPs based on the SEM micrographs is impeded by the presence of bulk agglomerates. Nevertheless, it is hypothesized that the size falls within the nanometer scale, specifically ranging from 100 to 300 nm. SEM image of pure graphene revealed a graphitic structure that is very porous and composed of many layers. This structure consists of hundreds of individual graphene nanosheets [80].

Furthermore, the graphene sheets exhibited a surface shape that was smooth and planar. Additionally, the SEM images demonstrate that the graphene sheets exhibited a stacked configuration as a result of the low strength of van der Waals interaction [81]. This stacking arrangement renders the graphene sheets vulnerable to the introduction of external doping components. The utilization of $CuFe_2O_4$ nanoparticles facilitates the effective elimination of substantial interlayer gaps existing among the graphene sheets, resulting in the creation of a nanocomposite material that is uniformly spread.

13.3.4 Energy Dispersive X-Ray Spectroscopy (EDX)

Figure 13.7 ensures the existence of the respective components in the processed samples, exhibiting varying ratios. The presence of iron, copper, carbon, and oxygen has been observed in all samples of $CuFe_2O_4$ and G/$CuFe_2O_4$-NP composite [82]. The spectroscopic results obtained from EDX analysis demonstrate that the copper, oxygen, carbon, and iron are evenly present within the nanoparticles of $CuFe_2O_4$. This uniform distribution significantly facilitates the development of spinel nanoparticles, hence preventing the occurrence of any secondary phases [83].

The spectra of $CuFe_2O_4$-NPs and its nanocomposites did not exhibit any indications of additional impurities, indicating a lack of contamination and confirming the complete sintering of the produced nanocomposites. Additional EDX analyses indicate that the nanoparticles exhibit full homogeneity in terms of elemental distribution. Furthermore, the composites of G/$CuFe_2O_4$-NPs also exhibited a notable level of composition uniformity.

13.3.5 Magnetic Properties

Figure 13.8 depicts the hysteresis patterns of each sample that were created. The presented graphs illustrate the magnetization hysteresis loops of three distinct materials: pure $CuFe_2O_4$ ferrites, pure graphene, and different ratios of graphene to copper ferrites. These graphs demonstrate a characteristic s-shaped trend, which signifies an excellent magnetic response to alterations in the applied magnetic field [84]. This behavior is attributed to the influence of an external magnetic field, namely, the 20S field. This implies that the G/$CuFe_2O_4$-NPs can be conveniently removed from the liquid procedure by utilizing an external magnet, which is crucial in practical applications. The saturation magnetization of the pure $CuFe_2O_4$-NP material was 16 emu/g. The data presented in Table 13.2 demonstrate a negative correlation between the loading of composite graphene and its corresponding trend. The observed behavior can be attributed to the diamagnetic properties exhibited by graphene. Furthermore, there is a potential correlation between the absence of collinearity in the magnetic moment on the surface of nanoparticles and the decrease in saturation magnetization as the graphene content increases [85]. Table 13.2 presents the magnetic aspects of all the samples.

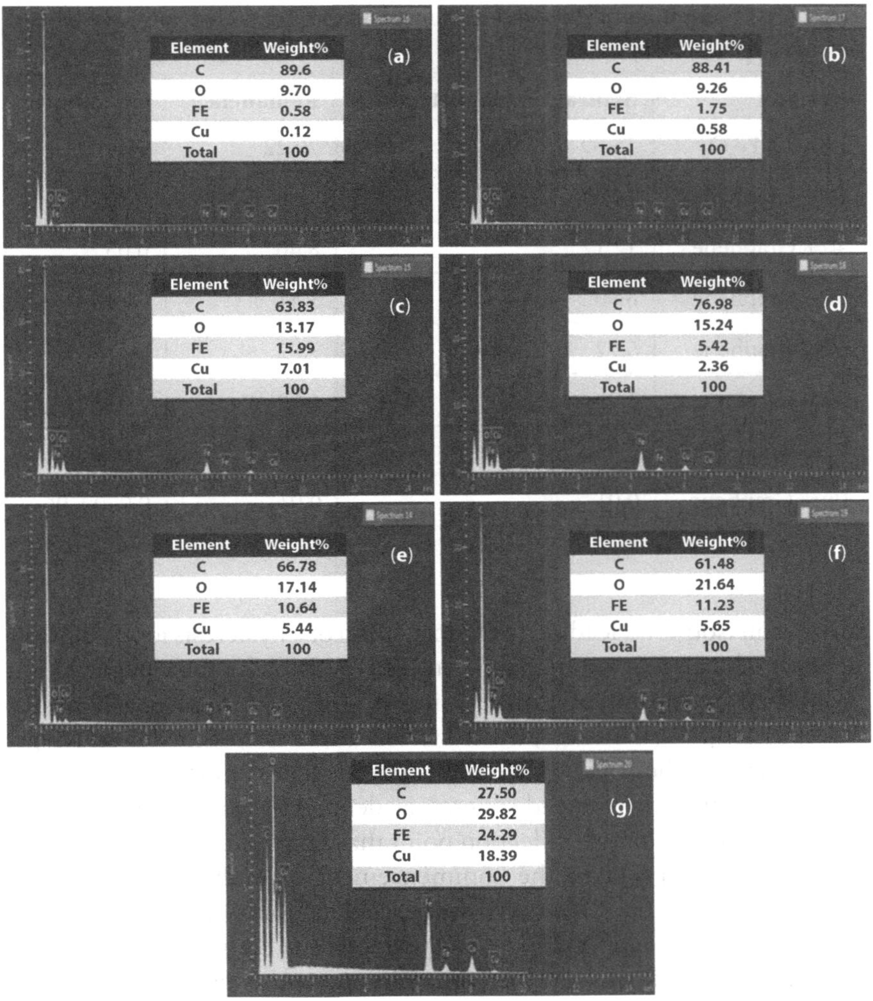

Figure 13.7 EDX graphs of graphene/$CuFe_2O_4$: (a) 80% graphene 20% $CuFe_2O_4$, (b) 70% graphene 30% $CuFe_2O_4$, (c) 60% graphene 40% $CuFe_2O_4$, (d) 50% graphene 50% $CuFe_2O_4$, (e) 40% graphene 60% $CuFe_2O_4$, (f) 20% graphene 80% $CuFe_2O_4$, and (g) Pure $CuFe_2O_4$.

13.3.6 Brunauer-Emmett-Teller Method for Textural Features

Nitrogen (N_2) adsorption or desorption isotherms have been performed at a temperature of 150°C to evaluate the behavior of $CuFe_2O_4$ nanoparticles and the produced nanocomposites. The results are illustrated in Figure 13.9. The graph illustrates that the process of desorption deviates from the

Table 13.2 Saturation magnetization, remanence, and coercivity of $CuFe_2O_4$ and G/$CuFe_2O_4$-NPs composites.

Samples	Saturation magnetization	Remanence	Coercivity
Pure $CuFe_2O_4$	16.69	29.22	0.10
20% Graphene	14.42	9.56	0.26
40% Graphene	12.12	8.40	0.05
50% Graphene	6.40	3.41	0.14
60% Graphene	5.72	3.02	0.015
70% Graphene	1.29	0.70	0.012
80% Graphene	2.33	1.22	0.008
Pure Graphene	0.01	0.01	0.0000196

adsorption isotherm, leading to the formation of a hysteresis loop. The designation of such adsorption isotherms as type IV is by the categorization established by the International Union of Pure and Applied Chemistry (IUPAC) [86]. Furthermore, the hysteresis curve is classified as H1 type, suggesting that the studied material is porous and composed of agglomerates with uniformly shaped spheres, as defined by the IUPAC [86]. The isotherm plot exhibits an inflection point that signifies the transition from monolayer adsorption to the commencement of multilayer adsorption [87]. All samples that have been synthesized exhibit a significant specific surface area. The compound $CuFe_2O_4$ exhibits the highest crystallite size, measuring approximately 53 nm while simultaneously possessing the lowest specific surface area, which amounts to 10.88 m^2/g. The observed BET surface area of $CuFe_2O_4$-NPs aligns with the literature for various metal oxides exhibiting comparable surface shapes [88, 89]. The BET-specific surface area of the 80% graphene composite is reported as 288.5 m^2/g in Table 13.3. This value is found to be 28 times more than that of pristine $CuFe_2O_4$-NPs. The primary factor responsible for the significant particular surface area of the $CuFe_2O_4$-graphene composite is the presence of graphene. Graphene facilitates the creation of large solid-electrolyte interphase (SEI) structures, which in turn enables the composite to achieve enormous capacities [90]. The produced nanocomposites possess a significant specific surface area, which enables enhanced adsorption capabilities. The range of pore sizes was determined by applying the desorption

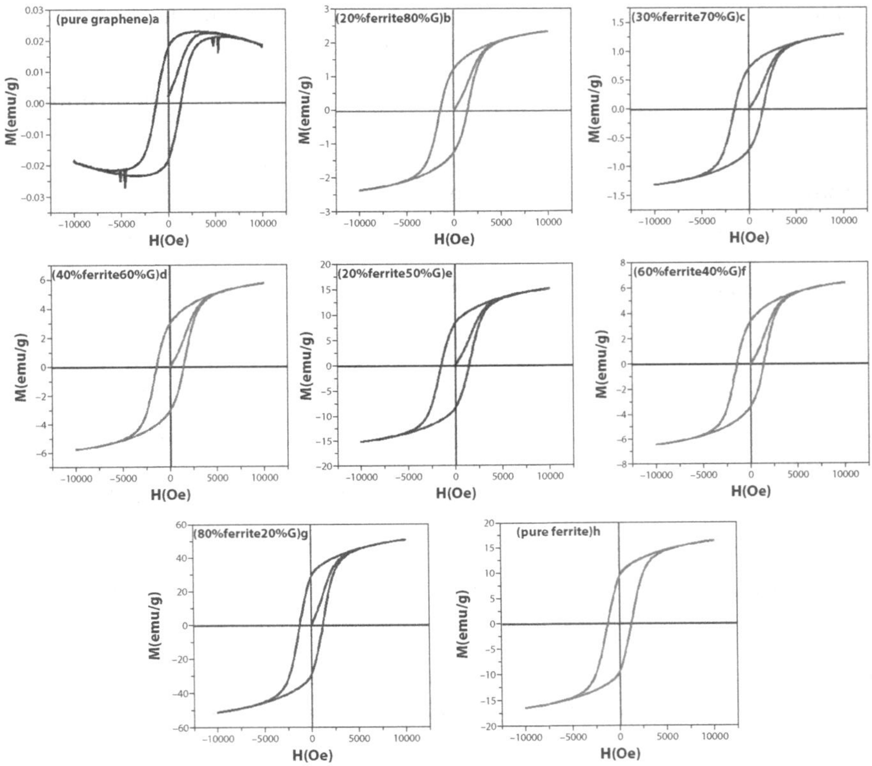

Figure 13.8 VSM graphs of graphene to $CuFe_2O_4$: (a) pure graphene, (b) 80% graphene 20% $CuFe_2O_4$, (c) 70% graphene 30% $CuFe_2O_4$, (d) 60% graphene 40% $CuFe_2O_4$, (e) 50% $CuFe_2O_4$ 50% graphene, (f) 40% graphene 60% $CuFe_2O_4$, (g) 20% graphene 80% $CuFe_2O_4$, and (h) pure $CuFe_2O_4$.

section of the isotherm, employing the BJH method, in conjunction with the measurement of BET surface area. The outcomes obtained in this study reveal the presence of a significant number of mesopores between 2 and 10 nm in size, and the highest possible pore size of 3.1 nm, as presented in Table 13.3. The adsorption volume of the pores for both pure ferrite and pure graphene is 0.01 and 0.052 cm^3/g, respectively. Furthermore, this volume increases as the graphene percentage rises, ultimately achieving a peak value of 0.088 cm^3/g. The porosity and specific surface area can be enhanced by immobilizing $CuFe_2O_4$ nanoparticles onto the graphene nanoflakes' surface [88]. This characteristic exhibits constructive attributes in relation to adsorption and catalytic performance [91]. Table 13.3 provides additional parameters for the as-synthesized samples, including BET

Table 13.3 Surface area, pore size, adsorption pore volume, and desorption pore volume of $CuFe_2O_4$ and G/$CuFe_2O_4$-NPs composites.

Samples	Surface area (m^2/g)	Pore size (nm)	Adsorption pore volume (cm^3/g)	Desorption pore volume (cm^3/g)
Pure $CuFe_2O_4$	10.88	3.0	0.10	0.003
20% Graphene	68.42	2.8	0.020	0.008
40% Graphene	133.8	3.1	0.042	0.019
50% Graphene	186.4	2.6	0.064	0.029
60% Graphene	230.7	2.7	0.080	0.038
70% Graphene	249.5	2.6	0.082	0.030
80% Graphene	288.5	2.9	0.088	0.044
Pure Graphene	174.3	2.7	0.052	0.024

pore volume, pore diameters, and surface area. Table 13.3 presents the various texture features of $CuFe_2O_4$ and G/$CuFe_2O_4$-NP composites.

13.3.7 Arsenic Removal

The composite materials that have been produced demonstrate a high potential for effectively removing arsenic and could be employed in the treatment of arsenic-contaminated water for consumption. In order to determine the saturation level, duration, and effectiveness of removing arsenic for the optimized composite sample [G (80 wt.%)/$CuFe_2O_4$-NPs (20 wt.%)], adsorption studies were conducted with initial concentrations of As(III) as well as As(V) at 20 mg/L at a temperature of 30°C and a pH of 8.0. Figure 13.10 illustrates the efficiency of removing As(III) and As(V). The observed values exhibit a significant rise, reaching 93.9% and 95.2% after 180 min. Subsequently, a gradual decrease in the rate of As removal was noted, and eventually reached saturation owing to adsorption after 380 min, achieving the highest removal percentage for As(III) of about 95.5% and 96.7% for As(V), in accordance with prior studies [92, 93]. The high efficacy in removing arsenic species can be attributed to the physical adsorption facilitated by the extensive surface area and porosity of the G (80 wt.%)/$CuFe_2O_4$-NP composite with 20 wt.%, as well as interactions between ions of arsenic individuals and $CuFe_2O_4$.

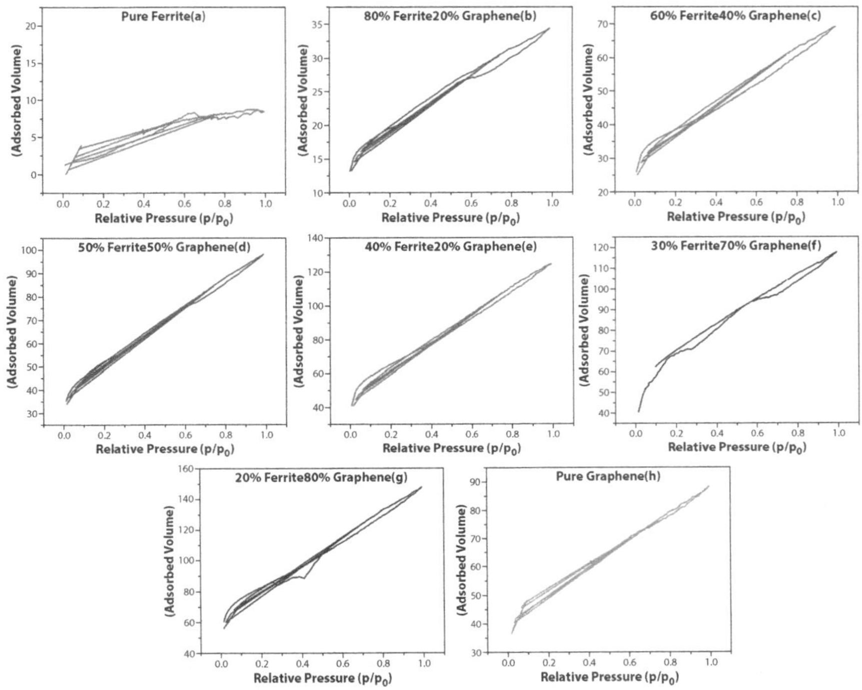

Figure 13.9 BET graphs of $CuFe_2O_4$ to graphene: (a) pure ferrite, (b) 80% $CuFe_2O_4$ 20% graphene, (c) 60% $CuFe_2O_4$ 40% graphene, (d) 50% $CuFe_2O_4$ 50% graphene, (e) 40% $CuFe_2O_4$ 60% graphene, (f) 30% $CuFe_2O_4$ 70% graphene, (g) 20% $CuFe_2O_4$ 80% graphene, and (h) pure graphene.

13.4 Conclusion

This research highlights the hazards of arsenic-containing wastewater in industrial regions of different areas across Pakistan as well as arsenic removal from wastewater The comprehensive study of as-synthesized composites was conducted using several characterization methods. The synthesis of composites consisting of G/$CuFe_2O_4$-NPs and pure $CuFe_2O_4$-NPs was successfully achieved using the sol-gel method. The XRD analysis confirmed the spinel structure of $CuFe_2O_4$-NPs. The obtained results indicated a crystallite size of 34 nm for the G (80 wt.%)/$CuFe_2O_4$-NP composites with 20 wt.%. The determined crystallite size of the $CuFe_2O_4$-NP sample was 53 nm. The observed phenomenon exhibited a positive correlation with the augmentation of ferrite concentration in the composites of G/$CuFe_2O_4$-NPs. Additionally, the results obtained from the VSM analysis demonstrate encouraging magnetic characteristics for the synthesized

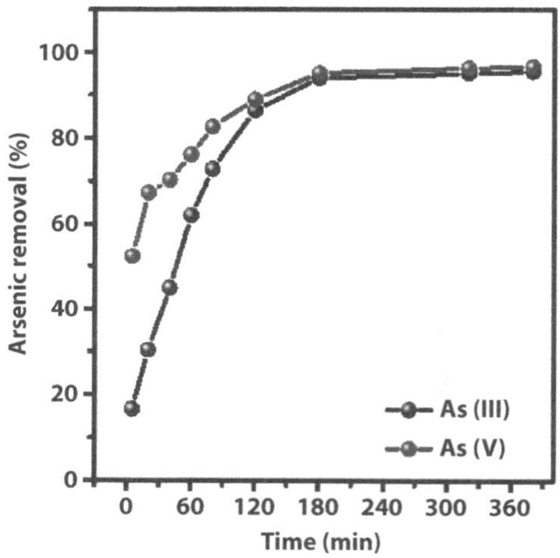

Figure 13.10 The influence of the duration of interaction on the behavior of As(III) and As(V) (80 wt.%)/$CuFe_2O_4$-NP (20 wt.%) composite adsorption with 20 mg L^{-1}As(III)/As(V), temperature 30°C, and pH 8.

composites. Notably, the magnetization saturation value of approximately 16 emu/g was observed for the pristine $CuFe_2O_4$-NPs. The observed trend indicates a steady drop in this value as the graphene content in the G/$CuFe_2O_4$-NP composites increases. This phenomenon can be related to the diamagnetic properties shown by graphene.

In addition, the BET surface area exhibits a positive correlation with the graphene concentration in the as-prepared composites. Furthermore, the BJH approach demonstrates that the pore size falls within the range of 2.6-3.1 nm, hence indicating the presence of mesopores in the composites that were created. Furthermore, it is noteworthy to observe that the pore volume for adsorption surpasses the desorption pore volume in all samples, indicating that the materials produced in this study have favorable absorption properties. The experimental results demonstrated that the composite material consisting of 80 wt.% G and 20 wt.% $CuFe_2O_4$-NPs exhibited high effectiveness in removing As(III) as well as As(V), making them suitable for cleaning wastewater applications. The point of adsorption saturation occurs after 380 min, exhibiting a peak removal efficiency of 95.5% for As(III) and for As(V) was found to be 96.7%. Hence, these composites demonstrate significant potential in the context of removing arsenic from the water we consume, making them interesting candidates for environmental remediation purposes.

References

1. Suter II, G. W. (2016). *Ecological risk assessment*. CRC press.
2. Molden, D. (2020). Scarcity of water or scarcity of management?. *Int. J. Water Resour. Dev.*, *36*(2-3), 258-268.
3. Gude, V. G. (2018). Desalination of deep groundwater aquifers for freshwater supplies–Challenges and strategies. *Groundw. Sustain. Dev.*, *6*, 87-92.
4. Shao, M., Tang, X., Zhang, Y., & Li, W. (2006). City clusters in China: air and surface water pollution. *Front. Ecol. Environ.*, 353-361.
5. Wu, J., Zhang, Y., & Zhou, H. (2020). Groundwater chemistry and groundwater quality index incorporating health risk weighting in Dingbian County, Ordos basin of northwest China. *Geochemistry, 80*(4), 125607.
6. Wang, H., Gu, H., Lan, S., Wang, M., & Chi, B. (2018). Human health risk assessment and sources analysis of nitrate in shallow groundwater of the Liujiang basin, China. *Hum. Ecol. Risk Assess.: Int. J.*, *24*(6), 1515-1531.
7. Şener, Ş. (2023). Groundwater quality, heavy metal pollution, and health risk assessment using geospatial techniques and index methods in Eber wetland and surroundings (Afyonkarahisar/Turkey). *Environ. Sci. Pollut. Res.*, *30*(18), 51387-51411.
8. Bolisetty, S., Peydayesh, M., & Mezzenga, R. (2019). Sustainable technologies for water purification from heavy metals: review and analysis. *Chem. Soc. Rev.*, *48*(2), 463-487.
9. Smith, A. H., Lingas, E. O., & Rahman, M. (2000). Contamination of drinking-water by arsenic in Bangladesh: a public health emergency. *Bull. World Health Organ.*, *78*, 1093-1103.
10. Naujokas, M. F., Anderson, B., Ahsan, H., Aposhian, H. V., Graziano, J. H., Thompson, C., & Suk, W. A. (2013). The broad scope of health effects from chronic arsenic exposure: update on a worldwide public health problem. *Environ. Health Perspect.*, *121*(3), 295-302.
11. Taheri, M., Gharaie, M. H. M., Mehrzad, J., Afshari, R., & Datta, S. (2017). Hydrogeochemical and isotopic evaluation of arsenic contaminated waters in an argillic alteration zone. *J. Geochem. Explor.*, *175*, 1-10.
12. Ali, W., Mao, K., Zhang, H., Junaid, M., Xu, N., Rasool, A., ... & Yang, Z. (2020). Comprehensive review of the basic chemical behaviours, sources, processes, and endpoints of trace element contamination in paddy soil-rice systems in rice-growing countries. *J. Hazard. Mater.*, *397*, 122720.
13. Polak-Juszczak, L., & Richert, J. S. (2021). Arsenic speciation in fish from Baltic Sea close to chemical munitions dumpsites. *Chemosphere*, *284*, 131326.
14. Santoro, S., Timpano, P., Avci, A. H., Argurio, P., Chidichimo, F., De Biase, M., & Curcio, E. (2021). An integrated membrane distillation, photocatalysis and polyelectrolyte-enhanced ultrafiltration process for arsenic remediation at point-of-use. *Desalination*, *520*, 115378.
15. Mochizuki, H. (2019). Arsenic neurotoxicity in humans. *Int. J. Mol. Sci.*, *20*(14), 3418.

16. Wu, Lian-Kui, *et al.* "Highly porous copper ferrite foam: a promising adsorbent for efficient removal of As (III) and As (V) from water." *J. Hazard. Mater.* 347 (2018): 15-245.
17. Wu, Lian-Kui, *et al.* "Graphene oxide/CuFe2O4 foam as an efficient absorbent for arsenic removal from water." *Chem. Eng. J.* 334 (2018): 1808-18195.
18. Mohammad, A. W., Teow, Y. H., Ang, W. L., Chung, Y. T., Oatley-Radcliffe, D. L., & Hilal, N. (2015). Nanofiltration membranes review: Recent advances and future prospects. *Desalination, 356*, 226-254.
19. Visvanathan, C., Aim, R. B., & Parameshwaran, K. (2000). Membrane separation bioreactors for wastewater treatment. *Crit. Rev. Environ. Sci. Technol., 30*(1), 1-48.
20. Rengaraj, S., Yeon, K. H., & Moon, S. H. (2001). Removal of chromium from water and wastewater by ion exchange resins. *J. Hazard. Mater., 87*(1-3), 273-287.
21. Cseri, L., Topuz, F., Abdulhamid, M. A., Alammar, A., Budd, P. M., & Szekely, G. (2021). Electrospun adsorptive nanofibrous membranes from ion exchange polymers to snare textile dyes from wastewater. *Adv. Mater. Technol.*, 2000955.
22. Vourch, M., Balannec, B., Chaufer, B., & Dorange, G. (2008). Treatment of dairy industry wastewater by reverse osmosis for water reuse. *Desalination, 219*(1-3), 190-202.
23. Chen, G. (2004). Electrochemical technologies in wastewater treatment. *Sep. Purif. Technol., 38*(1), 11-41.
24. Cooney, D. O. (1998). *Adsorption design for wastewater treatment*. CRC press.
25. Chai, W. S., Cheun, J. Y., Kumar, P. S., Mubashir, M., Majeed, Z., Banat, F., & Show, P. L. (2021). A review on conventional and novel materials towards heavy metal adsorption in wastewater treatment application. *J. Clean. Prod.*, 126589.
26. Qasem, N. A., Mohammed, R. H., & Lawal, D. U. (2021). Removal of heavy metal ions from wastewater: A comprehensive and critical review. *npj Clean Water, 4*(1), 1-15.
27. Rashid, R., Shafiq, I., Akhter, P., Iqbal, M. J., & Hussain, M. (2021). A state-of-the-art review on wastewater treatment techniques: the effectiveness of adsorption method. *Environ. Sci. Pollut. Res.*, 1-17.
28. Nazir, G., Rehman, A., Hussain, S., Afzal, A. M., Dastgeer, G., Rehman, M. A., & AlObaid, A. A. (2021). Heteroatoms-doped hierarchical porous carbons: Multifunctional materials for effective methylene blue removal and cryogenic hydrogen storage. *Colloids Surf., A*, 127554.
29. Nazir, G., Rehman, A., & Park, S. J. (2021). Valorization of shrimp shell biowaste for environmental remediation: Efficient contender for CO2 adsorption and separation. *J. Environ. Manage., 299*, 113661.
30. Rehman, A., & Park, S. J. (2019). Environmental remediation by microporous carbon: An efficient contender for CO2 and methylene blue adsorption. *J. CO2 Util., 34*, 656-667.

31. Reddy, D. H. K., & Lee, S. M. (2013). Application of magnetic chitosan composites for the removal of toxic metal and dyes from aqueous solutions. *Adv. Colloid Interface Sci.*, *201*, 68-93.
32. Tang, W. W., Zeng, G. M., Gong, J. L., Liang, J., Xu, P., Zhang, C., & Huang, B. B. (2014). Impact of humic/fulvic acid on the removal of heavy metals from aqueous solutions using nanomaterials: a review. *Sci. Total Environ.*, *468*, 1014-1027.
33. Ahmed, M. J. K., & Ahmaruzzaman, M. (2016). A review on potential usage of industrial waste materials for binding heavy metal ions from aqueous solutions. *J. Water Process Eng.*, *10*, 39-47.
34. Zhang, L. C., Jia, Z., Lyu, F., Liang, S. X., & Lu, J. (2019). A review of catalytic performance of metallic glasses in wastewater treatment: Recent progress and prospects. *Prog. Mater Sci.*, *105*, 100576.
35. Yang, Y., Zhang, C., & Hu, Z. (2013). Impact of metallic and metal oxide nanoparticles on wastewater treatment and anaerobic digestion. *Environ. Sci. Processes Impacts*, *15*(1), 39-48.
36. Upadhyay, R. K., Soin, N., & Roy, S. S. (2014). Role of graphene/metal oxide composites as photocatalysts, adsorbents and disinfectants in water treatment: a review. *RSC Adv.*, *4*(8), 3823-3851.
37. Xu, P., Zeng, G. M., Huang, D. L., Feng, C. L., Hu, S., Zhao, M. H., & Liu, Z. F. (2012). Use of iron oxide nanomaterials in wastewater treatment: a review. *Sci. Total Environ.*, *424*, 1-10.
38. Luo, H., Zeng, Y., He, D., & Pan, X. (2021). Application of iron-based materials in heterogeneous advanced oxidation processes for wastewater treatment: A review. *Biochem. Eng. J.*, *407*, 127191.
39. Feng, L., Cao, M., Ma, X., Zhu, Y., & Hu, C. (2012). Superparamagnetic high-surface-area Fe3O4 nanoparticles as adsorbents for arsenic removal. *J. Hazard. Mater.*, *217*, 439-446.
40. Ren, Z., Zhang, G., & Chen, J. P. (2011). Adsorptive removal of arsenic from water by an iron–zirconium binary oxide adsorbent. *J. Colloid Interface Sci.*, *358*(1), 230-237.
41. Wu, L. K., Wu, H., Liu, Z. Z., Cao, H. Z., Hou, G. Y., Tang, Y. P., & Zheng, G. Q. (2018). Highly porous copper ferrite foam: a promising adsorbent for efficient removal of As (III) and As (V) from water. *J. Hazard. Mater.*, *347*, 15-24.
42. Zhang, Y. U., Yang, M. I. N., Dou, X. M., He, H., & Wang, D. S. (2005). Arsenate adsorption on an Fe– Ce bimetal oxide adsorbent: role of surface properties. *Environ. Sci. Technol.*, *39*(18), 7246-7253.
43. Martinson, C. A., & Reddy, K. J. (2009). Adsorption of arsenic (III) and arsenic (V) by cupric oxide nanoparticles. *J. Colloid Interface Sci.*, *336*(2), 406-411.
44. Zhang, G., Ren, Z., Zhang, X., & Chen, J. (2013). Nanostructured iron (III)-copper (II) binary oxide: a novel adsorbent for enhanced arsenic removal from aqueous solutions. *Water Res.*, *47*(12), 4022-4031.

45. Kalaruban, M., Loganathan, P., Nguyen, T. V., Nur, T., Johir, M. A. H., Nguyen, T. H., & Vigneswaran, S. (2019). Iron-impregnated granular activated carbon for arsenic removal: application to practical column filters. *J. Environ. Manage.*, *239*, 235-243.
46. Su, H., Ye, Z., & Hmidi, N. (2017). High-performance iron oxide–graphene oxide nanocomposite adsorbents for arsenic removal. *Colloids Surf., A*, *522*, 161-172.
47. Oyewo, O. A., Elemike, E. E., Onwudiwe, D. C., & Onyango, M. S. (2020). Metal oxide-cellulose nanocomposites for the removal of toxic metals and dyes from wastewater. *Int. J. Biol. Macromol.*
48. Chen, B., Zhu, Z., Ma, J., Qiu, Y., & Chen, J. (2015). Iron oxide supported sulfhydryl-functionalized multiwalled carbon nanotubes for removal of arsenite from aqueous solution. *ChemPlusChem*, *80*(4), 740.
49. Ullah, R., Deb, B. K., & Mollah, M. Y. A. (2014). Synthesis and characterization of silica coated iron-oxide composites of different ratios. *Int. J. Compos. Mater.*, *4*(2), 135-145.
50. Joshi, S., Sharma, M., Kumari, A., Shrestha, S., & Shrestha, B. (2019). Arsenic removal from water by adsorption onto iron oxide/nano-porous carbon magnetic composite. *Appl. Sci.*, *9*(18), 3732.
51. Idumah, C. I., & Hassan, A. (2016). Emerging trends in graphene carbon based polymer nanocomposites and applications. *Rev. Chem. Eng.*, *32*(2), 223-264.
52. Shen, J., Zhu, Y., Jiang, H., & Li, C. (2016). 2D nanosheets-based novel architectures: Synthesis, assembly and applications. *Nano Today*, *11*(4), 483-520.
53. Naresh, U., Kumar, N. S., Basha, D. B., Benerjee, P., Naidu, K. C. B., Kumar, R. J., & Boddula, R. (2020). Synthesis and Properties of Graphene-Based Materials. *Monoelements: Properties and Applications*, 57-72.
54. Kyzas, G. Z., Deliyanni, E. A., & Matis, K. A. (2014). Graphene oxide and its application as an adsorbent for wastewater treatment. *J. Chem. Technol. Biotechnol.*, *89*(2), 196-205.
55. Tabish, T. A., Memon, F. A., Gomez, D. E., Horsell, D. W., & Zhang, S. (2018). A facile synthesis of porous graphene for efficient water and wastewater treatment. *Sci. Rep.*, *8*(1), 1-14.
56. Zhou, Q. *et al.* Real-time management of groundwater resources based on wireless sensors networks. *J. Sens. Actuator Netw.* 7, 4 (2018).
57. Kumar, P. *et al.* Effect of land use changes on water quality in an ephemeral coastal plain: Khambhat City, Gujarat, India. *Water* 11, 724 (2019).
58. Preti, F., Errico, A., Caruso, M., Dani, A. & Guastini, E. Dry-stone wall terrace monitoring and modelling. *Land Degrad. Dev.* 29, 1806–1818 (2018).
59. Zhang, Q., Xu, P. & Qian, H. Assessment of groundwater quality and human health risk (HHR) evaluation of nitrate in the centralwestern Guanzhong basin, China. *Int. J. Environ. Res. Public Health* 16, 4246 (2019).
60. Adnan, S. & Iqbal, J. Spatial analysis of the groundwater quality in the Peshawar District, Pakistan. *Procedia Eng.* 70, 14–22 (2014).

61. Adnan, S., Iqbal, J., Maltamo, M. & Valbuena, R. GIS-based DRASTIC model for groundwater vulnerability and pollution risk assessment in the Peshawar District, Pakistan. *Arab. J. Geosci.* 11, 458 (2018).
62. Adnan, S. *et al.* A simple approach of groundwater quality analysis, classification, and mapping in Peshawar, Pakistan. *Environments* 6, 123 (2019).
63. Zhang, Q., Xu, P. & Qian, H. Groundwater quality assessment using improved water quality index (WQI) and human health risk (HHR) evaluation in a semi-arid region of northwest China. *Expo. Health* 12, 487–500 (2020).
64. Zhang, Q., Xu, P., Qian, H. & Yang, F. Hydrogeochemistry and fluoride contamination in Jiaokou Irrigation District, Central China: Assessment based on multivariate statistical approach and human health risk. *Sci. Total Environ.* 741, 140460 (2020).
65. Melly, B. L., Gama, P. T. & Schael, D. M. Spatial patterns in small wetland systems: Identifying and prioritising wetlands most at risk from environmental and anthropogenic impacts. *Wetlands Ecol. Manage.* 26, 1001–1013 (2018).
66. Majumdar, S. & Pan, N. D. Combining opensource GIS and meta-analysis to link rainfall trend and human activity: case study on Gumti and Khowai drainage systems, Tripura, India. *Spat. Inf. Res.* 28, 287–298 (2019).
67. Schaub, S. Salient to whom? The positioning of German political parties on agricultural pollutants in water bodies. *Water* 11, 2278 (2019).
68. Yimer, F., Messing, I., Ledin, S. & Abdelkadir, A. Effects of different land use types on infiltration capacity in a catchment in the highlands of Ethiopia. *Soil Use Manage.* 24, 344–349 (2008).
69. Barlow, M. & Clarke, T. *Blue Gold: The Battle Against Corporate Theft of the World's Water* (Routledge, 2017).
70. Shahid, S. U., Iqbal, J. & Hasnain, G. Groundwater quality assessment and its correlation with gastroenteritis using GIS: A case study of Rawal Town, Rawalpindi, Pakistan. *Environ. Monit. Assess.* 186, 7525–7537 (2014).
71. Rasul, G. & Sharma, B. The nexus approach to water–energy–food security: An option for adaptation to climate change. *Clim. Policy* 16, 682–702 (2016).
72. Ahmad, W., Iqbal, J., Nasir, M. J., Ahmad, B., Khan, M. T., Khan, S. N., & Adnan, S. (2021). Impact of land use/land cover changes on water quality and human health in district Peshawar Pakistan. *Sci. Rep.*, *11*(1), 16526.
73. Shaheen, A., Iqbal, J., & Hussain, S. (2019). Adaptive geospatial modeling of soil contamination by selected heavy metals in the industrial area of Sheikhupura, Pakistan. *Int. J. Environ. Sci. Technol.*, *16*, 4447-4464.
74. Chandel, M., Moitra, D., Makkar, P., Sinha, H., Hora, H. S., & Ghosh, N. N. (2018). Synthesis of multifunctional CuFe 2O4–reduced graphene oxide nanocomposite: an efficient magnetically separable catalyst as well as high performance supercapacitor and first-principles calculations of its electronic structures. *RSC Advances*, *8*(49), 27725-27739.
75. Dhanda, R., & Kidwai, M. (2016). Magnetically separable CuFe 2O4/reduced graphene oxide nanocomposites: as a highly active catalyst for solvent free oxidative coupling of amines to imines. *RSC Advances*, *6*(58), 53430-53437.

76. Zhang, H., Gao, S., Shang, N., Wang, C., & Wang, Z. (2014). Copper ferrite–graphene hybrid: a highly efficient magnetic catalyst for chemoselective reduction of nitroarenes. *RSC Advances*, *4*(59), 31328-31332.
77. Sabari Girisun, T. C., Saravanan, M., & Venugopal Rao, S. (2019). Femtosecond nonlinear absorption and optical limiting action in nanoplatelet CuFe2O4-decorated rGO nanocomposites. *SN Appl. Sci.*, *1*(5), 1-14.
78. Peymanfar, R., Azadi, F., & Yassi, Y. (2018). Preparation and characterization of CuFe2O4 nanoparticles by the sol-gel method and investigation of its microwave absorption properties at Ku-band frequency using silicone rubber. In *Multidisciplinary Digital Publishing Institute Proceedings* (Vol. 2, No.
79. Tang, M., Li, X., Gao, C., Li, X., & Qiu, H. (2017). Adsorption performance of CuFe2O4/rGO nanocomposites towards organic dye. *Mater. Chem. Phys.*, *185*, 114-121.
80. Bashir, B., Shaheen, W., Asghar, M., Warsi, M. F., Khan, M. A., Haider, S., ... & Shahid, M. (2017). Copper doped manganese ferrites nanoparticles anchored on graphene nanosheets for high performance energy storage applications. *J. Alloys Compd.*, *695*, 881-887.
81. Li, Q., Li, S., Liu, Q., Liu, X., Shui, J., & Kong, X. (2021). Iodine cation bridged graphene sheets with strengthened interface combination for electromagnetic wave absorption. *Carbon*, *183*, 100-107.
82. La, D. D., Nguyen, T. A., Jones, L. A., & Bhosale, S. V. (2017). Graphene-supported spinel CuFe2O4 composites: novel adsorbents for arsenic removal in aqueous media. *Sensors*, *17*(6), 1292.
83. Israr, M., Iqbal, J., Arshad, A., Sadaf, A., Rani, M., Rani, M., & Jabeen, S. (2021). CuFe2O4/GNPs nanocomposites for symmetric supercapacitors and photocatalytic applications. *J. Phys. D: Appl. Phys.*
84. Singh, R., Ladol, J., Khajuria, H., & Sheikh, H. N. (2017). Nitrogen doped graphene nickel ferrite magnetic photocatalyst for the visible light degradation of methylene blue. *Acta Chim. Slov.*, *64*(1), 170-178.
85. Kumar, A., Rout, L., Achary, L. S. K., Dhaka, R. S., & Dash, P. (2017). Greener route for synthesis of aryl and alkyl-14H-dibenzo [aj] xanthenes using graphene oxide-copper ferrite nanocomposite as a recyclable heterogeneous catalyst. *Sci. Rep.*, *7*(1), 1-18.
86. Cychosz, K. A., & Thommes, M. (2018). Progress in the physisorption characterization of nanoporous gas storage materials. *Engineering*, *4*(4), 559-566.
87. Lowell, S., Shields, J. E., Thomas, M. A., & Thommes, M. (2012). *Characterization of porous solids and powders: surface area, pore size and density* (Vol. 16). Springer Science & Business Media.
88. Rezlescu, N., Rezlescu, E., Doroftei, C., Popa, P. D., & Ignat, M. (2012, October). Semiconducting spinel ferrite powders prepared by self-combustion method for catalyst applications. In *CAS 2012 (International Semiconductor Conference)* (Vol. 2, pp. 287-290). IEEE.

89. Fu, Y., Chen, Q., He, M., Wan, Y., Sun, X., Xia, H., & Wang, X. (2012). Copper ferrite-graphene hybrid: a multifunctional heteroarchitecture for photocatalysis and energy storage. *Ind. Eng. Chem. Res.*, *51*(36), 11700-11709.
90. Nirumand, L., Farhadi, S., & Zabardasti, A. (2018). Magnetically separable Ag/$CuFe_2O_4$/reduced Graphene oxide ternary Nanocomposite with high performance for the removal of Nitrophenols and dye pollutants from aqueous media. *Acta Chim. Slov.*, *65*(4), 919-931.
91. Reddy, M. V., Cherian, C. T., Ramanathan, K., Jie, K. C. W., Daryl, T. Y. W., Hao, T. Y., ... & Chowdari, B. V. R. (2014). Molten synthesis of $ZnO.Fe_3O_4$ and Fe_2O_3 and its electrochemical performance. *Electrochim. Acta*, 118, 75-80.
92. Peng, Y., Azeem, M., Li, R., Xing, L., Li, Y., Zhang, Y., & Zhang, Z. (2022). Zirconium hydroxide nanoparticle encapsulated magnetic biochar composite derived from rice residue: Application for As (III) and As (V) polluted water purification. *J. Hazard. Mater.*, *423*, 127081.
93. Weerasundara, L., Ok, Y. S., & Bundschuh, J. (2021). Selective removal of arsenic in water: A critical review. *Environ. Pollut.*, *268*, 115668.

14

Managing Construction and Demolition Waste Illegal Dumping through GIS: A Case Study of Urban Metropolitan

Shubhi Nuna[1], Bharath H. Aithal[1]* and Brajesh Dubey[2]

[1]*Energy and Urban Research Group, Ranbir and Chitra Gupta School of Infrastructure Design and Management, Indian Institute of Technology Kharagpur, Kharagpur, West Bengal, India*

[2]*Department of Civil Engineering, Indian Institute of Technology Kharagpur, Kharagpur, West Bengal, India*

Abstract

Rapid urbanization and urban growth in developing countries due to the increased population at the city centers have increased construction activities, generating huge amounts of waste, both during the construction and demolition phases. The consequences of improper waste management are potentially alarming due to its unambiguous effects on the environment and human health. Nearly half of the generated construction and demolition waste (CDW) is illegally discarded in urban areas. One significant management barrier is ineffective collection-disposal networks, which eventually lead to high labor, operating, and transportation costs. The use of Geographic Information Systems (GIS) is essential for streamlining and implementing sustainable waste management. The initial sections of the chapter discuss current waste management practices and the problems with inefficiency in their planning. It also describes how GIS can be integrated with multi-criteria evaluation. Predictive mapping of probable illegal dumping sites was carried out by analyzing the spatial distribution of existing dumpsites and examining factors that influence their occurrence. The final sections address the use of GIS to optimize waste collection. A network of intermediate designated collection points was developed using mapping techniques for various decision-support system-based planning. This has been correlated with land use data obtained by analyzing satellite imagery.

**Corresponding author*: bharath@infra.iitkgp.ac.in

Disha Thakur, Sanjay Kumar, Har Amrit Singh Sandhu and Chander Prakash (eds.) Sustainable Development Using Geospatial Techniques, (327–364) © 2024 Scrivener Publishing LLC

Keywords: Illegal dumping, location-allocation, multi-criteria evaluation, network analysis, predictive modeling, spatial analysis, waste management

14.1 Introduction

Inadequate approaches to the integrated management of construction and demolition waste (CDW) and infrastructure lead to illegal dumping in rivers, forests, roads, vacant plots, etc. Landfill space in India is rapidly depleting and the cost of waste management is rising. Cities need to make setting aside land for temporary collection points a top priority. The Swachh Survekshan 2021 has given more importance to action on CDW. But, a push is needed for proper design of waste segregation, collection, recycling, and reuse.

The collection of trash is usually handled by the local authorities, which is also the main point of contact between people who generate waste and the system that takes care of it. It is gaining attention due to its potential for service optimization using innovative tools provided by spatial modeling methodologies and Geographic Information Systems (GIS). GIS is a computer-based system that aids in the manipulation of data about specific geographic locations and has a wide range of applications. It enables the gathering, storage, retrieval, modification, and analysis of data to gain information for decision-making purposes. The implementation of sustainable CDW management is made simpler with its utilization. This chapter aims to propose a method for optimizing the CDW collecting system. Based on real-world data, the concept is applied to the Municipal Corporation of Gurugram (MCG). The method entails replacing and reallocating waste collection facility sites using multi-criteria evaluation techniques and network analysis. The advantages are measured in terms of reducing the distance traveled by the routine developed.

14.1.1 CDW Statistics for India

There is no systematic evaluation and quantification of the quantity of CDW produced, used, and disposed of to make plans for effective waste management. The MoEFCC (Ministry of Environment, Forest and Climate Change of India) claims that there is no structured database for CDW estimates. According to estimates from the Indian government, we officially generate between 165 and 170 million tonnes (MT) of CDW annually, of which only a meager 1% (6500 tonnes per day) is recycled [1]. However, this is viewed as a significant underestimation. Based on Technology

Information, Forecasting and Assessment Council (TIFAC) estimates, CSE projected 530 MT of CDW generation from building-related activities in 2013, which is an increase of 44% over official figures [2]. TIFAC is an independent organization created in 1988 by the Department of Science and Technology to anticipate technological developments, evaluate technological trends, and foster innovation. Data quality and accessibility issues arise as a result of the insufficiency of official statistics and the ensuing underestimating of this phenomenon. According to TIFAC, demolition creates 300–500 kg of waste for every square meter of building, whereas new construction creates 40–60 kg and rehabilitation creates 40–50 kg [3]. For most of our cities, managing CDW continues to be a challenge, as there are no official building and demolition permits issued, which can help the urban local bodies (ULB) to have a waste estimation. Typically, developers hire local contractors for the demolition process, who pay for recoverable items with immediate secondary market value. Doors, windows, glass panels, partitions, and furnishings are salvaged by the informal sector and bulky materials, such as concrete, masonry, and mortar, are discarded as rubble [4]. An estimated 1.3% of the CDW is recycled, 10–30% is used for backfilling, and the rest is dumped illegally [2, 5]. Waste disposal that is done improperly and illegally, where it does not belong, and/or where environmental harm is probable is known as "Illegal Dumping." This frequently takes place because the offender does not want to pay the disposal cost, does not take the time to prepare the waste for appropriate disposal or because correct disposal is too "inconvenient" (Figure 14.1).

In response to the aforementioned difficulty, a new waste management paradigm has arisen, moving the focus from waste prevention to appropriate disposal. Assuming that natural resources are cheap, numerous, and

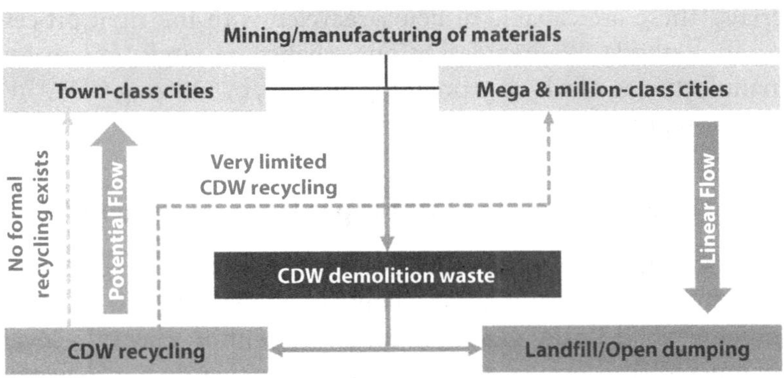

Figure 14.1 Circular flow of construction materials [6].

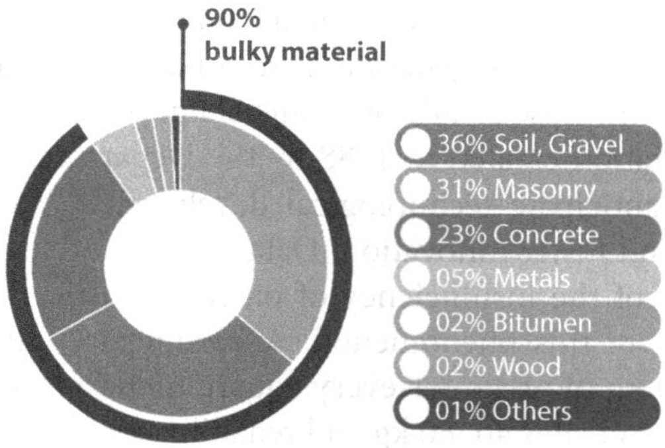

Figure 14.2 CDW composition in India [3].

easy to obtain and dispose of, the linear "take-make-consume-dispose" economic model is based on this assumption. However, it is not sustainable. The goal of the circular economy, which is restorative in nature, is to keep materials useful for as long as possible while still maintaining their value. This reduces fresh energy and material demands while also easing environmental constraints brought on by resource extraction, emissions, and waste management. CDW should be treated as a resource to be recycled. A considerable proportion of construction debris is recyclable and repurposed for use in the industry [2]. Current CDW management (CDWM) techniques are linear, which depletes nonrenewable and resource-intensive resources. Materials moving circularly can lessen the need to mine nonrenewable resources and the ensuing environmental harm.

Based on the CDW composition as reported by TIFAC (Figure 14.2), bulky materials, such as concrete, bricks, stones, soil, and gravel, majorly contribute to 90% proportion of the total waste [3]. Besides, the greatest recycling potential in the construction industry is represented by these materials. These are capable of being recycled with the right processing tools and methods. We may lessen our reliance on virgin raw materials, the manufacturing of which is polluting by itself, by utilizing CDW in new construction projects.

14.1.2 Negative Impacts of Improper CDW Management

Large construction projects frequently leave heaps of CDW on the sides of the road, in waterways, and in the vicinity of public spaces in our country. If municipal solid waste (MSW) is combined with it, energy recovery is also hampered. This occurs because most cities either do not have enough adequately labeled CDW collection spots or the public is unaware of where they are situated, which leads to illegal dumping.

Social Impacts:

- Health hazards have an impact on people's health and societal life.
- Land depletion takes public space, with negative consequences for workplace safety.
- Dumping MSW on the CDW dump creates a mix of waste, making the situation even more unhygienic.
- Frequently obstructs traffic and pedestrians, causing traffic congestion and sometimes accidents.

Environmental Impacts:

- Clog open drains and water channels, creating flooding in metropolitan areas during rainstorms.
- Hydrology is disrupted, and the aquatic ecology is destroyed.
- Leachate and tiny chemical particles from CDW erode the soil, resulting in land and groundwater contamination, particularly from hazardous components such as paints, oil, and asbestos sheets.
- CDW accumulates over time, forming fine dust that pollutes the air and reduces visibility.

Economic Impacts:

- Depletion of natural resources and excessive fuel use in transportation.
- CDW is difficult to sort once combined with MSW, resulting in higher processing costs and lower efficiency.
- Imposes major expenses on local government agencies, such as the cost of clearing CDW-clogged sewers.
- A large volume of CDW fills landfills, necessitating the construction of additional landfills or alternative dump sites.

Resource Shortage:

- High rates of primary resource extraction for building materials.

- In the building industry, sand, soil, stone, and limestone are all important materials. Due to mining bans and limitations, they are already experiencing supply interruptions and price increases.

14.1.3 CDW Management Barriers in India

CDWM currently has a poor status in India, as the most suitable procedures from waste collection to disposal are not being employed. Low motivation and a lack of creativity have prevented the adoption of innovative technology that may revolutionize waste management in India. Another significant obstacle to increasing CDWM in India is public attitudes.

Inadequacy in Governance:

- The slow process of installing facilities: Even with the extra pressure from the Bombay High Court to establish a CDW recycling plant, Mumbai still is lacking.
- Finances and business case concerns: Cities endeavor to safeguard their financial interests while enforcing regulations. ULBs have suggested that contractual parties responsible for the collection and transportation of CDW cover the ULB's expenses rather than receiving a tipping charge per tonne of CDW. Profits from the sale of recycled items are split between the recycling plant and the ULB.
- Low level of participation from state government agencies: ULBs should foster collaboration among City Development Authorities, Housing Development Boards, Public Sector Utility Companies, and other entities with significant involvement in construction and demolition activities to ensure the appropriate disposal of their CDW, implement *in situ* reuse of CDW in their projects, and create policies to purchase recycled goods.

Challenges from the Construction Industry:

- Low level of knowledge and a lukewarm response from the construction industry: The construction industry, as the primary producer of CDW, is the most important stakeholder in the execution of the rules (apart from natural calamities). It is expected that they would make sure that the waste they produce is disposed of appropriately and that they will use more recycled materials in their building procedures. But because of inertia and haphazard application of the rules and standards, progress has been slow.

- High pricing of recycled CDW products: To enhance the adoption rate of recycled CDW products, it is advisable to reduce their prices relative to traditional building materials. At this time, manufactured goods are taxed at 18%, whereas recycled aggregates are taxed at 5%; this disparity renders the use of recycled products uneconomical for the majority of consumers [2]. The GST should be reduced on CDW recycled products including tiles, bricks, sand, paver blocks, and aggregates.

Lack of Public Awareness:

- People continue to be concerned about the quality of recycled products. Therefore, the Central Pollution Control Board (CPCB), National Building Code (NBC), Central Public Works Department (CPWD), Bureau of Indian Standards (BIS), and Building Materials and Technology Promotion Council (BMPTPC) have adopted many guidelines for material reuse.
- Certain projects have been constructed by agencies using CDW materials in an effort to promote the use of recycled resources in the building industry. Pre-cast cement concrete blocks of Grade M10 constructed of recycled CDW procured from Delhi's Burari plant were utilized in the construction of the New Supreme Court complex. It employed about 1.8 million blocks as non–load-bearing walls. The Delhi Development Authority (DDA) initiative for the building of the prototype 100-m-wide and 10-km-long Express Route in North Delhi between NH-10 and Bakkarwala is employing recycled aggregates. It is estimated that 600,000 tonnes of recycled CDW will be utilized.

Inadequacy in Practice and Infrastructure:

- Constraints of finding land for recycling facilities: Any form of waste recycling facility needs substantial land area, and cities are sometimes hampered by older zoning restrictions and citizen attitude of *"not in my backyard"*. The state environment officials banned recycling facility locations proposed by Gurugram and Kolkata in the Aravalli Hills and wetland because it violated regulations governing environmentally sensitive areas.

- Challenges of intermediate collection and transportation of CDW: In Ahmedabad and Chennai, there is a CDW collection site in every municipal zone. These have adequate areas to gather waste; however, the general public is unaware of their presence. Furthermore, transporting waste to these collection stations is sometimes too expensive for small generators, and they may be fined unofficial fees for depositing CDW at these pickup stations, which discourages them even more. Kolkata offers emergency waste collection services. Based on payment, waste producers can select when they want their waste to be picked up. This is a slightly better alternative, but only if it covers all the possible city locations and has sufficient service capacity [2].

14.1.4 Proposed CDW Management Model

Figure 14.3 depicts a standardized CDWM model that has been created by GIZ—an international sustainable development organization and development alternatives for cities with CDW processing units [4].

- The ULB will charge generators upfront fees when approving their CDWM plan, which is related to the building or demolition permit.
- Generators will dispose of their waste at the closest approved intermediate collection location.
- CDW will be transported from these locations to the processing plant by the contractual agency.
- The amount of CDW entering each disposal site will be immediately recorded by a weighbridge and compared to the tracking serial number linked to each demolition permit. This will ensure that all of the generated waste is arriving at the authorized points and is not being disposed of unlawfully.

We can see that designated collection points are the key links between the CDW generator and the processing unit. It has been observed that

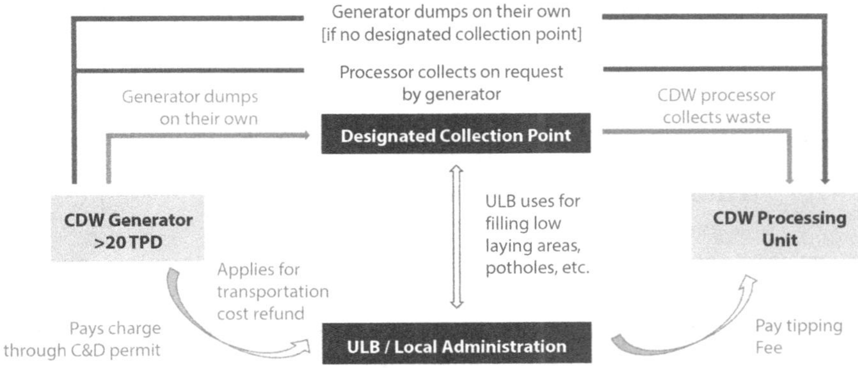

Figure 14.3 Material and cash flow in standardized CDWM [7].

these are either inadequate or their placement is inappropriate. Multiple factors should be considered while deciding their spatial location, such as proximity to roads, population density, urban development pattern, geographical characteristics land use and land cover, etc. To map these multiple criteria, GIS is effective. It is a computer-based spatial analytic approach that is used in CDWM to help decision-makers make better decisions by integrating alternatives and their preferences. GIS technology facilitates an effective environment for data gathering, processing, and presentation, hence contributing to the optimization of CDWM.

Global literature identifies two primary categories of waste management applications that employ GIS. In the first scenario, a geographical model is generated using GIS to forecast the likely placement of illicit dumpsites, considering comparable socioeconomic and physiographic factors. Systematically analyzing the interaction effects of geographical and temporal variables at the case level and social variables at the ward level is essential and this forms an important objective of this study. It does this by building a theoretical framework based on past research and applying it to the Municipality of Gurugram. Better outcomes might be possible with more research that considers a wider range of factors, including behavior, legislation, and other socioeconomic concerns. The second group of waste management apps that make use of GIS is more complex and includes programs that deal with garbage collection. Reallocating trash cans and collecting areas in their entirety has multiple applications. Reducing the amount of time and/or distance that the collection vehicle fleet has to travel is the main objective of these applications. To find the best place for CDW disposal infrastructure, our study used geographical analysis to solve the

CDW handling problem. An MCE approach and a network analysis were both applied using GIS.

14.2 CDWM Using GIS Tools and Multivariate Analysis Techniques

In order to optimize waste management outcomes, it is critical to consider technological solutions such as monitoring illegal garbage places and geographic mapping, but the steps done in India to control CDW are still in their infancy. People and government usually treat CDW as inert waste and it is not as much in focus as MSW. However, illegal dumping is not particular to our country; many countries are struggling with proper waste management due to unauthorized waste disposal. According to several writers, improving waste management efficiency can be accomplished by becoming more aware of the particular elements that influence the patterns of illegal dumpsites (ILs) in a given area [8–11].

14.2.1 Predicting Probable Locations of Illegal Dumpsites (ILs)

In the past 20 years, there have been limited studies on identifying IL spatial locations, due to the lack of monitoring, uniform database, incomplete government statistics, and high-resolution satellite or aerial imagery. These researchers attempted to identify and map the ILs and study the spatial factors that affect their locations, through remote sensing (RS), aerial infrared photography, crowdsourcing, and on-the-ground surveys [8–10]. The extracted factors responsible for the occurrence pattern of ILs include spatial, as well as, non-spatial factors. The factors that are most commonly addressed are socioeconomic characteristics, accessibility to roads, distance to industrial land use, land use, and proximity to authorized landfills. By utilizing RS tools, GIS, and multi-criteria decision analysis (MCDA) methodologies, geo-informatics technologies were applied in various studies to map already-existing illegal landfills and study variables impacting their occurrence to predict probable IL occurrences and delimit future occurrences [11–14].

14.2.2 Approaches to Spatial Analysis for Optimal Facility Locations

Numerous real-world issues have been resolved using facility location problems. Due to its direct correlation with transportation costs, the placement of collection points maintains the CDWM system's financial viability. Multi-criteria evaluation (MCE) and network analysis are two methodologies that are typically employed when addressing CDW. MCE approaches have been frequently applied with positive outcomes, such as for locating landfills [15, 16], transfer stations [17, 18], or treatment locations [19, 20]. MCE examines how much land should be allotted for a certain application based on a number of characteristics that the selected areas contain. It assists in the development of a land suitability map through exclusion and preference studies. In contrast, network analysis is used for smaller infrastructure allotment and is performed to determine service areas based on travel distance and locate the closest facility [14, 21, 22].

14.2.3 Location-Allocation Problems

In the 1970s, location-allocation models were first utilized to locate service infrastructure. These methods have been employed for the assessment of infrastructure in solid waste management, such as storage facilities or transfer stations along a transportation network [21, 23, 24]. It is a combinatorial optimization problem. Combinatorial optimization is a branch of mathematical optimization that deals with selecting the optimal solution from a finite collection of objects, where the set of viable solutions is discrete or may be reduced to a discrete set. Seven location-allocation models are offered by ArcGIS and along with exact, heuristic, and meta-heuristic approaches are also used to solve them. In the exact method, the search continues until you have found the optimum—over 17 trillion different possible solutions exist for even a simple task like selecting 10 solutions from 100 solution spaces.

$$\binom{N}{P} = \frac{N!}{P!(N-P)!} \tag{14.1}$$

Despite the fact that the outcomes of combinatorial studies are much simplified, it is still customary to utilize heuristic mechanisms, which rely on trial-and-error methods to find the optimal or nearly solution continuously [25]. Meta-heuristics are also capable of avoiding local optima. The three main factors considered by the location-allocation models are (i) demand points, (ii) candidate facilities, and (iii) a distance-time matrix. The demand points in this study are the locations where CDW is generated, i.e., high occurrence probability regions. The candidate facilities are those that match the emplacement requirements. The distance-time matrix is then used to determine how long it would take to travel between demand sites and candidate facilities. There are seven different categories of location-allocation problems that may be resolved with the application of ArcGIS version 10.8 and some input data [26]. These issues include minimizing impedance and facilities and maximizing coverage, attendance, capacitated coverage, and market share.

We will discuss the first four problem types since the others are employed when the candidate facilities provide distinct services (Figure 14.4). According to the problem type and criteria defined, the location-allocation solution selects the best candidate facilities to allocate demand to in the most effective manner. Facilities are the objects whose spatial location is determined by the model which takes into account interactions with other existing objects. Demand points are locations that indicate the objects or people that need access to the services that the facilities offer.

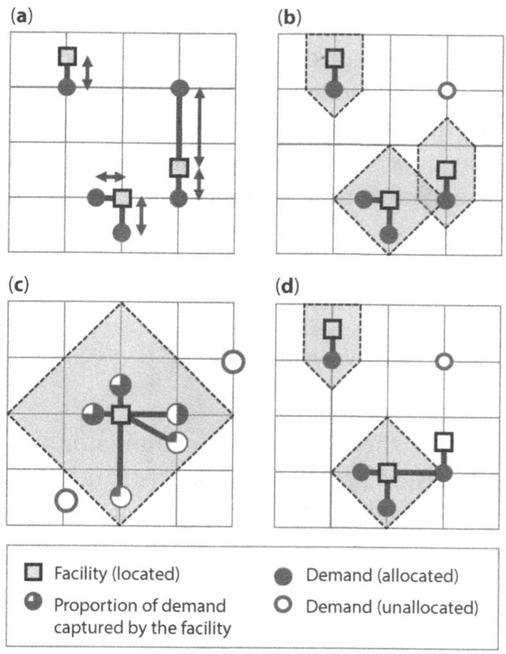

Figure 14.4 (a) Minimize impedance; (b) maximize coverage; (c) maximize attendance; (d) minimize facilities.

14.2.3.1 Minimize Impedance (MI)

This model was developed in 1964 and is also referred to as the "P-Median Problem" [27, 28]. It is utilized to determine the location of warehouses in order to reduce the total cost of transportation, be it distance or time. It has the following objective function [29]:

$$Minimize\ Z = \sum_{i=1}^{n} \sum_{j=1}^{m} c_{ij} x_{ij} \tag{14.2}$$

subject to

$$\sum_{j=1}^{m} x_{ij} = 1, i = 1, \ldots, n$$
$$x_{jj} - x_{ij} \geq 0, \quad i = 1, \ldots, n; \qquad j = 1, \ldots, m; \qquad i \neq j$$
$$\sum_{j=1}^{m} x_{ij} = P$$

where $\{1, \ldots, n\}$ is the set of demand points; $\{1, \ldots, n\}$ is the set of candidate facility points; c_{ij} is the shortest distance between the demand point i and the candidate facility j;

$$x_{ij} = \begin{cases} 1, & \text{if demand point } i \text{ is allocated to facility at } j \\ 0, & \text{otherwise} \end{cases}$$
$$x_{jj} = \begin{cases} 1, & \text{if a facility is located at site } j \\ 0, & \text{otherwise} \end{cases}$$

P is the number of facilities to be located.

14.2.3.2 Maximize Coverage (MC)

This location-allocation problem was created in 1974 and is used to allocate emergency services like fire stations as facilities are chosen based on the greatest demand possible within an impedance cutoff [30]. Its objective function is as follows:

$$Maximize\ Z = \sum_{i \in I} a_i y_i \tag{14.3}$$

subject to

$$\sum_{j \in N_i} x_j \geq y_i \text{ for all } i \in I$$
$$\sum_{j \in J} x_j = P$$

where a_i is the quantity of demand at point i; I is the set of demand points; J is the set of candidate facilities; N_i is the set of candidate facilities covering the demand of point i;

$$x_j = \begin{cases} 1, & \text{if a facility is located at site } j \\ 0, & \text{otherwise} \end{cases}$$
$$y_i = \begin{cases} 1, & \text{if demand point } i \text{ is covered by a facility within specified impedance } S \\ 0, & \text{otherwise} \end{cases}$$
S = distance or maximum service time

14.2.3.3 *Minimize Facilities (MF)*

Many contributions have been made to the placement of public and private services to the minimized facility model [31, 32]. It is similar to the maximize coverage problem, with additional requirements to minimize the number of solution facilities. Its objective function is as follows:

$$Minimize\ Z = \sum_{j=1}^{m} x_j \qquad (14.4)$$

subject to

$$\sum_{j \in N_i} x_j = 1, \quad i = 1, \ldots, n$$

where x_j is a Boolean variable and everything else is defined like the equations above.

14.2.3.4 *Maximize Attendance (MA)*

This location-allocation problem was developed in 1972 [33]. This maximizing problem implies that the facilities' ideal locations will be found close to the majority of demand sites. Its objective function is as follows:

$$Maximize\ Z = \sum_{i=1}^{n} \sum_{j=1}^{n} a_i \left(S - c_{ij}\right) x_{ij} \tag{14.5}$$

where everything is defined in a manner similar to the previous equations.

14.3 The Case Study of Gurugram Municipality

South of Delhi is the fast-growing urban enclave of Gurugram. A significant financial and industrial hub, Gurugram is located in India's National Capital Region (NCR). Over the past 30 years or so, the city has expanded rapidly. The city's population increased from 8.8 lakh to 18.9 lakhs between 2011 and 2021, which indicates the growth rate. The district has had incredible growth in all areas of development, but especially in business and urbanization. Multinational corporations, industrial goliaths, contact centers, software firms, retail centers, and skyscrapers are now concentrated there. To satisfy this insatiable demand, new buildings spring from the cityscape like plants in an overgrown wilderness, while old ones are demolished at a startling rate. This has made controlling CDW a huge task. The district of Gurugram is governed by the Municipal Corporation Gurugram (MCG), which oversees all administrative and civic functions. There are now 35 municipal wards and 16 communities and villages in the vicinity of the Dwarka Expressway and the Golf Course Extension Road that were added to the MCG's purview last year. There are designated 15 CDW collection points [34].

14.3.1 Current CDW Collection System

MCG and IL&FS Environmental Infrastructure and Services Limited (IEISL) signed a 20-year contract in 2016. IEISL received a 4-acre site in 2017, and the 300 TPD facility went into operation in December of that same year (and a design capacity of 1800 TPD). Specialized contractors that bring their staff and equipment and transfer the leftover garbage typically carry out demolition work. Based on the recoverable worth of recycled materials, the property owners pay a charge to the demolition contractors. Ideally, the remaining materials are properly disposed of using science. At a price set by IEISL authorities, output materials are sold to various bulk customers, most of whom are contractors. Bulk purchasers have two options for removing items from the plant storage area: either using their trucks or paying IEISL to carry them. MCG had employed two different concessionaires—the primary collection was used to be handled by Pragati AL Natural Resources Pvt Ltd and the secondary collection and processing fell under the purview of IEISL. The city made good strides in the primary collection, on-call removal, grievance settlement, and the development of a punitive procedure for non-compliance with CDW. But Pragati's contract was terminated in March of 2021 after an inquiry was opened against it, and CDW was once again being dumped around the city, covering roundabouts, residences, open spaces, and even the Aravallis [34].

14.3.2 Data Collection—Thematic and Spatial Database

The physical verification of a representative sampled 15 IL points previously identified through Google Earth imagery validated the points and showed that CDW is prominent, with cross-contamination with MSW at some places. The areas under construction were disposing of their waste at nearby empty plots and nearby vegetation (Figure 14.5). CDW was also

Figure 14.5 The examples of ILs located in the study area: along roadsides, cross-contaminated with MSW, near under-construction localities, and segregated CDW near the commercial area, widespread illegal dumping on the low-density areas away from the city center.

being subjected to unauthorized burning along with other waste. Some of these ILs were spread across 20,000–30,000 m^2 in clear sight from main roads.

This research has been carried out with the information collected through a representative sample of 310 ILs, distributed throughout MCG. For the IL localization process, potential ILs are identified using photo-interpretation of Google Earth imagery and UAV data for the years 2019 to 2022. A total of 310 ILs are detected and randomly sampled into calibration (70%) and validation (30%) data points (Figure 14.6). Various spatial and non-spatial factors studied in the literature review, including the socioeconomic, and physiographic factors, are mapped using QGIS. Principal components analysis (PCA) is performed to reduce the feature space and find the key factors.

Table 14.1 shows the 23 most correlated variables which were selected among the initially chosen 27 variables according to the sample size (217 points). Variables such as slope, distance to authorized collection points, distance to public semi-public areas, and urban built-up were removed through PCA.

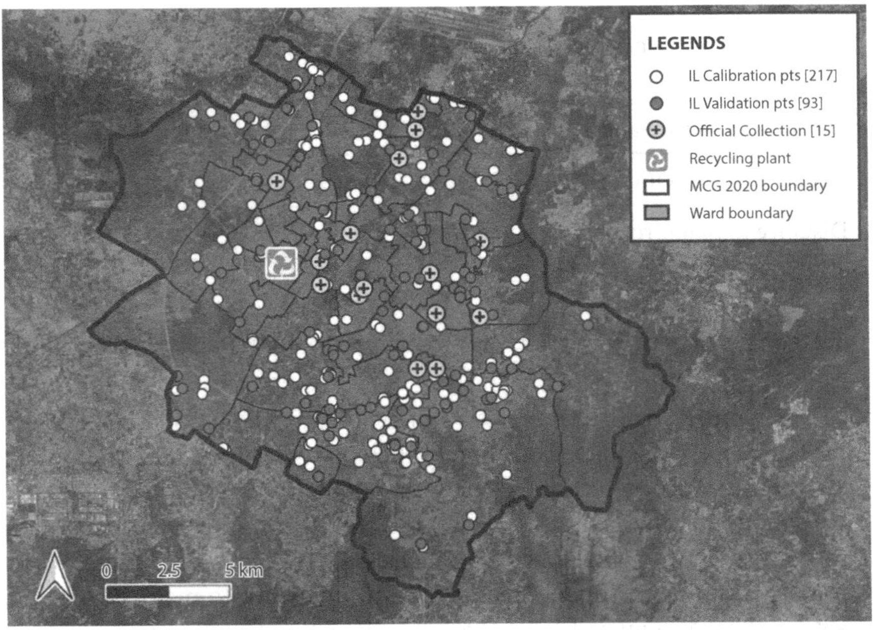

Figure 14.6 Map depicting the distribution of validation (93) and calibration (217) IL points in MCG.

Table 14.1 Selected significantly correlated 23 variables.

Distance attributes	Categorical/density attributes	Socioeconomic attributes (ward-wise)
Distance to industrial areas	Construction density	Population
Distance to commercial areas	Road density	Population density
Distance to residential areas	Elevation	Unemployment rate
Distance to city center	Visibility from rural roads	Literacy rate
Distance to ongoing construction sites	Normalized difference vegetation index (NDVI)	Illegal dumpsites coverage area
Distance to a recycling plant		Number of households
Distance to waterbody		
Distance to low-density areas		
Distance to vacant land		
Distance to vegetation		
Distance to major roads		
Distance to residential roads		

The required spatial database is not readily available for MCG. Ward boundaries were mapped according to the 2020 MCG jurisdiction boundary. The socioeconomic data, such as population, household, literacy rate, and unemployment rate, were collected through 2011 census data and rearranged for 2020 ward boundaries. Communication routes were extracted from OSM (open street map) and divided into major roads and residential roads. Figure 14.7 shows maps depicting (a) ward-wise population density (inhabitants per square kilometer); (b) communication routes;

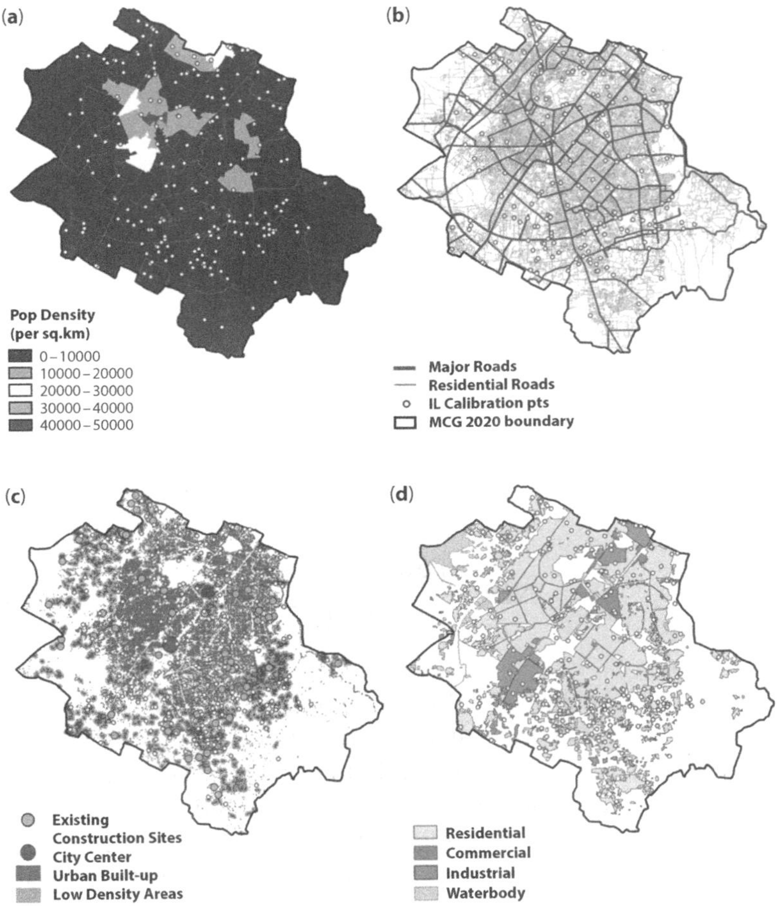

Figure 14.7 Maps depicting (a) ward-wise population density (inhabitants per sq. km); (b) communication routes; (c) locations of the city center, existing construction sites, urban built-up areas with low-density area clusters; and (d) land-use map.

(c) locations of the city center, existing construction sites, and urban built-up areas with low-density area clusters; and (d) land-use map.

NDVI (normalized difference vegetation index) and NBI (new built-up index) were calculated using Landsat-9 satellite images.

$$NDVI = \frac{(NIR - RED)}{(NIR + RED)} \tag{14.6}$$

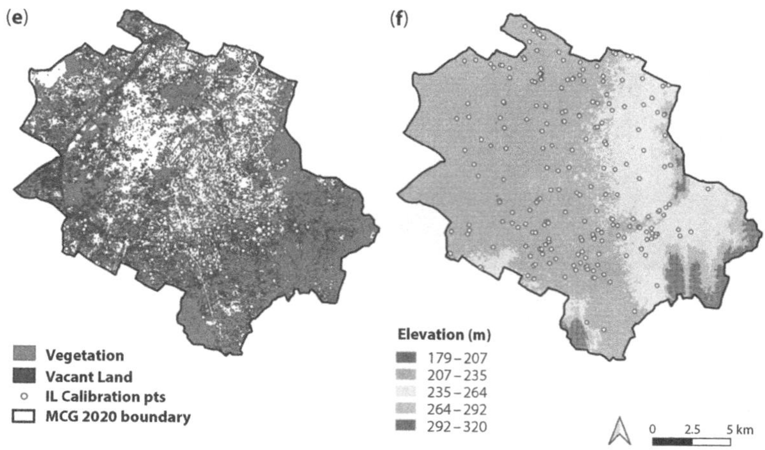

Figure 14.8 Maps depicting (e) vegetative and vacant land areas; and (f) elevation levels.

$$NBI = \frac{(RED - SWIR)}{(NIR)} \tag{14.7}$$

As NBI is considerably better for obtaining threshold values, it was used, instead of NDBI to map built-up areas. This index considerably decreases the confusion between vegetation and built-up by enhancing the vegetation land covers and applies to plain areas. It uses multiply and divide, producing a 3-digit number, stretching up its spectral value to the maximum. Low-density built-up area clusters are identified through cluster analysis. Waterbodies and canals were vectorized and a land-use map was developed for residential, commercial, and industrial use. Vegetation areas were extracted from the NDVI map and vacant areas from the NBI map. SRTM data were used to measure the elevation. Figure 14.8 shows maps depicting (e) vegetative and vacant land areas and (f) elevation levels of the study area.

14.4 Methodology

Figure 14.9 shows the framework of the adopted research methodology, it consists of four distinct stages. Mapping the current illegal CDW dumping sites throughout MCG and developing appropriate criteria for their occurrence constituted the initial phase. In sequence, the spatial database was created, with GIS layers constructed by the selected variables. Depending

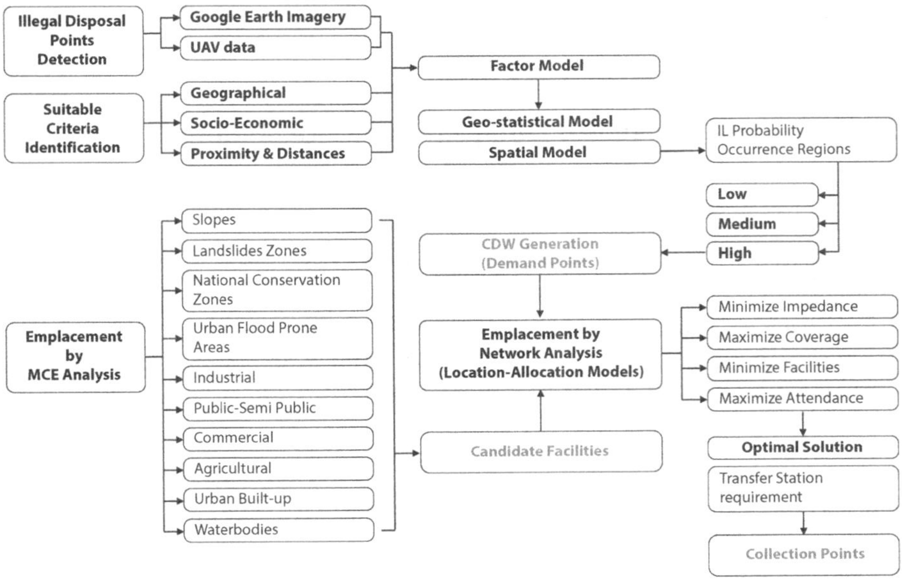

Figure 14.9 Framework of proposed methodology.

on their central tendency, the layers were then divided into binary classifications, which reflect the likelihood of an IL occurrence. The variables are classified into class-1 or class-0 representing the positive and null probability of IL occurrence respectively. The statistical parameter for a measure of central tendency was selected according to the type of variable and its distribution. Nominal variable types were classified using the mode, and ordinal variable types using the median. Interval/ratio variable types were classified according to the median and mean for skewed and normally distributed data respectively. Factor analysis was done and the variable weightage was computed through a percentage of explained variance by the nine components.

Spatial patterns related to each factor were created and overlaid to develop the final predictive probability occurrence map. The study region was thereafter divided into zones to uncover the most. The study region was thereafter divided into zones to uncover the most probable areas for future IL occurrence, which were considered the CDW generation zones. In the next step, the emplacement of collection points is carried out using two approaches: multi-criteria evaluation (MCE) and network analysis through four location-allocation problem types. Different impedance values were considered and the results were compared to select the most economical solution. Finally, the requirement for a transfer station was evaluated.

14.4.1 Probable Illegal Dumps Occurrence Mapping

Of all factor analyses carried out, the one that best met the requirements indicated was the one that offered nine factors (complex variables), which explains 75.47% of the total variance (Table 14.2), resulting in high communalities for almost all variables (Table 14.3). For the extraction method of PCA factors and rotation, we used the Quartimax method with Kaiser, resulting in a Kaiser-Meyer-Olkin (KMO) value of 0.533. Bartlett's test for sphericity produced a significance of 0.000, This demonstrates that the

Table 14.2 Total explained variance.

		Initial eigenvalues	**Rotation sums of squared loadings**	
Factor total		**% of variance**	**Cumulative %**	
F1	Urban development pattern	4.393	13.119	13.119
F2	Settlement	2.620	10.530	23.650
F3	Ward-specific characteristics	2.056	9.488	33.138
F4	Socioeconomics	1.866	8.214	41.352
F5	Land cover	1.647	7.785	49.138
F6	Geographical characteristics	1.395	7.734	56.871
F7	Distance to commercial areas	1.187	6.772	63.643
F8	Road accessibility	1.111	6.288	69.931
F9	Distance to residential areas	1.081	5.534	75.465

Table 14.3 Individual commonalities of the selected 27 variables calculated through PCA.

S.n.	Variables	Communality
1	Population (ward-wise)	0.932
2	Number of households (ward-wise)	0.922
3	Distance to a recycling plant	0.878
4	Road density	0.857
5	Construction density	0.817
6	Population density (ward-wise)	0.802
7	Unemployment rate (ward-wise)	0.798
8	Distance to industrial areas	0.788
9	Distance to city center	0.786
10	Elevation	0.773
11	Illegal dumpsite coverage area (ward-wise)	0.755
12	Distance to vegetation	0.749
13	Normalized difference vegetation index (NDVI)	0.728
14	Distance to low-density areas	0.728
15	Distance to commercial areas	0.724
16	Distance to residential areas	0.710
17	Literacy rate (ward-wise)	0.709
18	Distance to waterbody	0.694
19	Distance to ongoing construction sites	0.677
20	Distance to residential roads	0.658
21	Visibility from rural roads	0.657
22	Distance to major roads	0.608
23	Distance to vacant land	0.607

Table 14.4 Synthesis of the methodology developed.

Factor model			Geo-statistical model		Spatial model	
Factor	**Weight (%)**	**Variables**	**Variable loading**	**Statistical parameter**	**Spatial pattern**	**Map/final**
F1	17.38	Distance to city center	0.619	Mean	> 6.2 km	Spatial join analysis Zones with a probability of the presence of uncontrolled landfills
		Distance to a recycling plant	0.587	Mean	> 7.0 km	
		Visibility from rural roads	0.564	Median	< 12	
		Construction density	-0.816	Median	< 0.23 km^2/km^2	
		Road density	-0.886	Mean	> 17.5 km/km^2	
F2	13.95	Population (ward-wise)	0.950	Mean	Ward index	
		Number of households (ward-wise)	0.931	Mean		
F3	12.57	Distance to low-density areas	0.786	Median	< 202.7 m	
		Population density (ward-wise)	0.735	Mean	Ward index	
		Illegal dumpsite coverage area (ward-wise)	-0.604	Mean		
F4	10.89	Unemployment rate (ward-wise)	0.825	Mean	Ward index	
		Literacy rate (ward-wise)	0.782	Mean		
F5	10.32	Normalized difference vegetation index (NDVI)	0.832	Mode	199	

(Continued)

Table 14.4 Synthesis of the methodology developed. (*Continued*)

Factor model			**Geo-statistical model**		**Spatial model**	
Factor	**Weight (%)**	**Variables**	**Variable loading**	**Statistical parameter**	**Spatial pattern**	**Map/final**
		Distance to vacant land	0.610	Median	< 10.5 m	
		Distance to vegetation	-0.741	Median	< 21.0 m	
F6	10.25	Distance to industrial areas	0.787	Median	< 1.5 km	
		Distance to waterbody	0.697	Mean	< 1.3 km	
		Elevation	0.550	Median	< 228 m	
F7	8.97	Distance to commercial areas	-0.807	Median	< 0.45 km	
F8	8.33	Distance to ongoing construction sites	0.643	Median	< 1.2 km	
		Distance to major roads	0.642	Median	< 0.16 km	
F9	7.33	Distance to residential roads	0.752	Median	< 26 m	
		Distance to residential areas	0.723	Median	< 40 m	

correlation matrix does not conform to the characteristics of an identity matrix.

The nine components (Table 14.4) that represent the elements of the region associated with the highest likelihood of an IL occurrence are as follows:

1) Factor 1 comprises 17.38% of the total variance and comprises five variables mainly geographical and distance attributes. The included variables are road density, construction

density, distance to the city center, to the recycling plant, and visibility of ILs from rural roads. It shows that ILs are most likely to be present far from the city center and recycling plant, with less dense construction, and easily visible from roads.

2) Factor 2 encompasses 13.95% of the total variance and groups socioeconomic variables: ward-wise population and household data. The higher the socioeconomic level of a ward, the greater will be the IL occurrence probability. The ILs occur mostly in moderately populated areas.
3) Factor 3 accounts for 12.57% of the total variance and is determined by two socioeconomic variables (ward-wise population density and illegal dumpsite coverage area) and one metric variable (distance to low-density built-up areas). The majority of the ILs are present in areas with low population density (below 10,000 persons per square kilometer).
4) Factor 4 accounts for 10.89% of the total variance and comprises two socioeconomic variables: ward-wise literacy rate and unemployment rate.
5) Factor 5 contributes 10.32% of the total variance and encompasses three geographical variables: NDVI value, distance to vacant land, and vegetation. The values for calculating this spatial pattern are a median distance of 10.5 and 21 m from vacant land and vegetation coverages, respectively.
6) Factor 6 is associated with elevation and distance to waterbodies and industrial areas. This contributes to 10.25% of the total explained variance. ILs are not present very near industrial areas and waterbodies. In the study area, the high-altitude locations are less likely to see any IL.
7) Factor 7 accounts for 8.97% of the total variance. According to this factor, most ILs are found near commercial areas. In calculating its spatial pattern, a median buffer distance is calculated with 0.45 km from commercial areas.
8) Factor 8 indicates that most ILs are found in proximity to ongoing construction sites and easily accessed from major roads. This factor contributes to 8.33% of the total explained variance.
9) Factor 9 encompasses 7.33% of the total variance and groups distance to residential roads and residential areas. The probability is high within 26 m from residential roads and within 40 m from residential areas.

Depending on the factor analysis, the weight of each one of the nine factors is determined by the percentage of the total variance explained by what each one contributes. We have recalculated the explained variance or spatial pattern by each factor, by using the 75.47% value as the maximum limit of probability from zero to one hundred percent. For other factors, related to distance patterns and categorical variables, individual polygonal coverages are spatially overlaid to obtain an overall entity. Ward indexes are calculated for the spatial factors with socioeconomic variables, the ward having the highest index value will be contributing to the maximum probability explained by that factor and the ward with the lowest

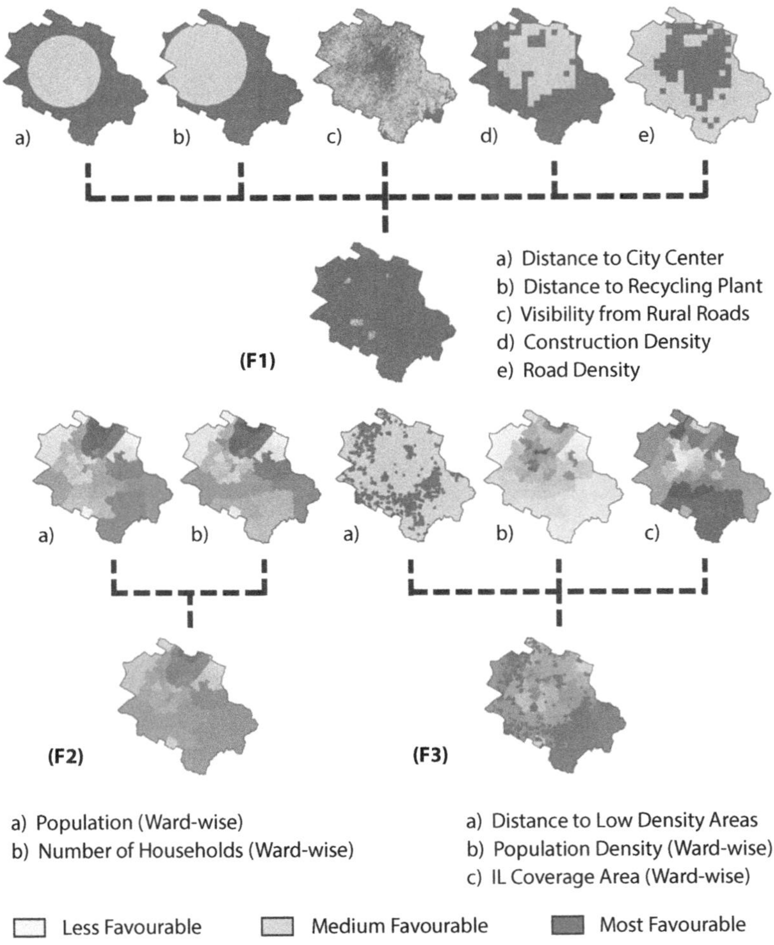

Figure 14.10(A) Spatial join analysis overlay of factors (F1 to F3).

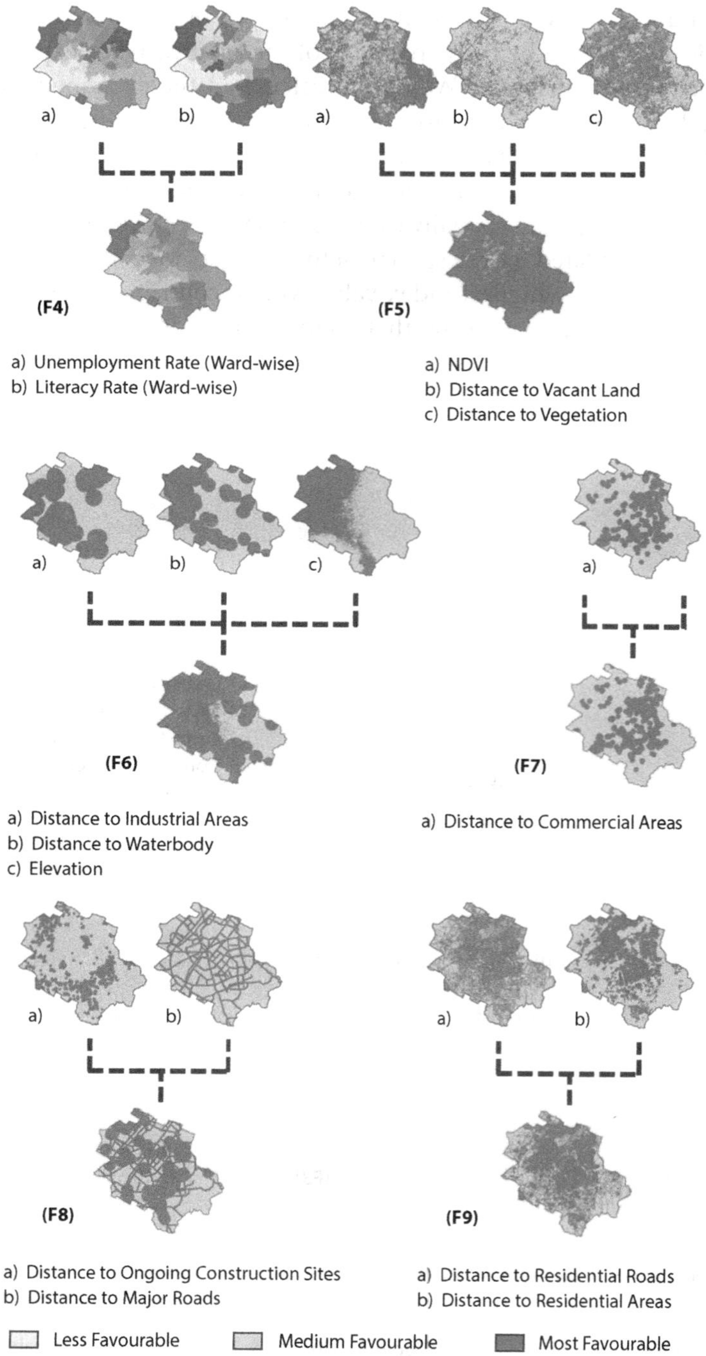

Figure 14.10(B) Spatial join analysis overlay of factors (F4 to F9).

index value will have 0% probability. For these factors, the socioeconomic variables are normalized and an average value is calculated, which when reclassified provides the index values. For other factors, related to distance patterns and categorical variables, individual polygonal coverages are spatially overlaid to obtain an overall entity.

After we get the spatial patterns of all nine factors [Figures 14.10(A and B)], we combine them using spatial join to produce the final predictive probability occurrence map (Figure 14.11). This map will depict the probability range from 0 to 100%. To reclassify this map into low, medium, and high probability areas, we take the cumulative total explained variance by the first three and first six factors (as we have nine factors). Table 14.5 depicts the three probability intervals: (1) low probability areas with values less than 43.91%, (2) medium probability areas with values between 43.91% and 75.36%, and (3) high probability areas with values over 75.36%.

Table 14.5 shows that more than 50% of the data points (both in calibration and validation data) are present in areas with over 75% probability of occurrence, which is equivalent to approximately three out of every four.

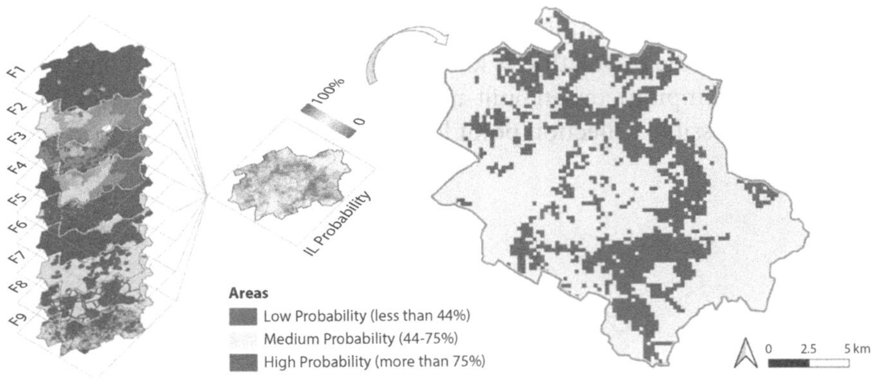

Figure 14.11 Final predictive probability occurrence map.

Table 14.5 Table showing the distribution of ILs across the three probability intervals.

Probability range		**Calibration**		**Validation**	
Low	0–43.91 %	3 Pts	1.4%	1 Pts	1.1%
Moderate	43.91–75.36%	100 Pts	46.1%	45 Pts	48.4%
High	75.36–100 %	114 Pts	52.5%	47 Pts	50.5%

This methodology has proven its validity by demonstrating demonstrates the distribution of a verified population of unauthorized dumps (47) is not random, but rather that 50.5% of the illegal dumps are found in the locations with the highest probability (above 75%). These high IL probability areas were converted into a point layer to generate CDW demand points (1707 points). The proposed methodology (FA and GIS) produces appropriate outcomes at the municipal level.

14.4.2 Emplacement by MCE Analysis

The spatial modeling provided by the MCE approach should be used if the decision-makers want to identify a single location where huge volumes of waste can be processed or disposed of (recycling plant) [22].

For the MCE approach, a zone of analysis was identified within the study area using some restrictive variables, where the required infrastructure could not be created. The variables used are slopes (associated with the deposit cells' stability), landslide zones, national conservation zones (related areas of natural relevance), areas prone to urban flooding (MCG is vulnerable to urban flooding), feasible land use variables—industrial, public semi-public, commercial and agricultural land use (to identify the regions where the population is least impacted and land use is compatible with waste dumping), urban built-up zones, and waterbodies. Figure 14.12 shows the possible collection point locations (point layer) developed from

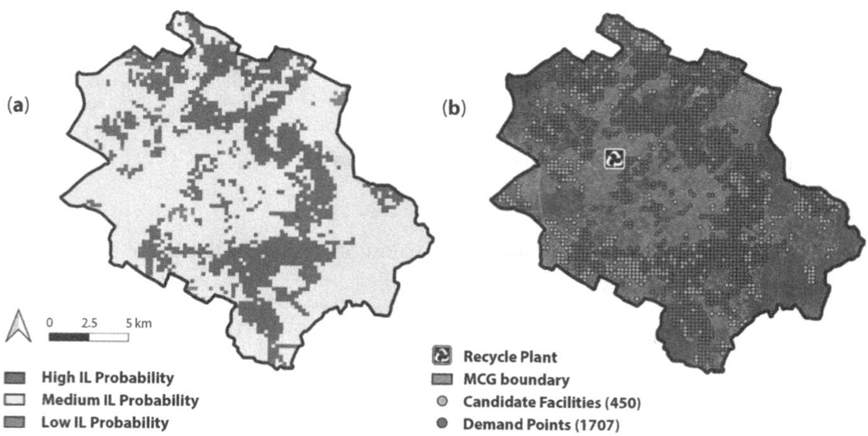

Figure 14.12 MCG CDW demand points and feasible collection points (candidate facilities).

the combination of all the restrictive variables. This resulted in 450 candidate facilities for collection point infrastructure.

14.4.3 Emplacement by Network Analysis

The spatial modeling provided by network analysis should be used if the decision-makers want to set up a system that is more focused on CDW treatment than eventual disposal (collection points and transfer stations). Unlike the MCE approach, it allows smaller surface area infrastructure [22]. The transport network used for analysis consists of MCG's major roads. It was built using ArcGIS 10.8.

Figure 14.13 demonstrates the results of the four location-allocation problems. The impedance is taken as 3 km, as it is advised by CSE to place collecting points no more than 3 km from the locations of the generating points [2]. Minimizing impedance and maximizing coverage (MI/MC) indicate similar outcomes. With 20 and 15 total collection point caps, the chosen facilities serve 95% and 86% demand points respectively, leaving 75 and 234 demand points unattended [Figures 14.13(a and b)].

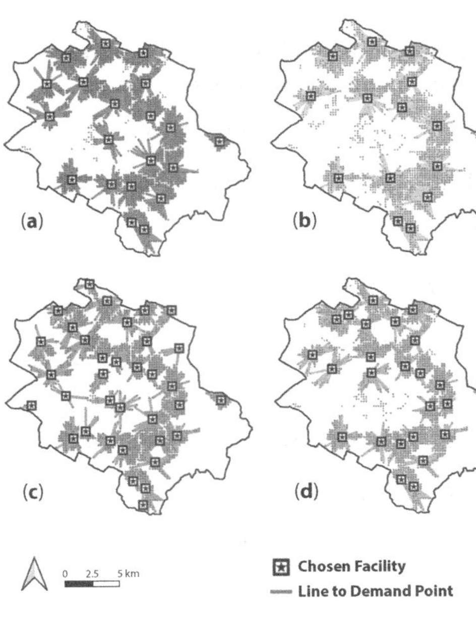

NO. OF FACILITY	(A) MI/MC	(B) MI/MC	(C) MF	(D) MA
35			6	
34			6	
33			6	
32			7	
31			15	
30			18	
29			21	
28			22	
27			24	
26			27	
25			38	36
24			39	40
23			40	40
22			43	40
21			45	43
20	40		48	45
19	40		49	49
18	43		50	58
17	46		55	59
16	58		57	61
15	63	40	58	63
14	65	45	60	63
13	68	66	61	63
12	70	69	62	68
11	74	73	65	68
10	79	89	67	68
9	88	91	68	71
8	92	92	68	73
7	94	107	75	79
6	107	116	77	81
5	107	119	78	84
4	113	120	82	85
3	114	142	84	88
2	132	145	85	88
1	136	159	86	93
SERVED	1629	1473	1692	1606
% SERVED	95%	86%	99%	94%

Figure 14.13 Results of different location-allocation models with 3-km impedance.

Minimum facilities (MF) serve almost all the demand points utilizing 35 collection facilities, only 15 points are unserved. This option handles more demand points than the others, but its construction and operation would be exceedingly expensive because 10 of the selected facilities would only be serving a small number of demand points [Figure 14.13(c)].

With 25 facility caps, the problem of maximum attendance (MA) can serve 94% of the demand, leaving 101 demand points unattended [Figure 14.13(d)]. Additionally, with facility caps of 20 and 30, MA serves 88% and 96% of the total demand, respectively.

14.5 Results and Discussion

Table 14.6 shows the findings of the MI/MC model's sensitivity analysis taking into account various impedance cutoffs and different facility caps. Figure 14.14 shows that a model with an impedance of 3 km and a total of 20 collection points, serving 95% of the demand, is an economical solution, as at present, there are officially 15 designated collection points.

Table 14.6 The findings of the MI/MC model's sensitivity analysis taking into account various impedance cutoffs.

Impedance cutoff (in km)	No. of facilities to choose from total of 450 facilities and unreachable demand points; percentage served						
		5	10	15	20	25	30
2.0 km	Unserved	-1300	-1014	-789	-596	-463	-360
	% Served	24%	41%	54%	65%	73%	79%
2.5 km	Unserved	-1089	-704	-453	-279	-171	-121
	% Served	36%	59%	73%	84%	90%	93%
3.0 km	Unserved	-946	-484	-234	-78	-40	-23
	% Served	45%	72%	86%	95%	98%	99%
3.5 km	Unserved	-750	-298	-85	-19	-9	-7
	% Served	56%	83%	95%	99%	99%	100%
4.0 km	Unserved	-580	-156	-24	-7	-7	-7
	% Served	66%	91%	99%	100%	100%	100%

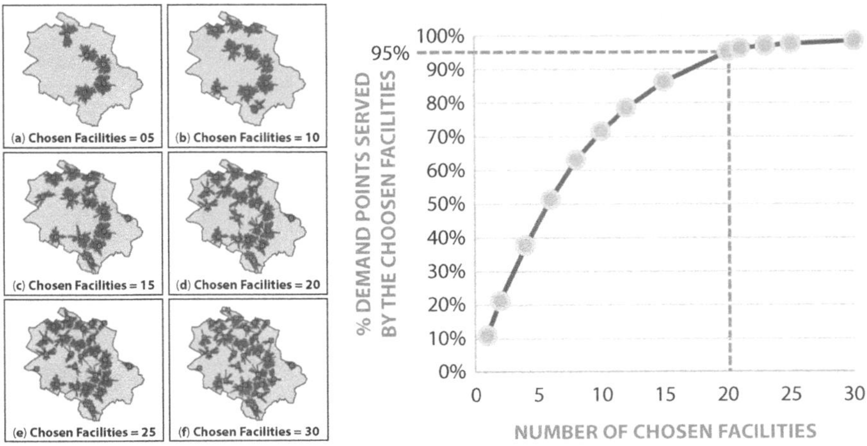

Figure 14.14 Sensitivity analysis of MI/MC model with 3-km impedance.

From the sensitivity analysis of the MI/MC model, 20 facilities within a 3-km distance from the demand points are taken as designated collection points (Figure 14.15). The CDW reception capacity of each chosen installation is calculated in Tables 14.7 and 14.8. Taking 1200 TPD CDW generation in MCG, each individual demand point generates 0.53 TPD waste.

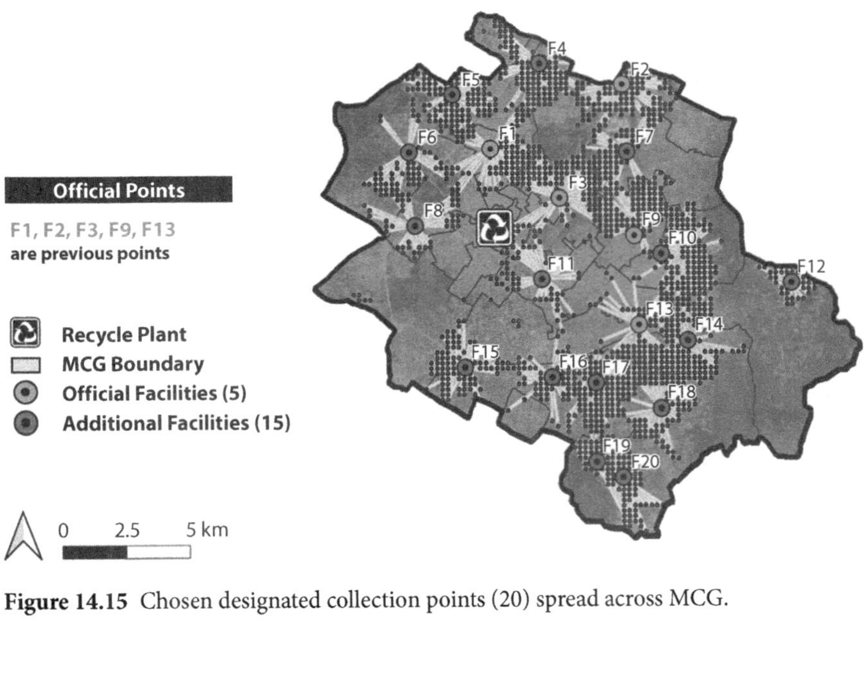

Figure 14.15 Chosen designated collection points (20) spread across MCG.

Table 14.7 CDW generation per demand point.

CDW generation in MCG		
MCG total generation	1200	TPD
Total demand points	1707	High
Total CDW generation by high probability areas (75%)	900	TPD
Total CDW generation by 1 demand point	0.53	TPD

Table 14.8 Individual facility reception capacity.

Facility name	**Demand count**	**Total length (in km)**	**TPD (tons per day)**	**Capacity (trucks/ day) 1 truck = 15 tonnes**
F1	79	156	42	3
F2	92	154	49	3
F3	94	184	50	3
F4	107	190	56	4
F5	113	202	60	4
F6	58	117	31	2
F7	74	131	39	3
F8	46	94	24	2
F9	107	189	56	4
F10	132	245	70	5
F11	40	78	21	1
F12	40	46	21	1
F13	65	111	34	2
F14	114	217	60	4
F15	88	164	46	3
F16	63	90	33	2
F17	136	219	72	5
F18	70	141	37	2
F19	43	75	23	2
F20	68	124	36	2
Total	1629	2927 km	859 TPD	57 trucks per day

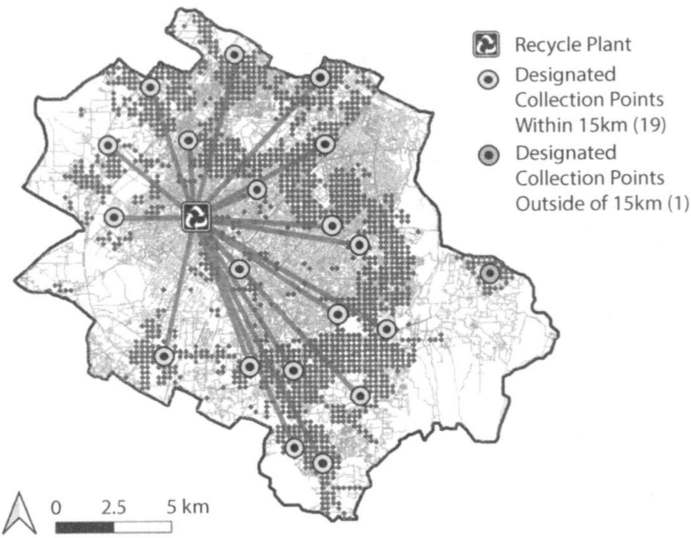

Figure 14.16 Map showing the distance of collection points from the recycling plant.

According to CSE, it is practical to build up transfer stations as tertiary storage/transfer stations in major cities since disposal sites are frequently greater than 15 km from the collecting point to reduce transportation costs [2]. Solving the minimum impedance problem with a 15-km distance and recycling plant as the facility, 19 out of 20 chosen collection points are reachable, excluding only one facility (F12) serving 40 CDW generation demand points (Figure 14.16).

14.6 Conclusion

CDW generation is rising globally, and the problem is exacerbated in developing countries due to inadequate infrastructure, financing, and political factors. Safe disposal of garbage has replaced waste reduction as the primary focus of the new waste management paradigm. As illegal dumping becomes increasingly apparent, it is important to understand how territorial and socioeconomic variables affect their spatial pattern. By testing on current illegal dumpsites in Gurugram, this study established that settlement patterns do indeed have a significant effect on predicting illegal dumps. However, the urban development pattern has the highest influence: illegal dumping increases as we move away from the city center, near rural areas, with less construction density. This suggests that, while catering to the dense city center, there is a larger requirement for waste collection points in the upcoming development areas. We used various location-allocation models to compare the demand served through candidate collection points: 20 chosen collection points model serving 95% of the demand within a 3-km travel distance was selected. Future research into CDW management should focus on establishing a clearer picture of how other related variables influence spatial prediction for illegal dumping. Finally, we hope the readers will get some insights and our approach might be a useful resource for them attempting in order to curb the practice of illegal dumping in urban regions.

Acknowledgment

We are thankful to the Indian Institute of Technology Kharagpur for the financial and infrastructure support.

References

1. S. Gupta and R. K. Malik, "The Impact of C&D Waste on Indian Environment: A Critical Review," *Civ. Eng. Res. J.*, vol. 5, no. 2, pp. 57-63, 2018.
2. A. Somvanshi and A. Verma, "*ANOTHER BRICK OFF THE WALL: Improving Construction and Demolition Waste Management in Indian Cities*," Centre for Science and Environment, New Delhi, 2020.
3. TIFAC, "Utilization of Waste from Construction Industry," Technology Information, Forcasting and Assessment Council, New Delhi, 2001.
4. NITI-Aayog, "Strategy on resource efficiency in construction and demolition sector," 2019.
5. M. Ramanathan and V. G. Ram, "Status of C&D Waste Recycling in India," *Sustainable Environmental Geotechnics*, Lecture Notes, vol. 89, pp. 95-106, 2019.
6. S. Jain, S. Singhal and N. K. Jain, "Construction and Demolition Waste Generation in Cities in India: An Integrated Approach," *Int. J. Sustain. Eng.*, vol. 12, no. 5, pp. 333-340, 2019.
7. A. R. Sekhar, Krishna Chandran and Vaibhav Rathi, "Training Manual on Construction and Demolition Waste Management in India for Cities and Towns," DEUTSCHE GESELLSCHAFT FÜR INTERNATIONALE ZUSAMMENARBEIT (GIZ) GmbH, New Delhi, 2017.
8. S. Silvestri and M. Omri, "A method for the remote sensing identification of uncontrolled landfills: Formulation and validation," *Int. J. Remote Sens.*, vol. 29, no. 4, pp. 975-989, 2008.

9. R. Jorda-Borrell, F. Ruiz-Rodríguez and Á. L. Lucendo-Monedero, "Factor analysis and geographic information system for determining probability areas of presence of illegal landfills," *Ecol. Indic.*, vol. 37, no. PART A, pp. 151-160, 2014.
10. K. Glanville and H. C. Chang, "Mapping illegal domestic waste disposal potential to support waste management efforts in Queensland, Australia," *Int. J. Geogr. Inf. Sci.*, vol. 29, no. 6, pp. 1042-1058, 2015.
11. G. Biotto, S. Silvestri, L. Gobbo, E. Furlan, S. Valenti and R. Rosselli, "GIS, multi-criteria and multi-factor spatial analysis for the probability assessment of the existence of illegal landfills," *Int. J. Geogr. Inf. Sci.*, vol. 23, no. 10, 2009.
12. A. L. Lucendo-Monedero, R. Jordá-Borrell and F. Ruiz-Rodríguez, "Predictive model for areas with illegal landfills using logistic regression," *J. Environ. Plann. Manage.*, vol. 58, no. 7, pp. 1309-1326, 2015.
13. L. C. Quesada-Ruiz, V. Rodriguez-Galiano and R. Jordá-Borrell, "Characterization and mapping of illegal landfill potential occurrence in the Canary Islands," *Waste Manage.*, vol. 85, pp. 506-518, 2019.
14. D. H. F. d. Paz, K. P. V. Lafayette and M. d. C. Sobral, "GIS-based planning system for managing the flow of construction and demolition waste in Brazil," *Waste Manage. Res.*, vol. 36, no. 6, pp. 541-549, 2018.
15. N. Alavi, G. Goudarzi, A. Babaei, N. Jaafarzadeh and M. Hosseinzadeh, "Municipal solid waste landfill site selection with geographic information systems and analytical hierarchy process: A case study in Mahshahr County, Iran," *Waste Manage. Res.*, pp. 98-105, 2013.
16. O. Demesouka, A. Vavatsikos and K. Anagnostopoulos, "Suitability analysis for siting MSW landfills and its multicriteria spatial decision support system: Method, implementation and case study," *Waste Manage.*, pp. 1190-1206, 2013.
17. R. Rafiee, N. Khorasani, A. Mahiny, A. Darvishsefat, A. Danekar and S. Hasan, "Siting Transfer Stations for Municipal Solid Waste Using a Spatial Multi-Criteria Analysis," *Environ. Eng. Geosci.*, pp. 143-154, 2011.
18. C. Bosompem, E. Stemn and B. Fei-Baffoe, "Multi-criteria GIS-based siting of transfer station for municipal solid waste: The case of Kumasi Metropolitan Area, Ghana," *Waste Manage. Res.*, pp. 1054-1063, 2016.
19. G. Tavares, Z. Zsigraiová and V. Semiao, "Multi-criteria GIS-based siting of an incineration plant for municipal solid waste," *Waste Manage.*, pp. 1960-1972, 2011.
20. J. Araiza, H. Nájera, R. Gutiérrez and M. Rojas, "Emplacement of solid waste management infrastructure for the Frailesca region, Chiapas, México, using GIS tools," *Egypt. J. Remote Sens. Space Sci.*, pp. 391-399, 2018.
21. A. Aremu, B. Sule, J. Downs and J. Mihelcic, "Framework to Determine the Optimal Spatial Location and Number of Municipal Solid Waste Bins in a Developing World Urban Neighborhood," *J. Environ. Eng. ASCE*, pp. 645-653, 2012.

22. J. A. Araiza-Aguilar, C. Gutiérrez-Palacios, M. N. Rojas-Valencia, H. A. Nájera-Aguilar, R. F. Gutiérrez-Hernández and R. A. Aguilar-Vera, "Selection of Sites for the Treatment and the Final Disposal of Construction and Demolition Waste, Using Two Approaches: An Analysis for Mexico City," *Sustainability*, 2019.
23. C. Valeo, B. Baetz and I. Tsanis, "Locations of recycling depots with GIS," *J. Urban Plan. D-ASCE*, pp. 93-99, 1998.
24. R. Vijay, A. Gautam, A. Kalamdhad, A. Gupta and S. Devotta, "GIS-based locational analysis of collection bins in municipal solid waste management systems," *J. Environ. Eng. Sci.*, pp. 39-43, 2008.
25. P. Densham and G. Rushton, "Strategies for solving large location-allocation problems by heuristic methods," *Environ. Plan. A.*, pp. 289-304, 1992.
26. E. S. R. Institute, "Location-Allocation Analysis," [Online]. Available: https://desktop.arcgis.com/en/arcmap/10.5/extensions/network-analyst/location-allocation.htm#ESRI_SECTION1_F8182D9F421E4EA4AEE11E7B360E1340.
27. S. Hakimi, "Optimum location of switching centers and the absolute centers and medians of a graph," *Oper. Res.*, pp. 450-459, 1964.
28. S. Hakimi, "Optimum distribution of switching centers in a communication network and some related graph theoretic problems," *Oper. Res.*, pp. 462-475, 1965.
29. H. Miller and S. Shaw, "*Geographic Information Systems for Transportation: Principles and Applications*," Oxford University Press: New York, NY, SA, p. 458, 2001.
30. R. Church and C. ReVelle, "The maximal covering location problem," *Pap. Reg. Sci.*, pp. 101-118, 1974.
31. C. Toregas, R. Swain, C. ReVelle and L. Bergman, "The location of emergency service facilities," *Oper. Res.*, pp. 1363-1373, 1971.
32. H. Rajagopalan, C. Saydam and J. Xiao, "A multiperiod set covering location model for dynamic redeployment of ambulances," *Comput. Oper. Res.*, pp. 814-826, 2008.
33. J. Holmes, F. Williams and L. Brown, "Facility Location under a maximum Travel Restriction: An example using Day Care facilities," *Geogr. Anal.*, pp. 258-266, 1972.
34. A. Biswas, K. Chaudhary, R. Singh, S. Tewari, S. Singh and S. Parida, "WASTE-WISE CITIES: Best practices in municipal solid waste management NITI Aayog," Centre for Science and Environment, New Delhi, 2021.

15

Assessment of Human Health Risk in Baitarani Basin, Odisha Using Water Quality Index (WQI), Cluster Analysis (CA), and Geographic Information Systems (GIS)

Abhijeet Das

Department of Civil Engineering, C.V. Raman Global University (CGU), Bhubaneswar, Odisha, India

Abstract

There has been a battle for decades between the health of natural ecosystems and artificially built systems. Globally, ecologists and environmentalists have become interested in riverfront development because of its socio-economic and ecologically sensitive features. All river basins, including Baitarani, have been contaminated by unexpected human activity. In the present study, water samples were gathered from 13 locations throughout PRM and POM, for its evaluation of spatial variations and its interpretation and also, assessed for suitability for drinking by using 22 parameters. Most of the parameters were discovered to fall within allowable limits of the Bureau of Indian Standards (BIS) among others, except for turbidity, DO, and TC. Water quality index (WQI) analysis was computed using five indexing methods: weighted arithmetic (WA) WQI, overall index of pollution (OIP), comprehensive pollution index (CPI), synthetic pollution index (SPI), and Nemerow pollution index (NPI). Geographical Information Systems (GIS) was employed to outline the spatial diagrams utilizing the inverse distance weighted (IDW) technique. Complex datasets about water quality were evaluated using multivariate statistical analysis techniques including cluster analysis (CA). Hence, to accomplish the goals outlined here, a study methodology that included lab tests, a questionnaire survey, and a mathematical model was developed. The findings of the WQIs indicate that excellent to poor (PRM) and good to poor (POM) in WA

Email: das.abhijeetlaltu1999@gmail.com; ORCID: https://orcid.org/0000-0003-4599-5462

Disha Thakur, Sanjay Kumar, Har Amrit Singh Sandhu and Chander Prakash (eds.) Sustainable Development Using Geospatial Techniques, (365–410) © 2024 Scrivener Publishing LLC

WQI, good to very poor (PRM/POM) in SPI, excellent to poor (PRM), and good to very poor (POM) in NPI, sub-clean to polluted water (PRM), and qualified to polluted (POM) in CPI and excellent to slightly polluted water in case of OIP. CA divides 22 water variables and 13 survey locations into three main clusters with associated surface water features using dendrogram plots. Based upon an integrated technique of various WQ indexes with GIS and CA, it is ascertained that sites from points S-8 represent poor-quality water. It was determined that the decreasing water quality was caused by an increase in the variety of anthropogenic activities, industrial waste dumping, and agricultural waste disposal on the river channel. Also, the rise in WQI values made it abundantly evident that the river's water quality had declined following channelization. Likewise, even after a significant amount of riverfront development, the primary problem of domestic sewage discharge into the river system with little to no treatment appears to remain unresolved. However, water at alerting stations that are susceptible to pollution needs to be treated first before being consumed.

Keywords: Anthropogenic, water quality index, GIS, CA, pollution

15.1 Introduction

Several researchers suggest that there is ample evidence from observational records and climatic forecasts that freshwater supplies are susceptible to significant impacts from climate change, which might have far-reaching effects on ecosystems and human societies. Generally, freshwater is a crucial resource for maintaining life, the environment, and human progress [1, 2]. Human survival depends on having access to safe water [3]. However, good water quality, therefore, promotes a healthier environment, which enhances human life [4, 5]. The primary supply of water for home, agricultural, and industrial uses is surface water bodies like rivers, lakes, and wetlands [6]. Most rivers have recently been subjected to high pollution loads brought on by the discharge of industrial effluents and untreated or only partially treated wastewater [7]. Runoff from agricultural activities and unprocessed large-scale mining waste are being dumped carelessly, significantly polluting the water bodies [8]. In addition to heavy metals, the wastewater from these sources also includes harmful organic and inorganic compounds [9] that pose substantial health risks and seriously harm the ecosystem, such as eutrophication [10]. Surface runoff from farmlands, which transports man-made pollutants, herbicides, and fertilizers, has also had an impact on the waterways [11]. The water quality of Indian rivers, including the river Baitarani, has been declining over the past few decades because of the continual discharge of partially/untreated industrial effluent, urban runoff, and sewage from point and non-point sources [12, 13]. When evaluating the overall water resource quality of any given location, both geogenic and anthropogenic influences become significant [14, 15]. Numerous research studies on various rivers in India have found that aquatic life is declining and the water quality is terrible [16]. The pathetic state of the rivers cannot be attributed to a single source; rather, the situation has reached these proportions because of the cumulative influence of many interrelated variables. Due to the highly dynamic character of the entire river system, development in a tiny portion of the river further alters the channel, reducing the likelihood of restoration. In order to address the rising demand and declining water quality, there are numerous water quality indices available for determining the quality state of every river body [17]. It has been reflected by many authors that the foremost WQI was created by Horton [18], which reflects arithmetic weighting along with multiplicative indicators [19]. To determine whether surface water is suitable for drinking [20], the weighted arithmetic water quality index method (WAWQI) was created, where water quality is categorized based on the most prevalent water quality characteristics and according to its degree of purity [21]. As a result of its many benefits over alternative techniques, this has been employed by numerous scientists [22]. One of the important objectives is that it can describe the suitability of both surface and groundwater for drinking purposes while being free from overestimation or underestimation [23], which is generally based on a mathematical equation that includes several water quality parameters without depending on specific parameters [24] and often considered to be bias-free [25]. Additionally, other alterations were produced in compliance with the characteristics, and multiple indexes were produced, including the comprehensive pollution index (CPI), the overall index of pollution (OIP), and, ultimately, the Nemerow pollution index (NPI) [26]. OIP promotes an understanding of the state of the surface water resources' water quality, particularly in Indian circumstances [27]. Furthermore, NPI was determined to assess how each variable affects the level of pollution source and its origin [28]. On the other hand, CPI was studied by Kamboj *et al.* [7] and Matta *et al.* [29] to categorize the current state of water quality. This research also examined the synthetic pollution index (SPI), yet also, considered as a valuable model, to validate the pollution level determined by WQI. It was created by Ma *et al.* [30] to evaluate how pollutants affected the quality of the water, and because of how straightforward its technique was, it was afterward used by many researchers [31]. Because of this, introducing a spatial approach, namely, Geographic Information Systems (GIS), could help in communicating to the public about the status of the river

Baitarani. GIS programs include geographical analysis tools with the capacity to handle enormous data [32, 33]. Techniques for surface interpolation help to anticipate concentrations in places, where samples could not be collected [34]. The method of IDW classification of spatial map interpolation was generating the parameters for the study area [35]. In IDW, the projected value is more influenced by points near the expected position than by points far from the predicted position [36]. The WQI and IDW interpolation method has been adopted by several researchers using GIS tools that are interested in analyzing surface water for drinking reasons. Using the WQI and GIS, Sarath Prasanth *et al.* [37] evaluated the surface water quality in the premise of Tamil Nadu region of India, and the results revealed that most of the surface water samples are fit for human consumption. This will improve the research output's accessibility and comprehension. For water quality monitoring, multivariate statistical methods are one of the top objectives when it comes to protecting surface water. For many years, they have been effectively employed in hydrochemistry. This aids in the understanding of water quality while interpreting intricate datasets. As a result, these methods are frequently used for different environmental data. So, cluster analysis (CA) was especially helpful in examining multiple random environmental factors at once, leading to the discovery of a new, smaller collection of uncorrelated variables that explained a significant amount of the variance in the original variables. This acts as a method for identifying pollution sources and aids the government in reducing water pollution [38]. CA is one of the techniques used to assess the degree of homogeneity in the observed parameters [39]. Moreover, using a preset selection criterion, each object in the cluster is classified using this method so that they can all be comparable to each other. The item clusters need to be loosely packed together within the cluster and homogeneous within the cluster. The objective is to compare data from related or comparable groups to one another and to data from other groups; the better or more distinct the clustering, the larger the difference across groups. Therefore, thorough investigations should be carried out to determine the causes of pollution and assess the water quality. To the best of the authors' knowledge, this reconnaissance research is the first to provide a thorough evaluation of the pollution situation throughout the whole length of the Baitarani River in Odisha. However, little effort has been made to evaluate the quality of the water. Very few researchers address the effects of declining water quality and not even a single piece of work has been done that integrates WQI, CPI, NPI, SPI, OIP, and CA to evaluate surface water quality. Consequently, there is a research void and more debate is necessary to understand the extent and causes of its decrease in safe water for consumption. Considering these factors, the impact of various physicochemical and biological parameters on the Baitarani River's water quality has been evaluated to discuss whether the water is suitable for drinking. This has been done using integrated approaches combining several water quality indices with GIS, along with multivariate statistical analysis, such as CA. In the ongoing work, this survey was performed for thirteen years (2010–2023) and designed to understand the seasonal variation, i.e., pre-monsoon (PRM) and post-monsoon (POM) seasons. Thus, the findings of this study will offer convincing scientific support for the need to control this river's water quality. Finally, using the previously indicated methods, the source and interdependency of several water quality measures were also evaluated.

15.2 Study Area

This study sheds light on the Baitarani River Basin, Odisha, which is placed between 85°0′0" to 86°30′0" E longitude and 21°0′0" to 22° latitude. Generally, its area encloses a value of 8645 km^2, and its elevations ranged between 32 and 1181 m above mean sea level (MSL). In the catchment, summertime temperatures range from 30°C to 36°C and wintertime lows are between 16°C and 17°C. The basin has a variety of seasons, including cool winters, scorching, dry summers, and monsoon seasons. The basin receives 1628 mm of rain on average year, with a predominance of subhumid weather. The topography of the basin is undulating, with an average slope of between 0% and 2%. Because agriculture predominates in this region, crops like rice, maize, green gram, wheat, groundnuts, and vegetables are grown all year round. The industrial sector, which includes the production of cars, leather goods, handicrafts, and stone carvings, is essential to the economy. Surface water is particularly significant in the study region since it is used largely for bathing, domestic uses (water for human consumption, cooking, dishwashing, and laundry), and watering and irrigating vegetable plants. To protect human health, it is essential to look at the quality of drinking water and document the management procedures. Therefore, sampling sites were chosen based on the population, probable anthropogenic activity, and socioeconomic significance of the area. The sampling stations' geographic coordinates, notable locations, and delineation are shown in Figure 15.1.

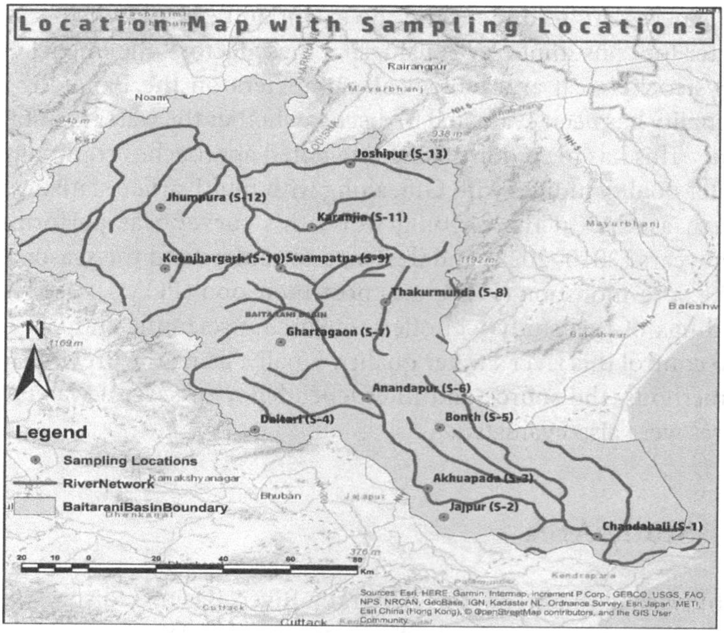

Figure 15.1 Index map of the study area.

15.3 Materials, Sampling, and Analysis

The water quality of the river Baitarani section in the State of Odisha is monitored by the Odisha State Pollution Control Board (OSPCB). The OSPCB database was used to collect the physicochemical monitoring data for the 13 years, i.e., 2010–2023. The surface water specimens were grouped in 1000-ml small plastic containers that were carefully labeled, rinsed, and stored in a cool box at 4°C and transported to the laboratory for analysis during pre-monsoon (PRM) (March-May) and post-monsoon (POM) (October-February) period following the standard methods recommended by American Public Health Association [40]. Specimens were taken every month at 13 monitoring stations in Odisha, and the raw data used in this study was represented as monthly average data. Additionally, care was taken to prevent sample agitation while being transported to the lab. At each location, a GPS receiver recorded the location of the sampling points. The methodology, which includes the software, tools, and analytical methods utilized to complete the job, is summarized in Table 15.1. Afterward, 13 sites were chosen for investigation of WQ

Table 15.1 Mathematical approaches implemented for the study.

Parameters	**Methods**
pH (mg/L)	Digital pH meter
Turbidity (NTU)	Digital nepheloturbidometer
Total dissolved solids [TDS (mg/L)]	TDS meter
Total suspended solids [TSS (mg/L)]	TSS meter
Electrical conductivity [EC (microsiemens/cm)]	Conductivity TDS meter
Dissolved oxygen [DO (mg/L)]	Modified Winkler's method
Alkalinity (mg/L)	Sulfuric acid titrimetric method
Biochemical oxygen demand [BOD (mg/L)]	Modified Winkler's method
Total hardness [TH (mg/L)]	EDTA titrimetric method
Bicarbonate (mg/L)	Titrimetric
Sulfate (mg/L)	Barium chloride
Nitrate (mg/L)	Ion selective electrode
Phosphate (mg/L)	Stannous chloride
Chloride [Cl (mg/L)]	Titration by sulfuric acid
Calcium [Ca (mg/L)]	EDTA titrimetric
Magnesium [Mg (mg/L)]	EDTA titrimetric
Sodium [Na (mg/L)]	Flame photometric
Potassium [K (mg/L)]	Flame photometric
Total coliform [TC (mg/L)] MPN/100 ml	Multiple tube fermentation
Fecal coliform [FC (mg/L)] MPN/100 ml	Multiple tube fermentation
Iron [Fe (mg/L)]	HACH colorimeter
Chromium [Cr (mg/L)]	Inductively coupled plasma mass spectroscopy
Computation of WQI	Drinking standards as per the Bureau of Indian Standards [43]
Spatial distribution maps	Inverse distance weighted (IDW) interpolation technique

(Figure 15.1). Twenty-two samples were examined utilizing the process described in the American Public Health Association [40]. To guarantee QA/QC protocols, about 500 cc of the quantity of materials was carefully collected and inspected in the laboratory. To confirm the reliability of the laboratory testing, material with an accuracy of less than 5% and charge balance faults were examined. IDW technique was chosen for creating interpolated maps using the spatial analysis tool from the toolbox. Additionally, the computer-generated base map for the area under study was developed by ArcGIS 10.5.

15.4 Methodology

15.4.1 Surface Water Quality (SWQ) Modeling

The SWQ of all 22 samples in this investigation was modeled using five indexing techniques, including WAWQI, SPI, NPI, CPI, and OIP. Those indexing models are deliberated in detail in the following sections.

15.4.1.1 WAWQI Model

The weighted arithmetic (WA) index method was used for the computation [41, 42] which typically uses several physical and chemical factors that are multiplied by a weighting factor before the final aggregate is calculated using the arithmetic mean. The weights allocated represented the importance of each parameter and its index was constructed by comparing water quality measures to regulatory requirements. The WQI calculations are broken down into three successive parts: The first step is to give each of the 22 characteristics a weight (w_i) based on how important they are to the quality of drinking water. The least significant parameter is given a weight of 2, while the most significant is given a weight of 5 (Table 15.2). By adopting the following equation $W_i = w_i / \sum_{i=1}^{n} wi$, the 2nd step is relative weight calculation. A quality rating scale (Q_i) for each parameter is computed by dividing each parameter's concentration in each water sample by its corresponding standard following the recommendations set out by the Bureau of Indian Standards [43], and then multiplying the result by 100 utilizing $Q_i = (C_i / S_i) * 100$. The water quality sub-index (SI_i) for each chemical parameter is finally established to compute the WQI, and the WQI is then calculated as $SI_i = W_i * Q_i$ and $WQI = \sum_{I=1}^{n} (SI)i$, where SI_i

Table 15.2 The relative weight of each selected parameter.

Parameters	Standard value BIS 10500:2012)	Weight (wi)	Relative weight (Wi)
pH	8.5	4	0.05882
Turbidity	5	4	0.05882
TDS	500	3	0.04412
TSS	500	2	0.02941
EC	2250	5	0.07353
DO	5	4	0.05882
Alkalinity	120	3	0.04412
BOD	5	3	0.04412
TH	300	2	0.02941
HCO_3^-	100	2	0.02941
SO_4^{2-}	250	3	0.04412
NO_3^-	45	5	0.07353
PO_4^{3-}	1.2	2	0.02941
Cl^-	250	4	0.05882
Ca^{++}	75	2	0.02941
Mg^{++}	50	2	0.02941
Na^+	200	2	0.02941
K^+	12	2	0.02941
TC	1000	5	0.07353
FC	300	2	0.02941
Fe^{++}	1	2	0.02941
Cr^{++}	0.5	2	0.02941

defines each parameter as a sub-index. Finally, this index has been distinguished into five groups: scores <50, excellent; between 50 and 100, good; 100 to 200, poor; 200 to 300, very poor; and finally, >300, unsuitable, as per Ramakrishnalah *et al.* [44].

15.4.1.2 SPI Model

After choosing the parameters and giving each parameter a unit weight, the SPI is obtained from the following expression, suggested by Ma *et al.* [30]. The following equation used is $\mathrm{SPI} = \sum_{i=1}^{n}\left(\frac{Vo}{Vs}\right) * Wi$, where V_o implies the analytical value of a variable and V_s illustrates as standard value, suggested by the Bureau of Indian Standards [43], which is highlighted in Table 15.3. By combining all the water quality parameters with the weightage, the critical value was found to be calculated as 1. It is further divided into five categories such as a value between 0 and 0.25 indicates excellent, values ranging from 0.25 to 0.50 imply good water, 0.50 and 0.70 belong under the poor zone, 0.75 to 1 denote very poor, and finally, scores >1 are referred as the worst quality of water.

15.4.1.3 NPI Model

This methodology is a composite index of pollution, which describes the cumulative effect of harmful pollutants for a certain use. Based on NPI [28], water quality pertains to 'excellent' when the score based on NPI is below 10; if it falls between 10 and 20, 20 and 30, 30 and 40, and >40, it signifies good, poor, very poor, and unsuitable category. According to Gauns *et al.* [45], it is determined as $PI_n = C_n/S_n$ and $NPI = \Sigma\ PI_n$. The variable PI_n signifies n^{th} parameters, while C_n talks about n^{th} observed value. Further, S_n depicts about n^{th} permissible reading.

15.4.1.4 CPI Model

CPI is performed to measure the overall level of water body pollution [46]. The CPI calculation formula [47] is expressed as $CPI = 1/n * \sum_{i=1}^{n} PIi$. The variable, namely, n, reflects about number of monitoring parameters, and $(PI)_i$ talks about the pollution index number i. It is noticed that $(PI)_i$ is constructed as $(PI)_i = Ci/S_i$, where C_i belongs to measured concentration; S_i pertains to permitted limitation as set by environmental standards. Additionally, it is divided into five categories, namely, the clean water group (suggests a score between 0 and 0.20), the value between 0.21 and 0.40 denotes sub-clean zone; reading between 0.41 and 1.00, refereed to be slightly polluted; a score from 1.01 to 2.00 belongs to polluted class; and finally, score >2.01, represents heavily polluted water.

15.4.1.5 OIP Model

The OIP index provided information regarding the quality of the water in India [27]. The following equation was used to estimate the final value as $OIP = (1/n) * \sum_{i=1}^{n} Pi$. The variable 'Pi' defines a value for pollution of the ith parameter, 'n' represents the quantity of various water variables, Pi = [V_n (experimental reading)/V_s (standard reading)]. As per its categorization, if the value of OPI is less than 1.9, it is said to be in an excellent zone and grouped to be in class C_1. Furthermore, if the score is around a value of 3.9, then the water refers to be 'acceptable' for consumption and it belongs in the class C_2 group. Hence, polluted (class C_3) ranged a value of >7.9, polluted (class C_4) spans a reading of >15.9, and finally, heavily polluted (class C_5) belongs to a value greater than 16, respectively.

15.4.2 Cluster Analysis (CA)

This is a type of approach that promotes the collection of multivariate strategies that primarily arrange variables or instances (observations or samples) into clusters based on their strong homogeneity within classes and high variability between classes [48]. Dendrograms, a two-dimensional chart that provides an understandable pictorial representation of the procedure, are commonly used to illustrate it [49]. To reduce the number, CA was used to classify all sampling sites into a cluster. Hierarchical CA (HCA) is used to quantify the separation of sample parameters. Indicators of a stronger relationship between samples/variables or a cluster of sampling sites/variables are branches with a linkage that is closer to one another. Clusters are formed by combining the points that are the most similar, and this mechanism is further elongated until all the points belong to the same cluster [50]. In this specific study, Ward's linkage model is incorporated, which makes use of squared Euclidean distances, and is applied for analysis.

15.5 Results and Discussions

The statistical data of surface water samples taken throughout the PRM and POM seasons are presented in Table 15.3 along with comparisons to data from the Bureau of Indian Standards (BIS 2012). Several parameters taken into consideration in this work are illustrated in Figures 15.2(a–v). The most important factor, i.e., water's pH, indicates how many hydrogen ions are present and whether it is acidic or basic. A score of 6.5 to 8.5 range is suggested for drinking water (BIS 2012). The value changes in the study region from 7.4 to 9 during the POM and from 7.3 to 9.7 during the PRM (Figure 15.2a). A few data points have pH readings higher than 8.5 and the majority of samples are within the threshold range. Its alkaline nature suggests that hard water minerals are present, and the discharge of water from farming lands may be a factor in alkalinity-influencing variables [51]. Higher pH was witnessed in S-6, S-9, and S-10 in PRM and S-10 in POM, indicating an increase in the productivity of the river. This could be a result of local industries' effluent seeping into the water and polluting it. The term "turbidity" refers to the cloudy look that particles, also referred to as suspended matter, generate in water [52]. Although it may only be measured with measuring tools, a white, black, or brown muddy suspension may be visible over 5 NTU. The values were observed between 8.2 and 25.20 NTU and 11.8 and 38.70 NTU in PRM and POM seasons, respectively (Figure 15.2b). All values obtained during both seasons were above the permissible limits (5 NTU) as per the Bureau of Indian Standards [43]. Inorganic substances like manganese and iron may be responsible for this deterioration. Such inorganic chemicals are deemed undesirable in excess because they have a negative impact on water distribution and sanitation systems [53]. TDS is made up mostly of inorganic salts that have been dissolved in surface water, including Ca^{2+}, Mg^{2+}, K^{+}, Na^{+}, HCO_3^{-}, Cl^{-}, and SO_4^{2-} [54]. These are frequent components of many effluents, but their toxicity or chemical makeup is often not well known [55]. The total dissolved mineral content in water serves as a broad indicator of the water's general appropriateness for a variety of functions. Customers frequently complained that drinking water with greater TDS had a bad taste [56]. In the present study, the value of TDS varies between 74 and 178 in PRM and 97 and 247 mg/L in POM (Figure 15.2c). All locations are under the optimal limit of 500 mg/L. TSS measures how many inorganic salts and how little organic materials are dissolved in water [57]. A low value denotes a bland flavor, whereas a high value might change the flavor and hardness of the water [58]. Higher values are explained by evaporation and decomposition processes that occur more quickly, as well as by the addition of surface runoffs after heavy precipitation [59]. The values vary between 30 and 121 mg/L in PRM and 43 and 127 mg/L in POM (Figure 15.2d). The appropriate TSS for human intake is taken at 500 mg/L. All samples are within the acceptable limits. Calculating conductance, which is directly connected to the ion concentration in the water, is a method for estimating the electric flow capacity of water. However, EC should not be greater than 2250 microsiemen/cm for drinking purposes. Lower levels are attributed to precipitation dilution [60]. Pollution may be indicated by a sudden increase or decrease in the conductivity of the water body. Anions like chloride, phosphate, and nitrate will be added, which will cause sewage discharge or agricultural runoff to have increased conductivity [61]. When water's EC is high, its TDS value is also high. Values observed in the current experiment during the PRM period varied from 96 to 318 and during the POM period from 121 to 393 and it is visible clearly in Figure 15.2e. All sites are within the prescribed limits. A water body's DO is influenced by a wide range of indicators, namely, photosynthetic activity, wind speed, temperature, and domestic waste [62]. It varies daily, seasonally, and with temperature changes [63]. The suggested concentration taken is 6 mg/L for human intake. The concentration varies from 4.78 to 8.01 in PRM and 5.03 to 7.69 in POM (Figure 15.2f). The river was pumped with one million gallons of the previous water during the previous month, which caused a slight increase in most of the tests [64]. Alkalinity reveals the water's capacity to serve as a buffer, which will help to keep its pH stable [65]. It gauges how effectively water can neutralize acids [66]. It is noted that dissolved CO_2, which is present in high quantities in many water channels, is the principal source of alkalinity. Due to the rapid rate of decomposition and subsequent release of CO_2, the addition of carbonate and bicarbonate ions cause its value to rise. Because of this, its value rises to be high [67]. The value should not exceed 120 mg/L. Regarding the research domain, the reported score of alkalinity varied from 43 to 99 mg/L in PRM and 69 to 99 mg/L during POM (Figure 15.2g). BOD is the volume of oxygen needed for microorganisms to successfully break down the organic matter in each water sample over a specified period and temperature. Low BOD values imply more toxicity in river water and the presence of home and industrial effluents in the water body; the more waste, the lower the need for oxygen [68]. The values varied from 0.86 to 4.23 mg/L and 0.88 to 4.54 mg/L in PRM & POM (Figure 15.2h). The results are within the BIS standard of 5.0 mg/L by a significant margin [69]. Although TH is referred to as water's ability to engage in a water reaction, hard water requires substantially more soap to produce a lather. High hardness can inescapably result from industrial waste handling untreated and improperly treated waste [70]. The permissible level of TH for human consumption is 300 mg/L, as reported by WHO norms [43]. In the recent study, the observed values ranged from 64-121 mg/L and 71-135 mg/L in PRM and POM, respectively (Figure 15.2i). Hence, in contrast to PRM, it is seen that there is a greater change in the hardness of the samples in POM because of the absorption of Ca^{2+} and Mg^{2+} salts in the recharging [71]. Each research site is significantly below the established limitations in both seasons. HCO_3^- concentration ranges from 41.92 to 87.55 mg/L during PRM and 55.64 to 91.46 mg/L during POM (Figure 15.2j). A higher HCO_3^- content could be a result of silicate weathering and disintegration. The fact that concentrations were marginally greater in POM than in PRM suggests that the weathering process contributed carbonate. Based on the Bureau of Indian Standards [43], the maximum for HCO_3^- is 100 mg/L. All samples are within the limits. There is a small change in the seasonal and spatial distribution that may be caused by the mixing of salt patches that are naturally present in the soil profile and by the rise in the water table [72]. Gypsum and other common minerals leach into water, causing the presence of SO_4^{2-} to accumulate naturally. Fertilizers can also be used to treat water to increase SO_4^{2-} levels [73]. Potential sources of sulfate in rocks include sulfur minerals, heavy metal sulfides that are typically present in rocks, limestone, and evaporites seen in sedimentary rocks, and intake because of volcanic eruptions, biochemical characteristics, human activity, and other variables [74]. However, in addition to using natural sources, fertilizers can be used to increase it. In the current research, the reported measurements varied between 2.40 and 6.87 mg/L in PRM and 2.31 to 7.16 mg/L in POM (Figure 15.2k). The measured levels were well below the BIS-recommended allowable range of 250 mg/L. Higher amounts of POM than PRM could be the result of leaching and anthropogenic activities that release sulfur gasses into the environment from industry and municipal utilities. The accumulation of nitrate in drinking water is a particular concern because of the potential negative impact of surface water use on human health. The maximum nitrate concentration in water for human consumption is 45 mg/L [43]. Higher concentrations can be caused by agricultural operations, septic tank leaks, unlined drainage and sewerage pipelines, home sewage, leaching from careless dumping of animal waste, and other factors. In addition to agricultural operations, a higher concentration may also be related to the disposal of animal dung, and mostly fertilizer bags that are cleaned by return flow irrigation, infiltration, and drinking practices when entering surface water [75]. In the present study, the values varied from 0.81 to 4.86 mg/L and 0.65 to 4.15 mg/L in PRM and POM, respectively (Figure 15.2l). All the samples in both seasons are within the limits. PO_4^{3-} is one of the elements limiting any water body's production. Agricultural runoff containing fertilizers and domestic wastewater are the principal sources of phosphate entering river ecosystems. Higher concentrations are a sign of pollution, and sewage, detergents, agricultural effluents, and fertilizers are the main sources of anthropogenic phosphorous. The concentration ranges from 0.25 to 1.04 in PRM and 0.31 to 1.17 in POM (Figure 15.2m). The limit as per the Bureau of Indian Standards [43] is 1.2 mg/L. All the sites are within the maximum allowable limit. More hazardous to human health is water with a greater Cl^- content [76]. The associated cation interferes with the locals' ability to taste the water that they drink, living beside the river channel. The rising Cl^- contents in surface water appear to be the result of geogenic or human activities. The principal sources include soil, animal waste, urban and industrial wastes, and inorganic salts like Na-Cl and K-Cl in the water. It is noticed to be an important indicator of water degradation [77]. The concentrations should not go above 250 mg/L. Chloride levels in these samples range from 7.87 to 28.18 in PRM and 8.72 to 28.86 in POM (Figure 15.2n). The fifth most common element in nature, Ca^{2+}, dissolves from rocks and soils and is the primary factor in water hardness. Inadequate calcium consumption can cause osteoporosis, malformed teeth, nephrolithiasis (kidney stones), rickets, hypertension, and stroke [78]. Possible causes of high calcium levels include sedimentary rocks and carbonate rocks. Ca^{2+} is crucial for human cell activity, hormone production, cancer, heart illness, blood pressure regulation, muscular contraction, neurological disease, and testicular descent [79]. The suggested upper limit for drinking water is 75 mg/L [43]. Calcium concentrations range from 14.83 to 28.72 mg/L in PRM and 14.03 to 29.74 mg/L in POM (Figure 15.2o). All locations are in the limits. Several types of rocks, sewage, and industrial garbage are the main links of Mg^{2+} in the water that is used for drinking [80]. High Mg^{2+} is caused by magnesium's greater solubility in water compared to calcium's corresponding molecules, which favors an increase in Mg concentration. Additionally, a high concentration in drinking water might produce unfavorable flavors that have laxative effects [81]. The range of magnesium vales in PRM is 1.58 to 4.63 mg/L, and in POM, it is 2.36 to 5.83 mg/L (Figure 15.2p). For drinking purposes, the necessary permissible limit is 50 mg/L. The majority of the samples are below the permitted level. Because sodium increases soil hardness and decreases permeability, it is crucial to consider the Na^+ concentration when assessing the quality of surface water for irrigation [82]. High salt concentrations have the potential to negatively impact soil permeability and structure, resulting in alkaline soils. Na^+ can contaminate natural water sources. The optimum concentration in potable water is taken as 200 mg/L. The value of Na^+ spans from 2 to 10.10 mg/L in PRM and 3.6 to 13.30 mg/L in POM (Figure 15.2q). All locations satisfy the BIS limits. Since K^+ salts are more soluble than sodium salts, they crystallize later during evaporation. It is a crucial component that all humans need. Furthermore, employing $KMNO_4$, which serves as an oxidant during the process of treating water, thereby helps in removing pollutants from drinking water [83]. K^+ is a vital nutrient, but when ingested in excess, it can have laxative effects. Potassium is less common than sodium in all water bodies, which may be a result of the mineral's limited ability to migrate and resilience to weathering-induced breakdown. In the ongoing area, the value fluctuates between 0.7 and 3.2 mg/L and 0.8 and 2.9 mg/L in PRM and POM, respectively (Figure 15.2r). The maximum limit prescribed by the Bureau of Indian Standards [43] is 12 mg/L. All samples in both seasons, satisfy the limits. In the normal gut flora of people and animals, TC and FC are abundant and typically do not harm [78]. Serious illnesses like bacteremia, meningitis, and urinary tract infections may be caused by it. These can prosper as well as develop systems for delivering water, especially in the context of biofilms [78]. The value of TC varies between 970 and 8000 MPN per 100 ml in PRM and 2500 and 11,000 MPN per 100 ml in POM (Figure 15.2s). The limit set by the Bureau of Indian Standards [43] is 1000 MPN/100 ml. All places exceed the limitations, suggesting the infiltration process and stormwater runoff from urban areas, indicating that the bacteriological quality of water exhibits poor grades and needs to be treated before consumption in these areas [84]. The range of FC varied between 70 and 360 in PRM and 90 and 510 MPN per 100 ml (Figure 15.2t). The limit prescribed is 300 MPN per 100 ml. Most of the samples satisfied well within the limits in both seasons except site 8. Higher levels are found at site 8 as a result of anthropogenic activities such as excessive nitrogen fertilizer use and poor waste disposal. Fe^{2+} is the core problem with rural drinking water [79]. Although a low iron content is necessary for human nutrition and plant metabolism and cannot cause significant harm, it promotes the growth of unfavorable bacteria inside waterworks and supply systems, which leads to the deposition of a slushy coating on the piping. In addition, high Fe^{2+} level >1 mg/L, leads to an overabundance that may cause nausea, stomach issues, vomiting, diabetes, hemochromatosis, and other health issues. The concentration varies from 0.19 to 1.08 mg/L and 0.13 to 1.43 mg/L in PRM and POM, respectively (Figure 15.2u). Except for site 8, the majority of the samples meet the requirements for both seasons. Cr^{2+} is a necessary component for the proper functioning of various cellular enzymes, including hydrolases, kinases, transferases, and decarboxylases, as well as for the metabolism of humans, animals, and plants. A surplus of it produces unpleasant flavors, odors, colors, stains, and rust [78]. The concentration varied between 0.05 and 0.17 mg/L in the case of PRM and 0.06 and 0.15 mg/L in POM (Figure 15.2v). The desirable limit set by the Bureau of Indian Standards [43] is 0.5 mg/L. All the sampling locations are within the safe limits. The degree to which the river water quality deviates from the required level for human consumption cannot be inferred explicitly from the values of these metrics [85]. This study's objective is to examine the water quality using the SPI, NPI, CPI, WQI, and OIP, as well as to identify the sources of pollution using multivariate statistical tools. This allowed us to suggest necessary remedial procedures taken to make the water fit for ingestion by people.

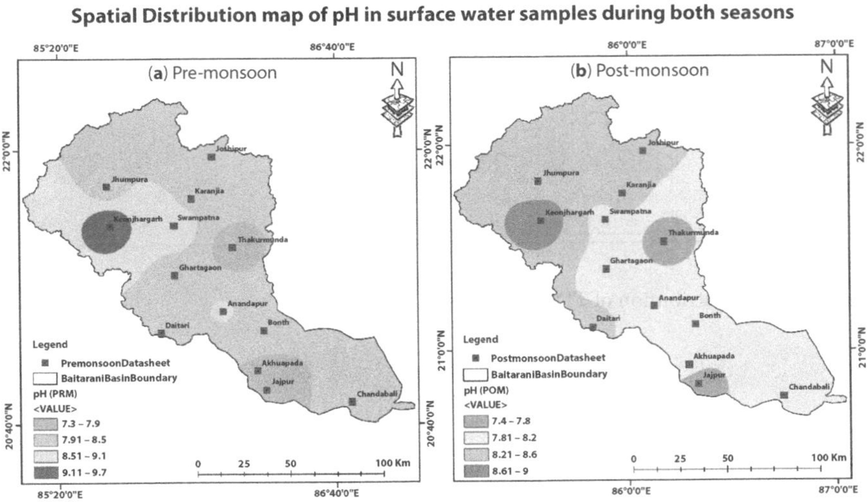

Figure 15.2a Distribution of pH for both seasons.

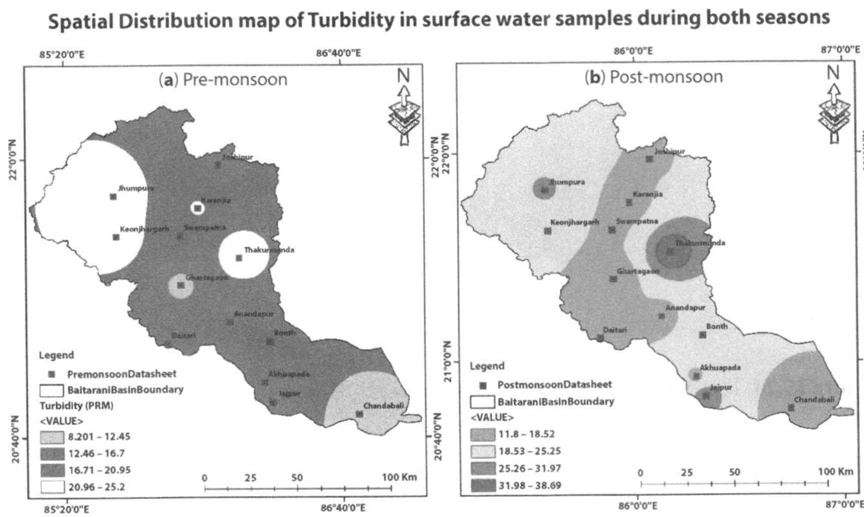

Figure 15.2b Distribution of turbidity for both seasons.

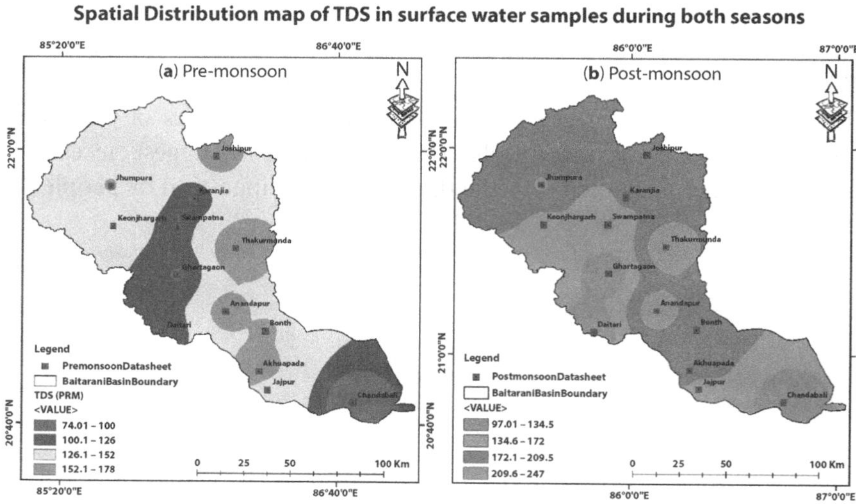

Figure 15.2c Distribution of TDS for both seasons.

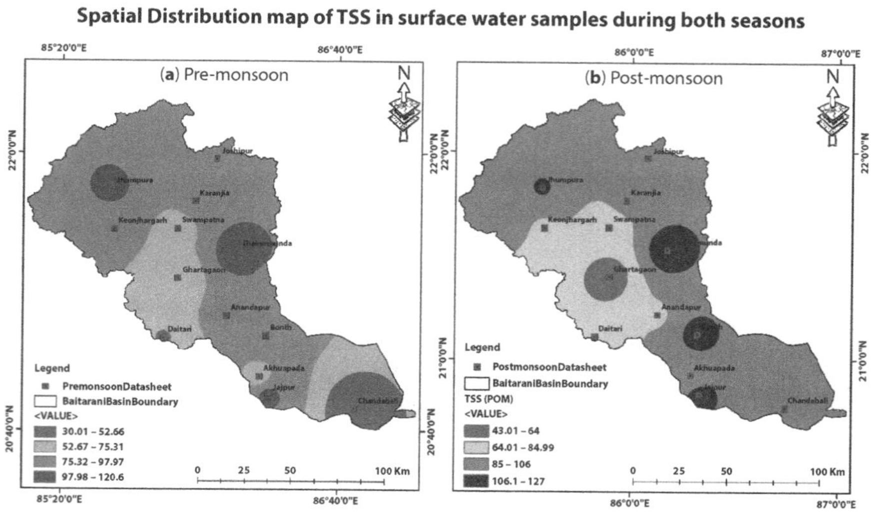

Figure 15.2d Distribution of TSS for both seasons.

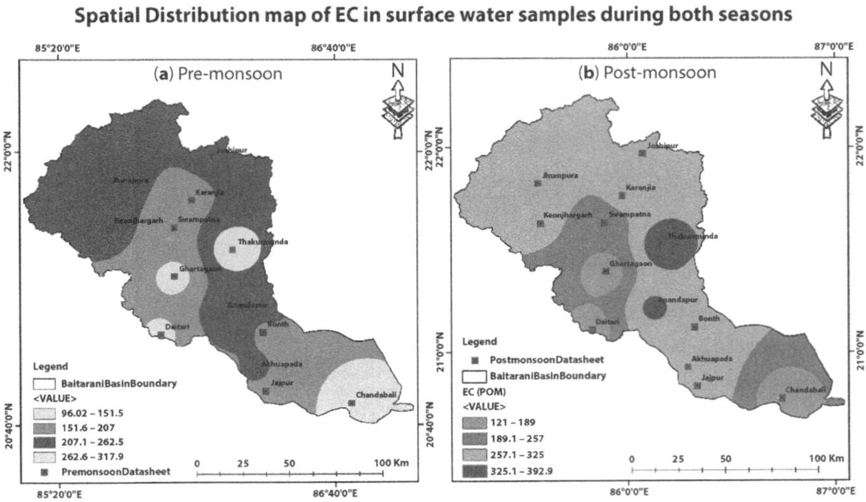

Figure 15.2e Distribution of EC for both seasons.

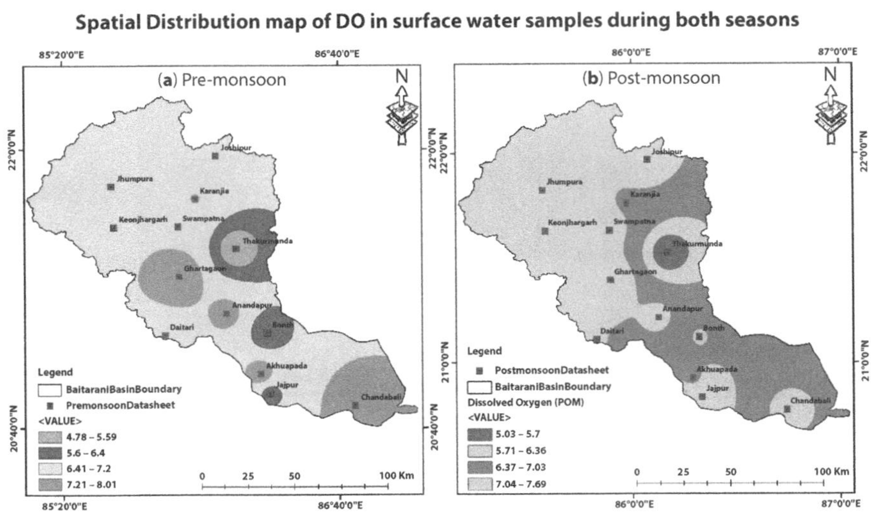

Figure 15.2f Distribution of DO for both seasons.

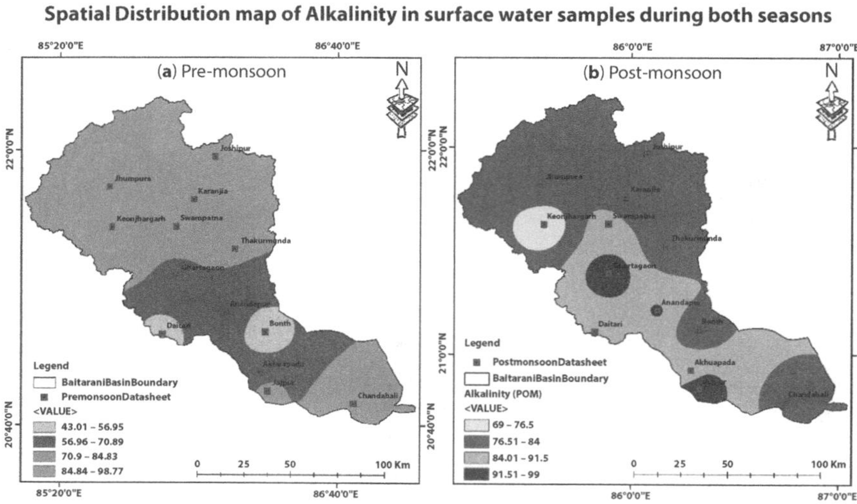

Figure 15.2g Distribution of alkalinity for both seasons.

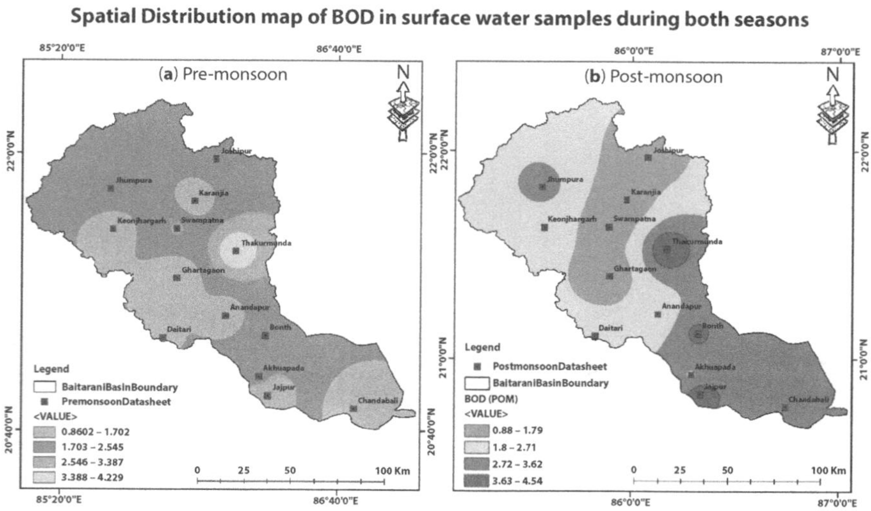

Figure 15.2h Distribution of BOD for both seasons.

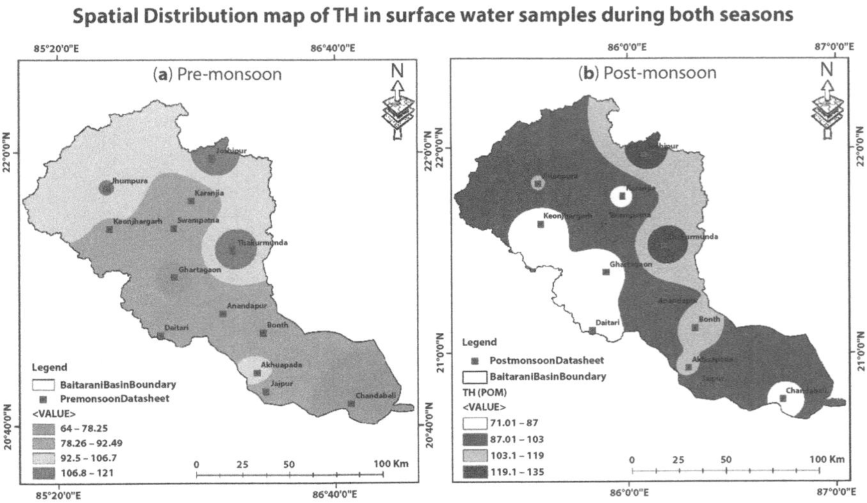

Figure 15.2i Distribution of TH for both seasons.

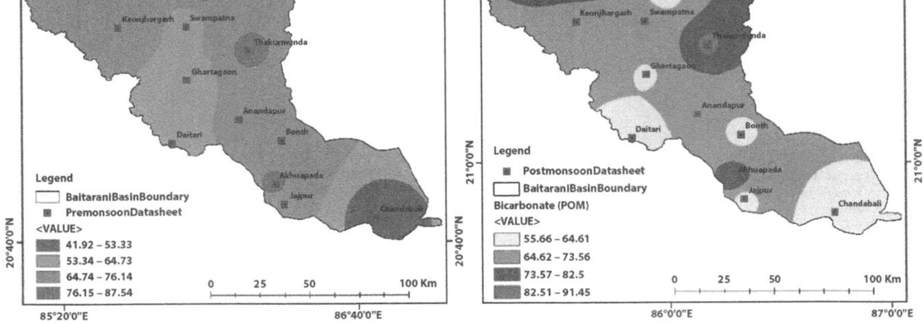

Figure 15.2j Distribution of HCO_3^- for both seasons.

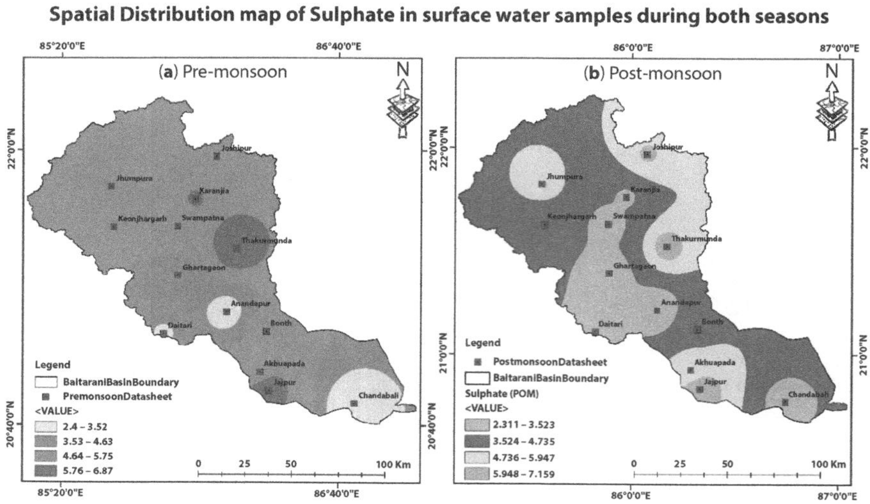

Figure 15.2k Distribution of SO_4^{2-} for both seasons.

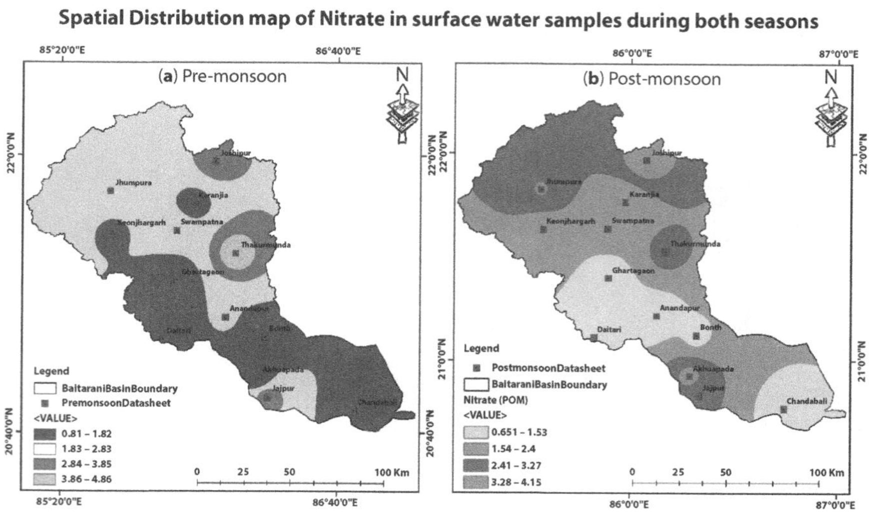

Figure 15.2l Distribution of NO_3^- for both seasons.

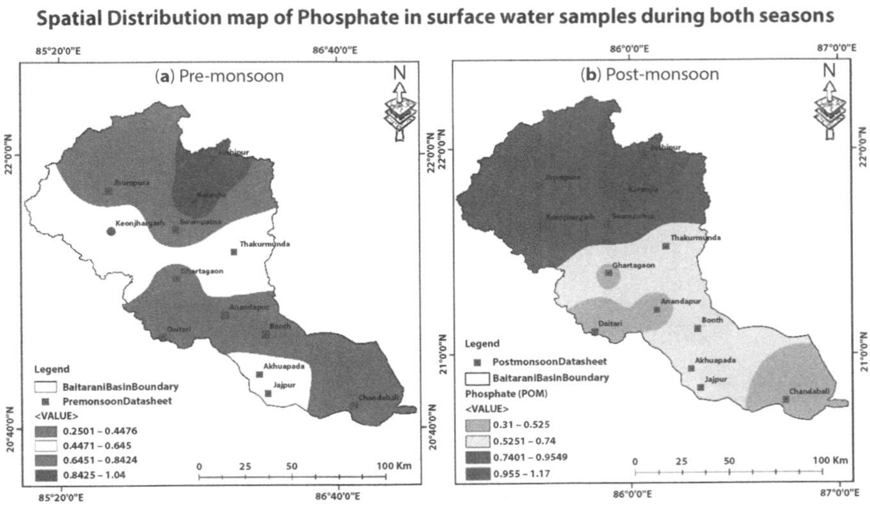

Figure 15.2m Distribution of PO_4^{3-} for both seasons.

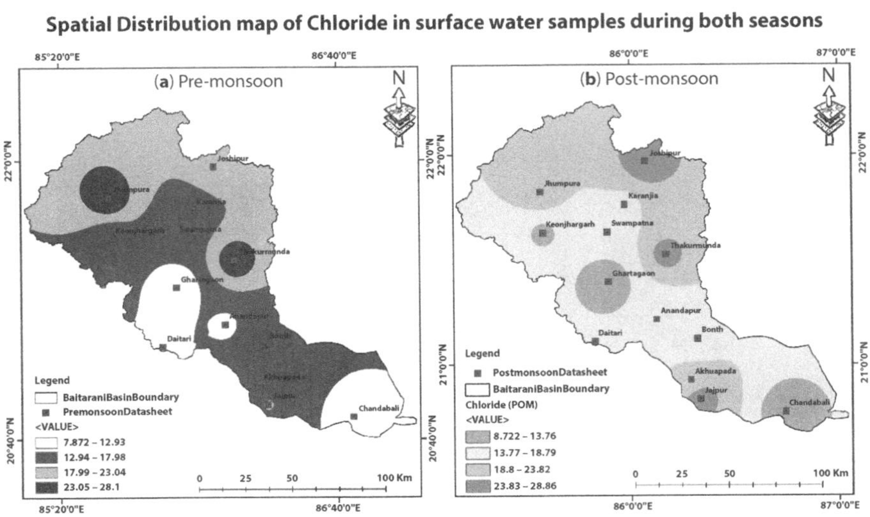

Figure 15.2n Distribution of Cl^- for both seasons.

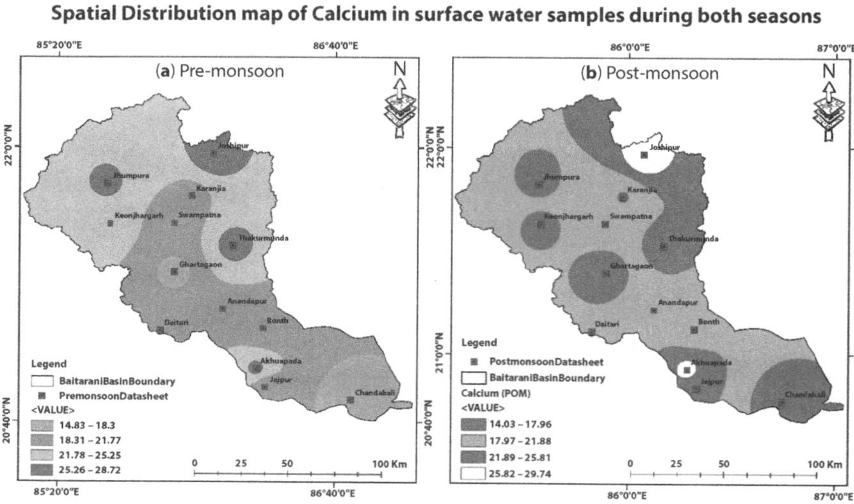

Figure 15.2o Distribution of Ca^{2+} for both seasons.

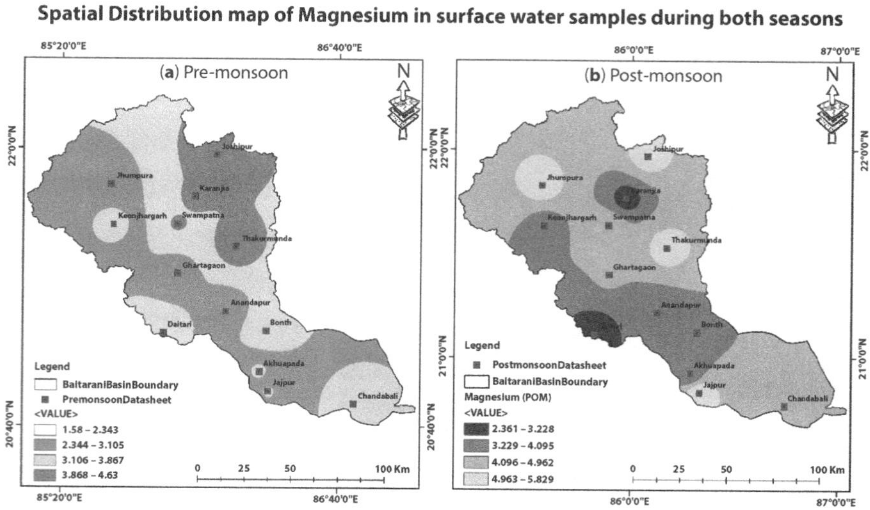

Figure 15.2p Distribution of Mg^{2+} for both seasons.

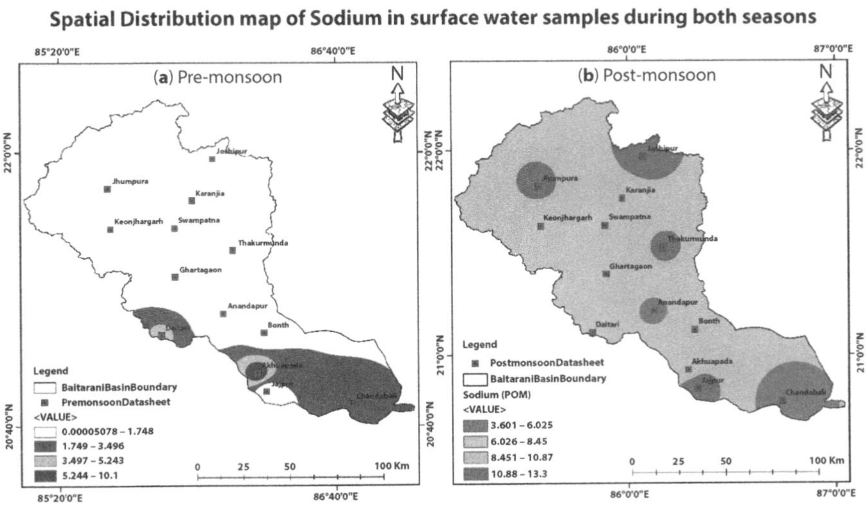

Figure 15.2q Distribution of Na^+ for both seasons.

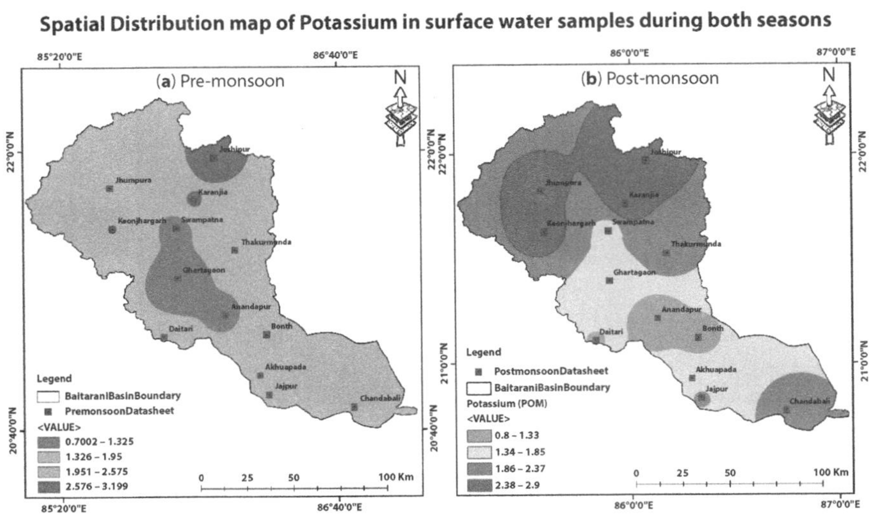

Figure 15.2r Distribution of K^+ for both seasons.

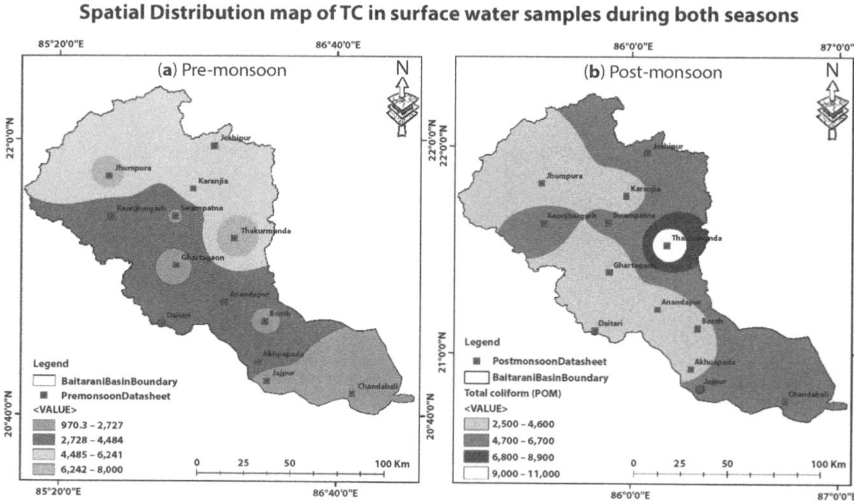

Figure 15.2s Distribution of TC for both seasons.

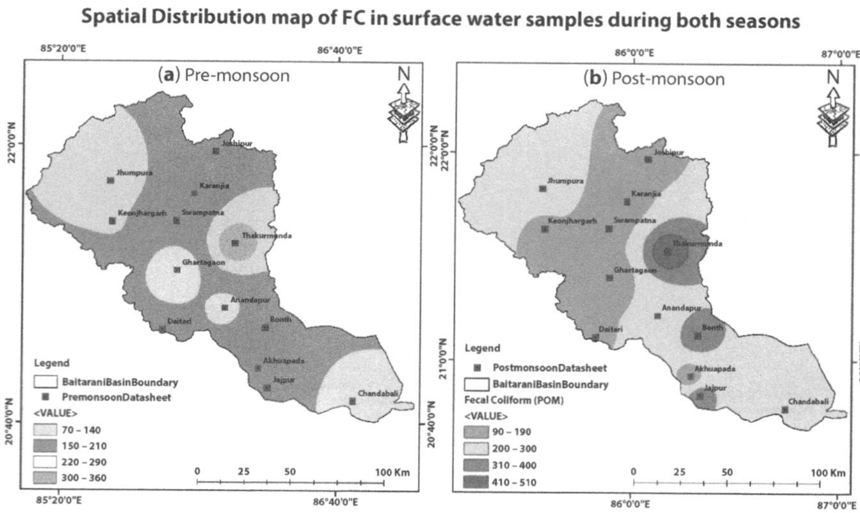

Figure 15.2t Distribution of FC for both seasons.

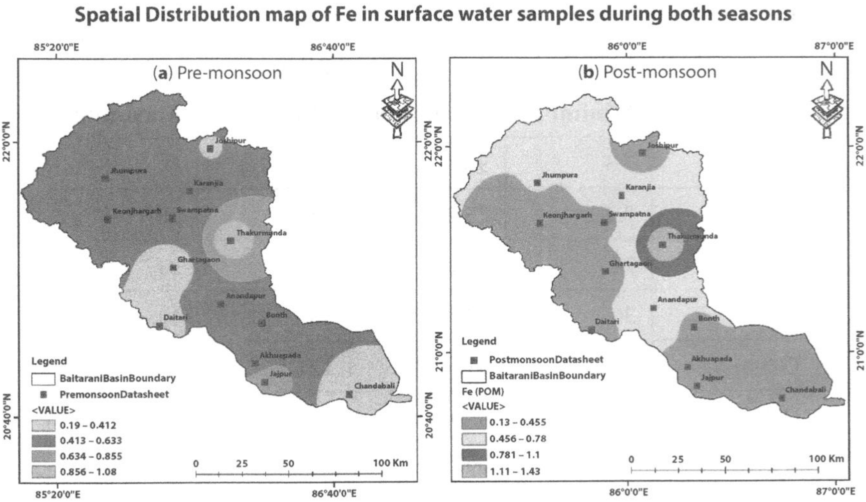

Figure 15.2u Distribution of Fe^{2+} for both seasons.

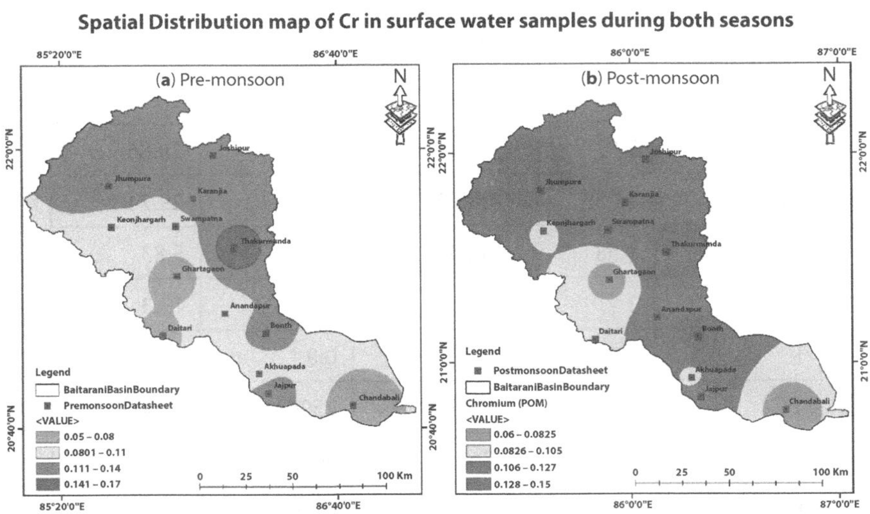

Figure 15.2v Distribution of Cr^{2+} for both seasons.

Table 15.3 Statistical analysis of analyzed physicochemical properties.

	PRM		POM	
Parameters	**Minimum**	**Maximum**	**Minimum**	**Maximum**
pH	7.3	9.7	7.4	9
Turbidity	8.2	25.2	11.8	38.7
TDS	74	178	97	247
TSS	30	121	43	127
EC	96	318	121	393
DO	4.78	8.01	5.03	7.69
Alkalinity	43	99	69	99
BOD	0.86	4.23	0.88	4.54
TH	64	121	71	135
HCO_3^-	41.92	87.55	55.64	91.46
SO_4^{2-}	2.4	6.87	2.31	7.16
NO_3^-	0.81	4.86	0.65	4.15
PO_4^{3-}	0.25	1.04	0.31	1.17
Cl^-	7.87	28.18	8.72	28.86
Ca^{++}	14.83	28.72	14.03	29.74
Mg^{++}	1.58	4.63	2.36	5.83
Na^+	2	10.1	3.6	13.3
K^+	0.7	3.2	0.8	2.9
TC	970	8000	2500	11000
FC	70	360	90	510
Fe^{++}	0.192	1.084	0.132	1.432
Cr^{++}	0.051	0.172	0.062	0.153

15.5.1 Blending Approach of Different WQIs with GIS

The WA WQI seems to be another widely accepted model used to enhance drinking water quality [86, 87]. The obtained WQI range is 42.75 to 123.86 in PRM and 60.26 to 167.24, suggesting excellent to poor and good to poor categories. Only 1 sample falls into the excellent category, 8 (PRM) and 12 (POM) samples fall in the good category, and 4 (PRM) and 1 (POM) samples fall under the poor class. The distribution of the percentages shows that 7.69% of sources were of excellent quality, 61.54% had good water quality, and 30.77% exhibited poor water quality in the case of PRM, while in POM, 92.31% of sources were of good quality and 7.69% belongs to poor quality, for both seasons is being represented in Figure 15.3. Variations of WA WQI throughout the sites can be seen in Figure 15.4. Variation of the interpolation map across the study area is represented in Figure 15.5. The map indicates that sites S-8, S-11, S-12, and S-13 in PRM and site S-8 in POM lie in the poor category. The high WQI values in surface water samples in sites S-8, S-11, S-12, and S-13 were mostly brought on by the presence of increased turbidity and coliform levels. All the sub-indices for each parameter were added up to create an overall SPI. The obtained SPI range is 0.39 to 1.3 in PRM and 0.35 to 1.4 in POM, suggesting a good to very poor category for both seasons (Figure 15.6). As per SPI value, out of 13 samples, 9 (PRM) and 12 (POM) fall into the good category and 4 (PRM) and 1 (POM) holds a very poor water status. Figure 15.3 displays the sample distribution in terms of percentage. The SPI map interpolation across the research area is shown in Figure 15.7. The map amply demonstrates that sites S-8, S-11, S-12, and S-13 in PRM and site S-8 in POM lie in the group of very poor zones. For the 22 samples, NPI was calculated based on the reported water quality metrics. The score of NPI is created by performing cumulative addition of all the contaminating indices for all variables, present at every sampling site (Figure 15.8). The desired NPI range is 8.24 to 21.99 in PRM and 11.15 to 31.50 in POM suggesting 'excellent to poor' and 'good to very poor' classes. Moreover, as per NPI readings, from 13 samples, 2 PRM samples belong to excellent, 9 PRM samples and 12 POM samples are good, and 2 PRM samples and 1 POM sample are poor. Figure 15.3 illustrates the % variation of the samples, which revealed that 'excellent water' is present in 15.38% of PRM, a 'good' zone of water is present in 69.23% of PRM and 92.31% of POM, and poor water is present in 15.38% of PRM and 7.69% of POM. The constructed spatial map can be noticed in Figure 15.9. The location of sites S-8 and S-12 in PRM and site S-8 in POM is widely manifested on the map and falls under the poor zone. Furthermore, the CPI is obtained to determine the total amount of contamination in the river water [88–90]. The value ranged between 0.37 and 1.2 in PRM and 0.51 and 1.32 in POM (Figure 15.10), reflecting that the water falls under the 'sub-clean to polluted' and 'qualified to the polluted category'. The highest CPI value was recorded in both seasons in site S-8.

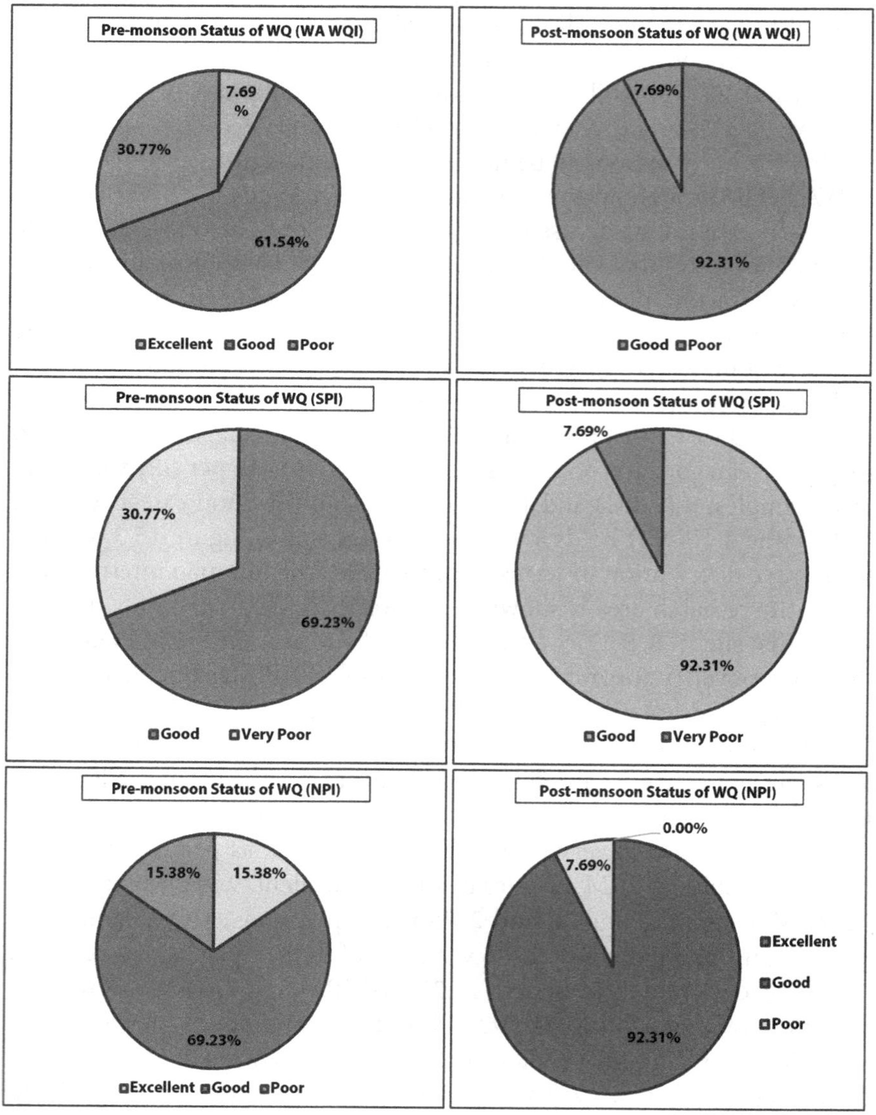

Figure 15.3 Percentage-wise variation due to WA WQI, SPI, NPI, CPI, and OIP. *(Continued)*

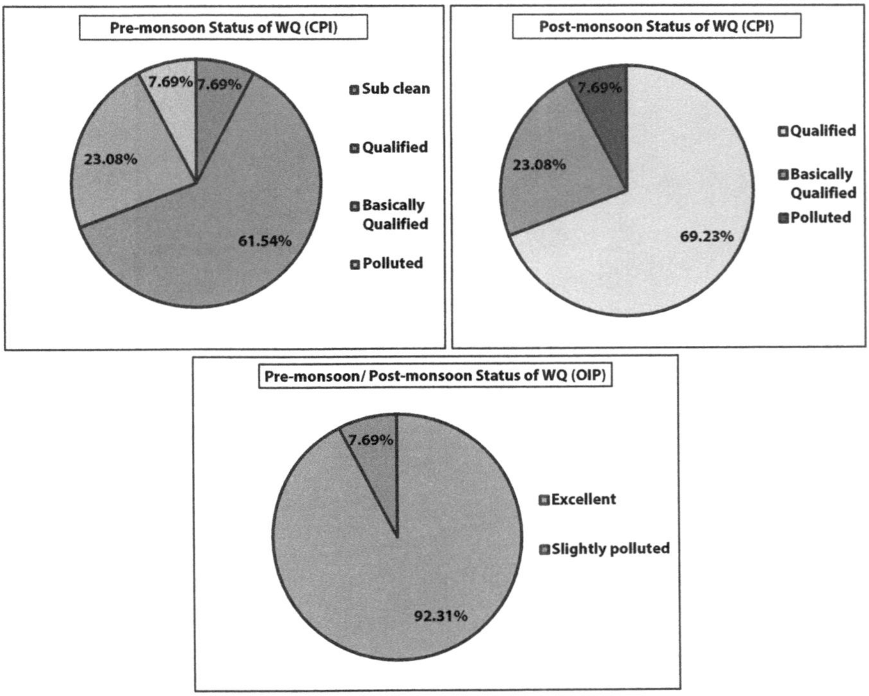

Figure 15.3 (Continued) Percentage-wise variation due to WA WQI, SPI, NPI, CPI, and OIP.

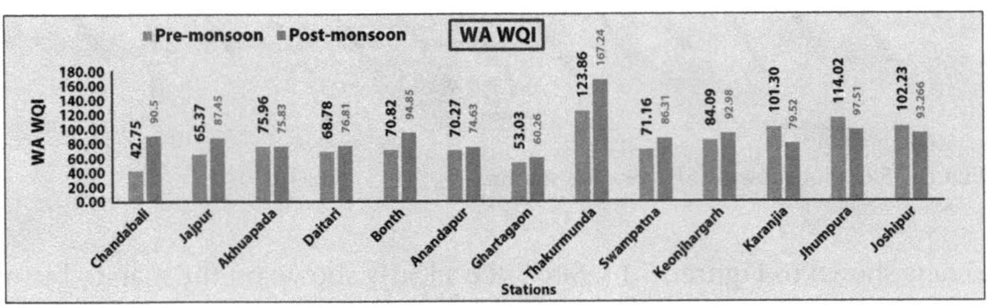

Figure 15.4 Plot depicting WA WQI for each station.

Greater values of turbidity, DO, and TC readings could be a valid reason for high CPI values. As per CPI ratings, out of 13 samples, 1 (PRM) falls into a sub-clean category, 8 (PRM) and 9 (POM) fall into a qualified category, 3 (PRM) and 3 (POM) fall into the qualified category, and only 1 fall into the polluted category for both seasons. Figure 15.3 illustrates the sample distribution in percentage form. The CPI map interpolation across the research

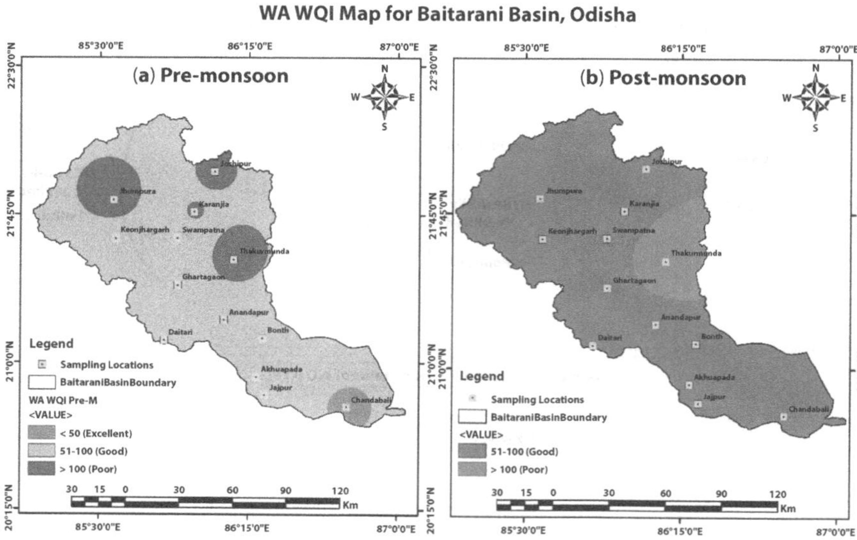

Figure 15.5 WA WQI map for Baitarani River.

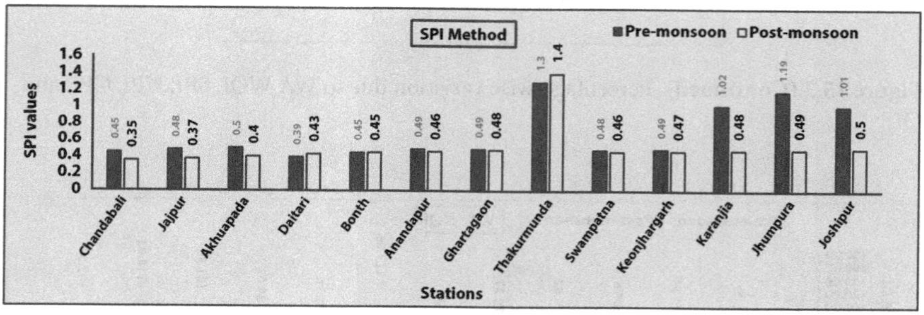

Figure 15.6 Plot showing SPI for each station.

area is shown in Figure 15.11. Site S-8 is clearly shown on the map to be in the polluted category for both periods. The calculated OIP score indicates excellent to moderate levels of pollution for both seasons, i.e., 0.37 to 4.40 in PRM and 0.5 to 4.2 in POM (Figure 15.12). According to the results, 12 samples in both seasons fall into the excellent category and 1 sample in both periods falls into the slightly polluted class. The overall allocation of samples in percentage shows that in both seasons, 92.31% belongs to excellent water and 7.69% signifies slightly polluted water, which can be seen in Figure 15.3. In Figure 15.13, you can see the completed spatial map. The interpolated feature diagram demonstrates that only site 8 belongs to

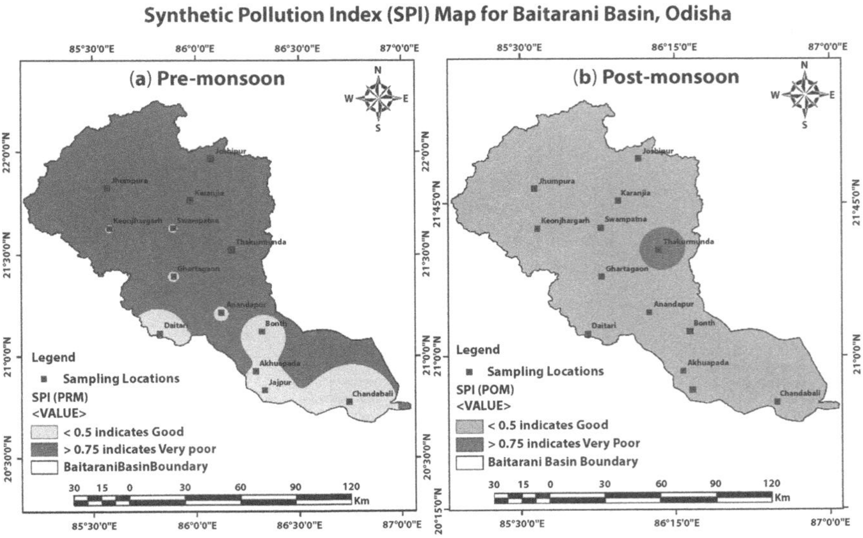

Figure 15.7 SPI map for Baitarani River.

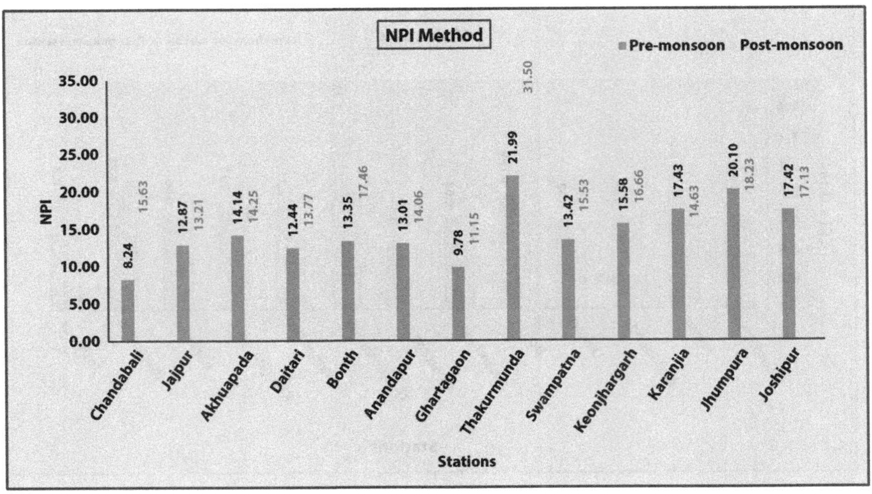

Figure 15.8 Plot showing WQI for each station.

the slightly polluted zone. The water quality index approach along with the pollution indices methodology has some limitations and uncertainty [91, 92]. Numerous other water quality factors that were absent from the index make it impossible for one figure to fully convey the narrative of

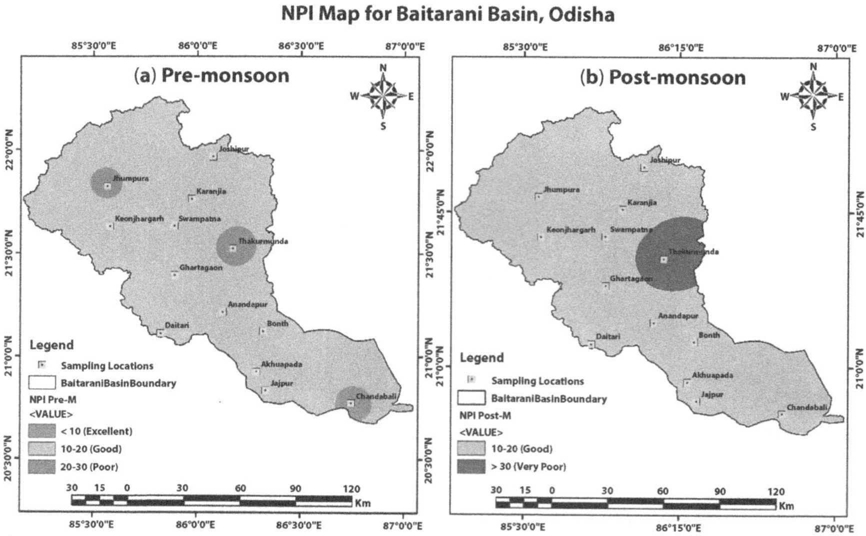

Figure 15.9 NPI map for Baitarani River.

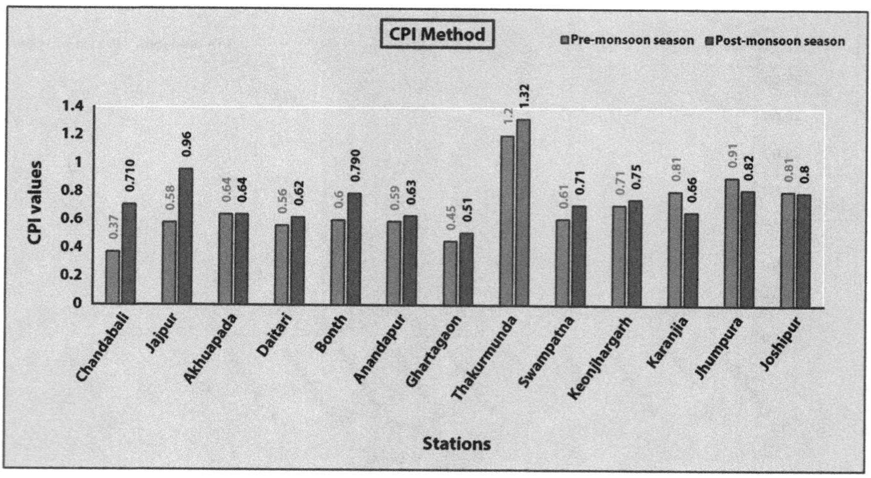

Figure 15.10 Plot showing CPI for each station.

water quality. Additionally, Ward's approach was applied to HCA to determine the association between multiple sampling sites (Euclidean distance as an indicator of resemblance). HCA was applied to the 13 sampling sites using 22 variables. Three main groups were visually picked from the

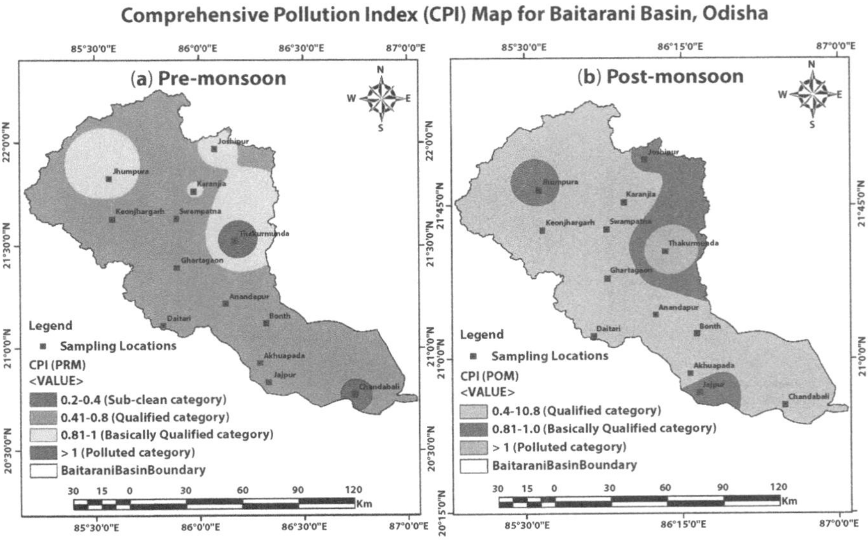

Figure 15.11 CPI map for Baitarani River.

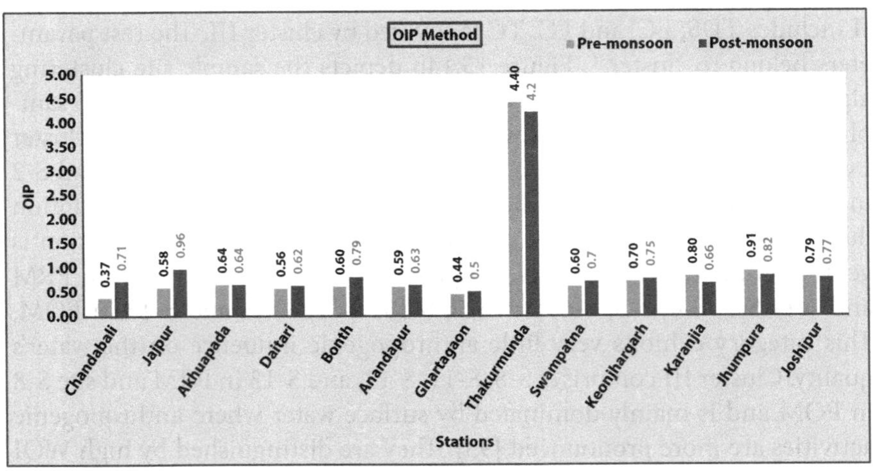

Figure 15.12 Plot showing OIP for each station.

dendrogram of the resulting HCA of the data. A dendrogram representation of the parameter using CA is drawn as an interpolated spatial map and it is visible in Figure 15.14a. Here, cluster I includes pH, turbidity, BOD, SO_4^{2-}, NO_3^-, PO_4^{3-}, Cl^-, Ca^{2+}, Mg^{2+}, Na^+, K^+, Fe^{2+}, and Cr^{2+}. TDS, TSS, EC, alkalinity, TH, HCO_3^-, and FC were the seven indicators formed within

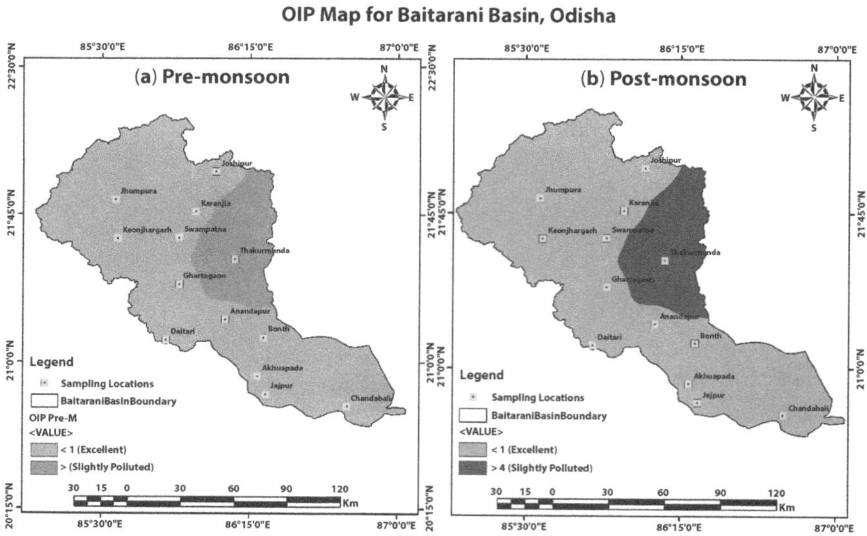

Figure 15.13 OIP map for Baitarani River.

cluster II. Cluster III encompasses TC in PRM. In the case of POM, cluster II includes TDS, EC, and FC. TC is covered by cluster III. The rest parameters belong to cluster I. Figure 15.14b depicts the sample site clustering algorithm from a dendrogram viewpoint. On the other hand, from 13 sampling locations, three clusters were generated by CA. The resulting cluster I comprises sites S-1, S-2, S-5, S-7, and S-9 in PRM and sites S-1 and S-2 in POM. These places could be defined as having low levels of pollution due to their low WQI values, and their main source of contamination is geological material. Cluster II comprises S-3, S-4, S-6, and S-10 in PRM and sites S-3, S-4, S-5, S-6, S-7, S-9, S-10, S-11, S-12, and S-13 in POM. This category exhibits very little anthropogenic influence on the water's quality. Cluster III comprises S-8, S-11, S-12, and S-13 in PRM and site S-8 in POM and is mainly dominated by surface water where anthropogenic activities are more pronounced [93]. They are distinguished by high WQI scores. Although only a small number of samples have high water quality, the bulk of samples have good water quality. Additionally, this demonstrates how anthropogenic activities are steadily affecting the surface water in these places. The increased agricultural runoff entering the river at these sampling places in both seasons, as well as the intense fishing activity there, particularly in October, can be largely blamed for cluster III.

Keeping this in mind as stated above in the context of CA, it displays the water quality in clusters, with cluster III being the best quality and cluster

I being the lowest. Portraying the whole stream, the contaminated survey sites (coming under cluster III) accounted for around (4 of 13 sampling points in total in PRM and 1 of 13 in total in POM). According to the CA findings (Figure 15.14c), the POM season's water quality was superior to that of the PRM season. It has been noted that the WA WQI approach also produced similar results.

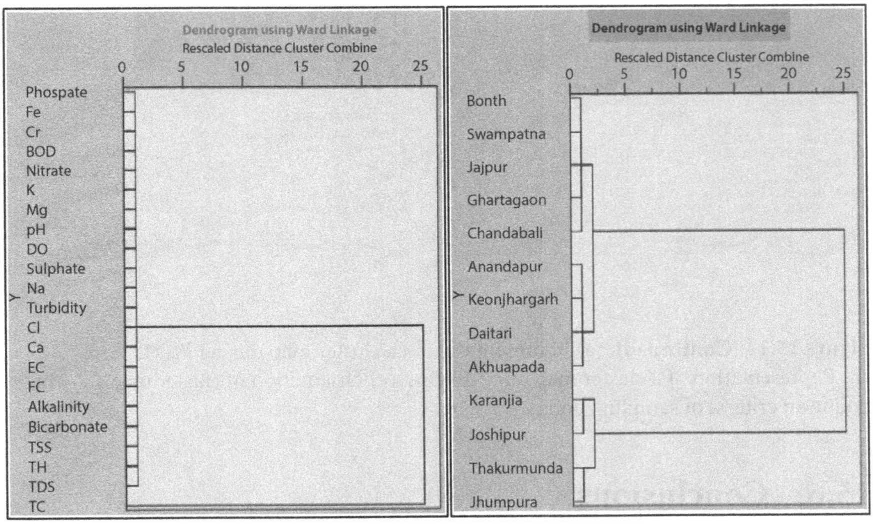

(a)

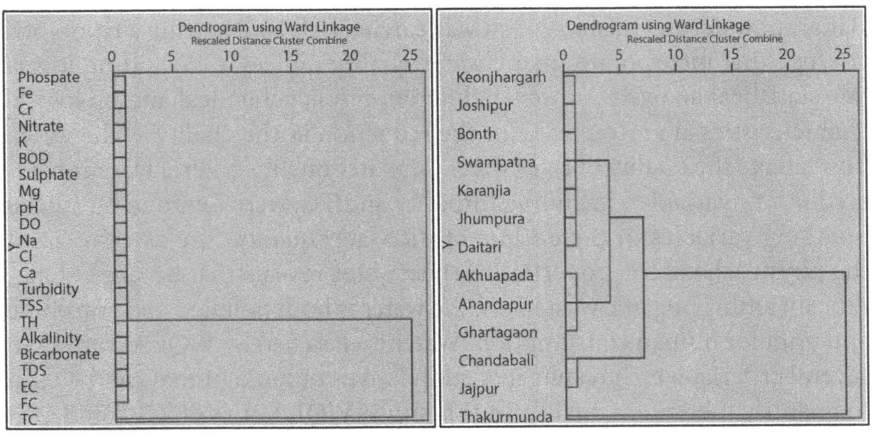

(b)

Figure 15.14 (a) Representation of a cluster map during PRM. (b) Representation of a cluster map during POM. (c) Orientation of cluster map showing pollution criteria of sampling points. *(Continued)*

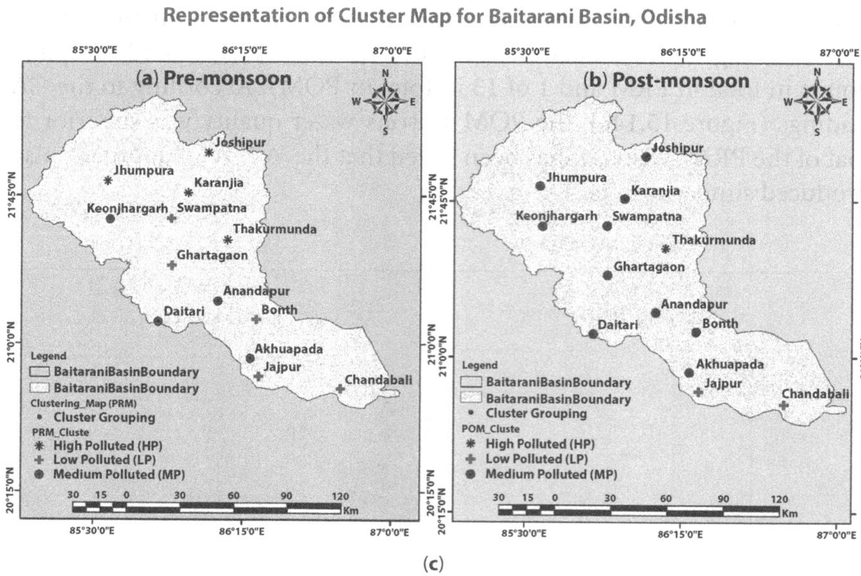

(c)

Figure 15.14 (Continued) (a) Representation of a cluster map during PRM. (b) Representation of a cluster map during POM. (c) Orientation of cluster map showing pollution criteria of sampling points.

15.6 Conclusions

The main role of this study is to identify and assess the pollution loads in river Baitarani water in Odisha, which was carried out, by adopting a combined strategy that incorporates many water quality indices, GIS, and multivariate statistical analysis. However, different physicochemical and biological characteristics are recognized for the reduction in the quality of the water. To evaluate the quality of river water, 22 water quality criteria were considered. A few variables, including turbidity and TC, were found to be crucial polluting variables in the decline of the water quality. The assessment of the physicochemical properties of river water reveals that because of significant anthropogenic waste flow, the water is both polluted and eutrophic and unfit for human consumption. Water is classified by WQI according to several criteria, which results in a cumulative component that can be used to quantify the water's quality. In this study, WQI, SPI, NPI, CPI, and OIP are calculated for further evaluation. According to CA, the presence of TC is caused by bacterial contamination, whereas turbidity is caused by silicate weathering from alluvial deposits and anthropogenic activity like the discharge of home sewage and industrial effluents. Additionally, CA divides its 13 sampling sites into three groups based on their water quality characteristics, which it refers to as groupings of low, medium, and high polluted zones. This study also proposed various treatment methods based on the level of the water quality and the source of polluting substances. Moreover, a GIS-based integrated strategy of several water quality indices outlines the most important locations, which are either appropriate or inappropriate for consumption by humans. Considering the results, site S-8 represents poor-quality water in all methods. In the later part, CA confirms the outcomes of the WA WQI approach. Hence, this innovative investigation also supports the use of multivariate statistical analysis, which includes CA, to examine geographical fluctuation and pinpoint the potential toxic sources in the area under investigation. The regulatory body can use the findings of this study and the suggested methods to document and frame efficient water quality management programs to address pollution.

Acknowledgments

The research reported in this paper was supported by C.V. Raman Global University (CGU), Bhubaneswar, Odisha, India. The author acknowledges the anonymous reviewers for their insightful comments that helped to improve the article. The author would like to express their gratitude to Miss Rani Das, Ruchika Square, Bhubaneswar, Odisha, for providing financial support, in carrying out this innovative work.

References

1. Srivastava PK, Gupta M, Mukherjee S (2012) Mapping spatial distribution of pollutants in groundwater of a tropical area of India using remote sensing and GIS. *Appl. Geomatics*, 4:21–32. https://doi.org/10.1007/s12518-011-0072-y
2. Eid, M.H., Elbagory, M., Tamma, A.A., Gad, M., Elsayed, S., Hussein, H., Moghanm, F.S., Omara, A.E.-D., Kovács, A., Péter, S., 2023. Evaluation of groundwater quality for irrigation in deep aquifers using multiple graphical and indexing approaches supported with machine learning models and GIS techniques, Souf Valley, Algeria. *Water*, 15, 182. https://doi.org/10.3390/w15010182.
3. Das, A. (2022). Multivariate statistical approach for the assessment of water quality of Mahanadi basin, Odisha. *Mater. Today: Proc.*, *65*, A1-A11.
4. Sharma ASC, Gupta S, Singh NR (2013) Studies on the physicochemical parameters in water of Keibul Lamjao National Park, Manipur, India. *J. Environ. Biol.*, 34:1019–1025

5. Singha, S., Pasupuleti, S., Singha, S.S., Singh, R., Kumar, S., 2021. Prediction of groundwater quality using efficient machine learning technique. *Chemosphere*, 276, 130265. https://doi.org/10.1016/j.chemosphere.2021.130265.
6. Mishra, S., & Kumar, A. (2020). Estimation of physicochemical characteristics and associated metal contamination risk in the Narmada River. *Environ. Eng. Res.*, 26(1), 1–11.
7. Kamboj, V., Kamboj, N., & Sharma, S. (2017). Environmental impact of riverbed mining-a review. *Int. J. Sci. Res. Rev.*, 7(1), 504–520.
8. Obiri S (2007) Determination of heavy metals in water from boreholes in Dumasi in the Wassa West District of the western region of Republic of Ghana. *Environ. Monit. Assess.*, 130(1–3):455–463.
9. Dawood, A.S. and Ahmed, A.N., 2016. Using GIS for Assess the Groundwater Quality in Southwest Side of Basrah City. *Muthanna J. Eng. Technol. (MJET)*, 4(2), pp.75-87.
10. Chavan AD, Sharma MP, Bhargava R (2009) Water quality assessment of the Godavari River. *Hydro. Nepal J. Water Energy Environ.*, 5:31–34.
11. Bhardwaj, R., Gupta, A., Garg, J. K, 2017 "Evaluation of heavy metal contamination using environ metrics and Indexing approach for river Yamuna, Delhi stretch, India", *Water Sci. J.*
12. Mishra, S., Kumar, A., & Shukla, P. (2020). Estimation of heavy metal contamination in the Hidon River. *Appl. Water Sci.*, 11(2), 1–9.
13. Das, A. (2023). Assessment of potability of surface water and its health implication in Mahanadi Basin, Odisha. *Materi. Today: Proc.*
14. Naylor R (2003) *Ghana an Oxfam country profile*. Oxfam, Oxford.
15. Das, A. (2022, August). Water Criteria Evaluation for Drinking Purposes in Mahanadi River Basin, Odisha. In *International Conference on Trends and Recent Advances in Civil Engineering* (pp. 237-260). Singapore: Springer Nature Singapore.
16. Shah, A. K., & Geeta, J. S. (2017). Evaluation of water quality index for River Sabarmati, Gujarat, India. *Appl. Water Sci.*, 7(3), 1349–1358.
17. Bui, D.T., Khosravi, K., Tiefenbacher, J., Nguyen, H., Kazakis, N., 2020. Improving prediction of water quality indices using novel hybrid machine-learning algorithms. *Sci. Total Environ.*, 721, 137612 https://doi.org/10.1016/j.scitotenv.2020.137612.
18. Horton RK (1965) An index-number system for rating water quality. *J. Water Pollut. Control Fed.*, 37:300–306.
19. Lkr, A., Singh, M. R. & Puro, N. 2020 Assessment of water quality status of Doyang River, Nagaland, India, using Water Quality Index. *Appl. Water Sci.*, 10 (1), 1–13. https://doi.org/10.1007/s13201-019-1133-3.
20. Brown RM, McClelland NI, Deininger RA, Tozer RG (1970) A water quality Index- do we dare. *Water Sew. Works*, 117:339–343.
21. Carpenter S, Caraco NF, Correll DL, Howarth RW, Sharpley AN, Smith VH (1998) Nonpoint pollution of surface waters with phosphorus and nitrogen. *Ecol. Appl.*, 8(3):559–568.

22. Islam ARMT, Ahmed N, Bodrud-Doza M, Chu R (2017) Characterizing groundwater quality ranks for drinking purposes in Sylhet district, Bangladesh, using entropy method, spatial autocorrelation index, and geo-statistics. *Environ. Sci. Pollut. Res.*, 24(34):26350–26374
23. Patel SV, Chavda P, Tyagi S (2019) 'carrying out assessment of groundwater quality of villages of Bhavnagar district of Gujarat, India. *Environ. Claims J.*, 31(1):79–92.
24. Sharma, D. & Kansal, A. 2011 Water quality analysis of River Yamuna using water quality index in the national capital territory, India (2000– 2009). *Appl. Water Sci.*, 1 (3–4), 147–157. https://doi.org/10.1007/s13201-011-0011-4.
25. Jothivenkatachalam K, Nithya A, Chandra Mohan S (2010) Correlation analysis of drinking water quality in and around Perur block of Coimbatore District, Tamil Nadu. India *Rasayan J. Chem.*, 3:649–654.
26. Poonam, T., Tanushree, B. & Sukalyan, C. 2013 Water quality indices – important tools for water quality assessment: a review. *Int. J. Adv. Chem.*, 1 (1), 15–28.
27. Sargaonkar A and Deshpande V (2003) Development of an overall index of pollution for surface water based on a general classification scheme in Indian context. *Environ. Monit. Assess.*, 89 43-67.
28. Numerow NL, Sumitomo H (1970) Benefits of water quality enhancement report no. 16110 DAJ, prepared for the US Environmental Protection Agency. Retrieved from Link
29. Matta G, Kumar A, Nayak A, Kumar P, Kumar A, Tiwari AK (2020) Water quality and planktonic composition of river Henwal (India) using comprehensive pollution index and biotic-indices. *Trans Indian Natl. Acad. Eng.*, https://doi.org/10.1007/s41403-020-00094-x
30. Ma, J., Ding, Z., Wei, G., Zhao, H. & Huang, T. 2009 Sources of water pollution and evolution of water quality in the Wuwei basin of Shiyang river, Northwest China. *J. Environ. Manage.*, 90 (2), 1168–1177. https://doi.org/10.1016/j.jenvman.2008.05.007.
31. Sunar, S., Tripathi, O. P. & Mishra, B. P. 2020 Water quality and pollution indices application in monitoring water quality of Serlui river impacted by Serlui-B dam, Mizoram, North East India. *Curr. Sci.*, 119, 1685. 10.18520/cs/v119/i10/1685-1689.
32. Madhloom HM, Alansari N (2018) Geographical information system and remote sensing for water resources management case study: the Diyala River, Iraq. *Int. J. Civ. Eng. Technol.*, 9:971–984.
33. Uddin, Md.G., Nash, S., Olbert, A.I., 2021. A review of water quality index models and their use for assessing surface water quality. *Ecol. Indic.*, 122, 107218 https://doi. org/10.1016/j.ecolind.2020.107218.
34. Venkatramanan S, Chung SY, Ramkumar T, Rajesh R, Gnanachandrasamy G (2015) Assessment of groundwater quality using GIS and CCME WQI techniques: a case study of Thiruthuraipoondi city in Cauvery deltaic region, Tamil Nadu, India. *Desalin. Water Treat.*, P 1-16. doi:10.1080/19443994.2015.1048740.

35. Selvam S, Iruthaya Jeba Dhana Mala R, Muthukakshmi V (2013b) A hydrochemical analysis and evaluation of groundwater quality index in Thoothukudi district, Tamilnadu, South India. *Int. J. Adv. Eng. Appl.*, 2(3):25–37.
36. Al-Madhlom Q, Nordell B, Chabuk A, Al-Ansari N, Lindblom J, Laue J, Hussain HM (2019) Potential use ofUTES in Babylon Governorate, Iraq. *Groundw. Sustain. Dev.*, 10:1–18.
37. Sarath Prasanth SV, Magesh NS, Jitheshlal KV, Chandrasekar N (2012) Evaluation of groundwater quality and its suitability for drinking and agricultural use in the coastal stretch of Alappuzha District, Kerala, India. *Appl. Water Sci.*, 2(3):165–175.
38. Saha, P. & Paul, B. 2019a Groundwater quality assessment in an industrial hotspot through interdisciplinary techniques. *Environ. Monit. Assess.*, 191 (2). https://doi.org/326.10.1007/s10661-019-7418-z.
39. Shrestha S, Kazama F (2007) Assessment of surface water quality using multivariate statistical techniques: a case study of the Fuji river basin. Japan. *Environ. Model. Softw.*, 22(4):464–75.
40. APHA (1995) *Standard methods for the examination of water and wastewater*, 19th edn. American Public Health Association, New York.
41. Abbasi, S. A. (2002). Water quality indices, state of the art report, National Institute of Hydrology, scientific contribution no. INCOH/SAR-25/2002, Roorkee: INCOH. pp: 73.
42. Lateef KH (2011) Evaluation of groundwater quality for drinking purpose for Tikrit and Samarra cities using water quality index. *Eur. J. Sci. Res.*, 58(4):472–481.
43. Bureau of Indian Standards (BIS) 10500 (2012) Specification for drinking water. Indian Standards Institution, New Delhi, pp. 1–5.
44. Ramakrishnalah CR, Sadas Hivalah C, Ranganna G (2009) Assessment of water quality index for the groundwater in Tumkur Taluk, Karnataka state, India. *E.-J. Chem.*, 6(2):523–530.
45. Gauns A, Nagarajan M, Lalitha R, Baskar M (2020) GIS-based assessment of groundwater quality for drinking and irrigation by water quality index. *Int. J. Curr. Microbiol. Appl. Sci.*, 9:2361–2370. 269.
46. Yang, Y.; Zhou, F.; Guo, H.; Sheng, H.; Liu, H.; Dao, X.; He, C. Analysis of spatial and temporal water pollution patterns in Lake Dianchi using multivariate statistical methods. *Environ. Monit. Assess.*, 2010, 170, 407–416. [CrossRef] [PubMed].
47. Thakur N, Rishi M, Sharma DA, Keesari T. Quality of water resources in Kullu Valley in Himachal Himalayas, India: perspective and prognosis. *Appl. Water Sci.*, 2018;8(1):1–3.
48. Massart, D.L., Vandeginste, B.G.M., Deming, S.N., Michotte, Y., Kaufman, L., 1988. *Chemometrices: A Textbook*. Elsevier, Amsterdam.

49. Jankowska J, Radzka E, Rymuza K (2017) Principal component analysis and cluster analysis in multivariate assessment of water quality. *J. Ecol. Eng.*, 18:92–96. https://doi. org/10.12911/22998 993/68141
50. George R, Martin GD, Nair SM *et al.* (2016) Geochemical assessment of trace metal pollution in sediments of the Cochin backwaters. *Environ. Forensics*, 17:156–171.
51. Dinka MO, Loiskandl W, Ndambuki JM (2015). Hydrochemical characterization of various surface water and groundwater resources available in Matahara areas, Fantalle Woreda of Oromiya region. *J. Hydrol. Reg. Stud.*, 3:444-456.
52. Kavanaugh MC and ART (1979) Measurement of turbidity In: *Advances in laboratory techniques for quality control* American Water Works Association, Denver, pp 41–53.
53. Akhtar, N.; Syakir, M.I.; Anees, M.T.; Qadir, A.; Yusuff, M.S. Characteristics and Assessment of Groundwater. In *Groundwater*; Chapter 3; Intech Open: London, UK, 2020; pp. 1–20.
54. Adimalla N, Venkatayogi S (2017) Mechanism of luoride enrichment in groundwater of hard rock aquifers in Medak, Telangana State. South India. *Environ. Earth Sci.*, 76(1):45. https ://doi. org/10.1007/s1266 5-016-6362-2
55. Chapman, D.V., 1996. *Water Quality Assessment-A Guide to Use Biota, Sediment and Water in Environmental Monitoring.* 2nd Edn., ISBN: 041921590, pp: 626.
56. Gorgoglione, A.; Bombardelli, F.A.; Pitton, B.J.L.; Oki, L.R.; Haver, D.L.; Young, T.M. Uncertainty in the parameterization of sediment build-up and wash-off processes in the simulation of water quality in urban areas. *Environ. Model. Softw.*, 2019, 111, 170–181. [CrossRef]
57. Harilal CC, Hashim A, Arun PR, Baji S (2004) Hydro geochemistry of two rivers of Kerala with special reference to drinking water quality. *J. Ecol. Environ. Conserv.*, 10(2):187–192.
58. Yadav KK, Gupta N, Kumar V, Choudhary P, Khan SA (2018) GIS- based evaluation of groundwater geochemistry and statistical determination of the fate of contaminants in shallow aquifers from different functional areas of Agra city, India: levels and spatial distributions. *RSC Adv.*, 8(29):15876–15889.
59. Tiwari, T.N. and Manzoor, A., 1988. Pollution of Subarnarekha River near Jamshedpur and the suitability of its water for irrigation. *Indian J. Environ. Prot.*, 8(7), pp.494-497.
60. Mandal DK (2013) River linking and flood management of Mahananda River basin in West Bengal. *J. Bengal Geogr.*, II(I):1–15 (ISSN 2319-619X)
61. Kaminsky LS, Mahoney MC, Leach J, Melius J, Jo Miller M (1990) Fluoride: benefits and risks of exposure. *Crit. Rev. Oral Biol. Med.*, 1:261–281.
62. Wavde PN, Arjun B (2010) Groundwater quality assessment at Malegaon region of Nanded in Maharashtra (India). *J. Environ. Sci. Eng.*, 52(1):57–60.
63. Victor H (2017). Hindustan Times (February, 13), 7 tube wells pumping water, but Sukhna's level is same as January 16. https://www.hindustantimes.

com/punjab/7-tube-wells-pumping-water-but-sukhna-s-level-is-same-as-january-16/story-kcTIZmXEohrsRJoH2ZJwsM.Html

64. Kodom K, Onoyinka AA. Mkude IT, Otu SA, Yeboah JO (2018). Physicochemical parameters assessment and seasonal variations of surface water quality of the volta lake in Ghana.
65. Lodh R, Paul R, Karmakar B, Das MK (2014) Physicochemical studies of water quality with special reference to ancient lakes of Udaipur City, Tripura, India. *Int. J. Sci. Res. Publ.*, 4(6):1–9.
66. Kumar SK, Logeshkumaran A, Magesh NS, Godson PS, Chandrasekar N (2014) Hydro-geochemistry and application of water quality index (WQI) for groundwater quality assessment, Anna Nagar, part of Chennai City, Tamil Nadu, India. *Appl. Water. Sci.*, 5:335–343.
67. Verma RK, Murthy S, Tiwary RK, Verma S (2019) Development of simplified WQIs for assessment of spatial and temporal variations of surface water quality in upper Damodar River basin, Eastern India. *Appl. Water Sci.*, 9(1):21.
68. Kalavathy S, Sharma TR, Suresh KP (2011) Water quality index of river Cauvery in Tiruchirappalli district, Tamilnadu. *Arch. Environ. Sci.*, 5:55–61.
69. Kumar, M. & Puri, A. 2012 A review of permissible limits of drinking water. *Indian J. Occup. Environ. Med.*, 16 (1), 40–44. https://doi.org/10.4103/0019-5278.99696.
70. McGowan W (2000) *Water processing: residential, commercial, light- industrial*, 3rd edn. Water Quality Association, Lisle, IL.
71. Kamble, Swapnil R., and Ritesh Vijay. 2011. "Assessment of Water Quality Using Cluster Analysis in Coastal Region of Mumbai, India."
72. Vasanthavigar M, Srinivasamoorthy K, Prasanna MV (2012) Evaluation of groundwater suitability for domestic, irrigational, and industrial purposes: a case study from Thirumanimuttar River basin, Tamil Nadu, India. *Environ. Monit. Assess.*, 184(1):405–420.
73. Hem JD (1970) *Study and interpretation of the chemical characteristics of natural water*. Geological survey water-supply paper 1473. United States government printing office, Washington.
74. Nikanorov AM, Brazhnikova LV, Sketches B (2009) Water chemical composition of rivers, lakes and wetlands. In: Khublaryan MG (ed) *Types and properties of water*, vol. 2. EOLSS Co. Ltd., pp 42–80. Available at http://www.eolss.net/sample-chapters/c07/e2- 03.pdf. Accessed December 2015.
75. Prasad M, Sunitha V, Sudharsan Reddy Y, Suvarna B, Muralidhara Reddy B, Ramakrishna Reddy M (2019) Data on Water quality index development for groundwater quality assessment from Obulavaripalli Mandal, YSR district, A.P India. *Data Brief*, 24:103846 https://doi.org/10.1016/j.dib.2019.103846
76. Pramanik, A. K., Majumdar, D., & Chatterjee, A. (2020). Factors affecting lean, wet season water quality of Tilaiya reservoir in Koderma district, India during 2013-2017. *Water Sci.*, 34(1), 85–97.

77. Podhade D, Lal SB, Singh S, Mehera B, Khare N, James A (2020) Evaluating the impact of wetland health on wildlife health by soil and water quality analysis. *Int. J. Curr. Microbiol. Appl. Sci.*, 9(9):839–849.
78. WHO (2011) *Guidelines for Drinking-water Quality*. Geneva.
79. Islam, M. S. & Mostafa, M. G. 2021f Groundwater quality and risk assessment of heavy metal pollution in Middle-West part of Bangladesh. *J. Earth Environ. Sci. Res.*, 3 (2), 1–15. doi.org/10.47363/JEESR/2021(3)143.
80. Ikram, R.M.A., Ewees, A.A., Parmar, K.S., Yaseen, Z.M., Shahid, S., Kisi, O., 2022b. The viability of extended marine predators algorithm-based artificial neural networks for streamflow prediction. *Appl. Soft Comput.*, 131, 109739 https://doi.org/10.1016/j.asoc.2022.109739.
81. Dimple, Dimple, Rajput, J., Al-Ansari, N., Elbeltagi, A., 2022b. Predicting irrigation water quality indices based on data-driven algorithms: case study in semiarid environment. *J. Chemother.*, 2022, e4488446 https://doi.org/10.1155/2022/4488446.
82. Rajput, J., Dimple, Singh, P.K., Kumar, D., Gaddikeri, V., Elbeltagi, A., 2023. Combination of discretization regression with data-driven algorithms for modeling irrigation water quality indices. *Ecol. Inf.*, 102093 https://doi.org/10.1016/j.ecoinf.2023.102093
83. Cooley, S.W., Ryan, J.C., Smith, L.C., 2021. Human alteration of global surface water storage variability. *Nature*, 591, 78–81. https://doi.org/10.1038/s41586-021-03262-3.
84. Vishwakarma, D.K., Pandey, K., Kaur, A., Kushwaha, N.L., Kumar, R., Ali, R., Elbeltagi, A., Kuriqi, A., 2022. Methods to estimate evapotranspiration in humid and subtropical climate conditions. *Agric. Water Manage.*, 261, 107378 https://doi.org/ 10.1016/j.agwat.2021.107378.
85. Jain, R., Kishore, P., Singh, D., 2019. Irrigation in India: status, challenges and options. *J. Soil Water Conserv.*, 18, 1–11.
86. Gao, J., Li, Z., Chen, Z., Zhou, Y., Liu, W., Wang, L., Zhou, J., 2021. Deterioration of groundwater quality along an increasing intensive land use pattern in a small catchment. *Agric. Water Manag.*, 253, 106953 https://doi.org/10.1016/j. agwat.2021.106953.
87. He, X., Li, P., Wu, J., Wei, M., Ren, X., Wang, D., 2021. Poor groundwater quality and high potential health risks in the Datong Basin, northern China: research from published data. *Environ. Geochem. Health*, 43, 791–812. https://doi.org/10.1007/ s10653-020-00520-7.
88. Das, A. (2023). Modelling of Surface Water Quality and Spatial Mapping in Mahanadi Basin, Odisha. *J. Mines Met. Fuels*, 71(9).
89. Das, A. (2023). Identification of Surface Water Contamination Zones and its Sources on Mahanadi River, Odisha Using Entropy-Based WQI and MCDM Techniques. *Engineering Research Transcripts*, 4, 67-92.
90. Das, A. (2022, December). Surface Water Quality Modelling Using Water Quality Index (WQI) and Geographic Information System (GIS) on the

Mahanadi Basin, Odisha. In *International Conference on Water Technologies* (pp. 21-46). Singapore: Springer Nature Singapore.

91. Laltu, A. D. (2023). Drinking Suitability and Source Apportionment of Physical Parameters in Surface Water in Mahanadi Basin, Odisha. Odisha (April 14, 2023).
92. Das, A. (2023). Anthropogenic Effects on Surface Water Quality Assessment in Baitarani River Basin, Odisha Using GIS and MCDM Techniques. *Engineering Research Transcripts*, 5, 37-64.
93. Das, A., Goya, A., & Soni, A. (2023, July). Use of water quality indices and its evaluation to verify the impact of Mahanadi River basin, Odisha. In *AIP Conf. Proc.* (Vol. 2721, No. 1). AIP Publishing.

16

Drone Mapping for Agricultural Sustainability: Applications and Benefits

Arun Bawa*, Gurjinder Baath, Pulkit Juneja and Jaiveer Brar

Texas A&M AgriLife Research, Blackland Research & Extension Center, Temple, TX, USA

Abstract

Drone mapping holds transformative potential for agricultural systems, driving improvements in efficiency, resilience, and sustainability. This chapter explores the integration of drone mapping into farming practices to optimize resource utilization, minimize environmental impact, and ensure the long-term viability of agricultural systems. It discusses advancements in remote sensing, focusing on various sensing platforms used for precision agriculture. Emphasis is placed on the utilization of drones to overcome the limitations of satellite and ground-based monitoring methods. Detailed explanations are provided for current applications of drone mapping, including crop monitoring, weed and pest detection, water and nutrient stress monitoring, and variable rate application. The chapter also addresses challenges associated with drone mapping, such as high costs, complex data processing, and regulatory constraints that hinder widespread adoption among farmers and crop consultants. The chapter concludes by summarizing the challenges and future directions of drone mapping in agriculture, highlighting opportunities for real-time data processing, predictive analytics, and precision applications. Furthermore, the combination of machine learning, artificial intelligence, and the Internet of Things (IoT) is expected to facilitate advanced applications, empowering decision-making, and enabling smart farming practices. This chapter overall presents useful information for readers seeking to understand drone applications in farming systems and their future implications.

***Keywords*:** UAV, UAS, remote sensing, vegetation indices, precision agriculture, crop monitoring

**Corresponding author*: arun.bawa@ag.tamu.edu

Disha Thakur, Sanjay Kumar, Har Amrit Singh Sandhu and Chander Prakash (eds.) Sustainable Development Using Geospatial Techniques, (411–436) © 2024 Scrivener Publishing LLC

16.1 Introduction

Agricultural sustainability is a global concern, as crop yields must improve to feed over 9 billion people by 2050 while minimizing negative environmental and climate impacts [1]. However, increasing food production can have significant environmental and climate change effects, including increased soil erosion, diminished water quality, and elevated greenhouse gas emissions [2]. For instance, increased food production through agricultural extensification, i.e., converting natural land to agricultural land, introduces detrimental impacts on the ecosystem such as biodiversity losses [3]. Conversely, increased food production through agricultural intensification, i.e., increasing crop productivity over existing croplands through modern agricultural practices, can also have negative implications on the ecosystem, such as depleting aquifers under increased irrigation practices [1]. Therefore, it is essential to promote sustainable agriculture, which aims to increase food production while maintaining farm profitability and minimizing environmental impacts.

As climate change awareness grows, the importance of agricultural sustainability in society has increased, leading to new initiatives such as the European Parliament's drive to agricultural sustainability, "produce more with less", under the European Union research program "Horizon 2020"[1] and United States' "partnership for climate-smart commodity" program[2] by the United States Department of Agriculture (USDA) to develop and promote climate-smart commodities. In this sense, precision agriculture is an effective tool that introduces a multidisciplinary approach to address the objectives of more production per unit of resource. Precision agriculture, often termed site-specific crop management, involves the strategic management of crop areas based on specific requirements, rather than applying a uniform management practice across the whole field. Within a given field, there can be significant variations in the crop's environment. Elements like soil nutrient levels, physical and chemical properties, moisture levels, and occurrence of abiotic/biotic stresses can differ from one location to another. By adapting crop management practices and inputs based on the specific conditions of each location, farmers can potentially save costs, enhance yields, and minimize undesirable environmental impacts such as agricultural nutrient or chemical losses to surface runoff.

---

[1]https://research-and-innovation.ec.europa.eu/funding/funding-opportunities/funding-programmes-and-open-calls/horizon-2020_en

[2]https://www.usda.gov/climate-solutions/climate-smart-commodities

The American National Research Council[3] defines precision agriculture as "a management strategy that uses information technology to bring data from multiple sources to bear on decisions associated with crop production". Precision agriculture leverages modern agricultural approaches, including drone mapping, information and communication technology, and geospatial techniques, to acquire valuable data for crop management. It takes into consideration the diversity and variability of landscape within agricultural fields.

With the growing importance of precision agriculture, there is a need for a multidisciplinary approach to upscale current point/plant scale agricultural data collection to field/canopy scales [4]. Remote sensing has emerged as an effective tool for collecting data at the field scale, but traditional approaches such as using remote sensors on towers come with limitations such as a fixed position from where data is collected. Furthermore, the use of aircraft or satellite datasets also has spatial and temporal data resolution limitations [5]. As such, the application of drone mapping or data collection using unmanned aerial vehicles (UAVs) is emerging as crucial in addressing the limitations of traditional remote sensing and introducing an era of on-demand remote sensing with fine spatial and temporal resolutions. Drone mapping has several advantages over traditional remote sensing techniques, including their ability to low altitude flights and high-resolution data with rapid turnaround times. These capabilities allow for more frequent and precise data collection, enabling agriculturists to make rapid informed decisions and improve agricultural sustainability.

Drone mapping can revolutionize site-specific management in agriculture by allowing the creation of high-resolution maps of agricultural land, which can help scout crop health, identify stress areas, and assess the effectiveness of management strategies. Site-specific management involves the use of data-driven approaches to optimize agricultural inputs, like water, fertilizers, and pesticides, at the subfield scale, for increased yields, reduced environmental impact, and improved economic viability. Drone mapping can offer a cost-effective and much more efficient way to collect this information. Drones can capture data on a wide range of crops and soil properties utilizing an array of sensors, including multispectral and thermal cameras. These data can then be processed using specialized software to create high-resolution maps of the field, which can help identify areas of variability and inform site-specific management decisions. Multispectral and thermal sensors can detect variations in plant health that may not be

[3]https://nap.nationalacademies.org/catalog/5491/precision-agriculture-in-the-21st-century-geospatial-and-information-technologies

visible to the naked eye, allowing farmers/crop consultants to make data-driven decisions that can improve crop health, reduce environmental impact, and increase economic viability. For instance, UAV-based multispectral imagery has proven to be effective in detecting and mapping the spatial distribution of wheat (*Triticum*) stripe rust at a sub-plot level [6]. Drone-based approaches are often more accurate in detecting the disease than ground-based surveys or satellite imagery. Su *et al.* [6] employed the sensitivity of bands and indices to evaluate and analyze rust quantitatively and to detect the yellow rust stage in wheat. Similarly, drone-based multispectral imagery is utilized to identify and map the spatial distribution of sugarcane aphids within sorghum fields [7], resulting in more accuracy in estimating the extent of infestation than traditional scouting methods. Furthermore, by combining drone mapping with ground-based scouting and sampling, site-specific management decisions can be further refined and optimized.

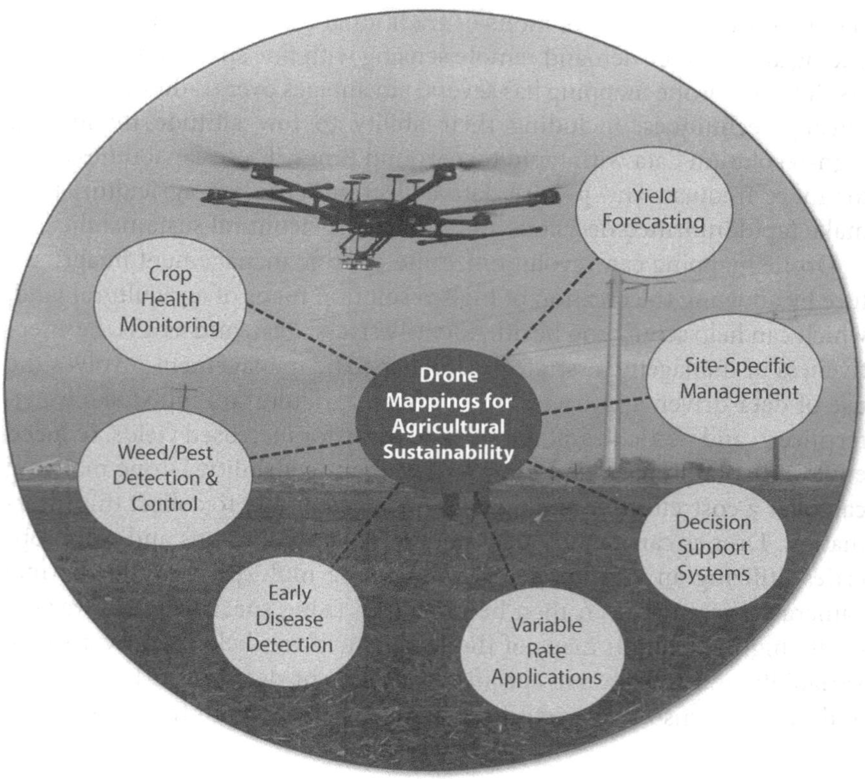

Figure 16.1 Drone mapping applications in agriculture.

As drone technology gets improved further and more accessible, the use of drone mapping in agriculture will become more widespread. This chapter presents an overview of the potential of drone mapping in the context of sustainable agriculture. It also explores the various applications of drone mapping in agriculture, including crop monitoring, yield forecasting, site-specific management, and targeted chemical applications for weed, pest, and disease management (Figure 16.1). The economic, social, and environmental benefits associated with the use of drone mapping in agriculture, as well as the challenges and limitations of the technology, are examined. Finally, the chapter highlights future opportunities and directions for drone mapping in promoting agricultural sustainability.

16.2 Agricultural Remote Sensing: Satellite and Drones

Sustainable agriculture has become a critical area of research and development, with technological advancements resulting in various tools and techniques aimed at increasing crop yields while maintaining environmental quality and long-term sustainability. Two prominent tools used for agricultural mapping and monitoring are drone and satellite imagery. Both have their unique features and applications and understanding their capabilities can assist crop consultants/farmers and researchers in making informed decisions on the selection of tools to use for specific tasks.

16.2.1 Satellite

Satellite remote sensing refers to obtaining Earth's surface information through the captured satellite data. Satellite remote sensing involves the use of several types of sensors that capture data at different wavelengths. These include visible, infrared, and microwave sensors, among others. Each type of sensor is sensitive to different types of information and can provide unique insights into the Earth's surface. This technology offers significant advantages, including nondestructive data collection across larger surface areas, ensuring systematic and comparable data acquisition [8].

The application of satellite remote sensing in agriculture gained attention in the 1970s and 1980s when researchers started to utilize remote sensing data for field mapping and monitoring. Since then, satellite remote sensing has revolutionized the way we monitor and manage agricultural systems, providing valuable information at large spatial and temporal scales. Major applications include predicting yields, identifying crops, monitoring flood damage, analyzing land use, assessing crop growth and phenology, estimating leaf area index (LAI), measuring plant height, managing nutrients, classifying weeds, and disease management [9]. The usefulness of such sensing methods has grown in popularity due to advancements in sensors for fine spatial resolution [10], frequent data collection [11], publicly available datasets [10], and the accumulation of data over several years [12]. Satellite imagery can cover vast geographical areas by spanning hundreds of hectares in a single image [13], hence enabling rapid data acquisition, facilitating early agronomic management decisions, and the identification of areas requiring intervention to prevent production losses. Its potential for sustainable agriculture continues to be explored, and ongoing advancements in satellite technology and data processing methods along with integrating with drone mapping (discussed in section 16.2.4) are likely to further enhance its capabilities in the future.

16.2.2 Drones

Drones have emerged as a promising tool to aid in precision agriculture, environmental monitoring, and natural resource management. Drone mapping involves the use of UAVs to collect data about agricultural land. Drones are equipped with different types of sensors that gather data on crops, soil, and the surrounding environment. The most common types of sensors used in drone mapping are optical sensors (Table 16.1), which include RGB (red, green, blue), multispectral, hyperspectral, and thermal sensors.

RGB sensors are widely used in drone mapping for their ability to capture images in the red, green, and blue bands. These sensors are commonly used for generating high-resolution maps of agricultural land. Since all three bands (red, green, and blue) are visible to the human eye, RGB images can be easily interpreted by humans. Moreover, RGB sensors are relatively affordable and user-friendly, making them a popular choice for agricultural monitoring by farmers or crop consultants. Although RGB imagery may not generate spectral detail as much as other options, several studies reported its effectiveness in estimating biomass and 3D modeling of field crops [59]. In the case of crop 3D modeling, RGB images can provide point clouds of crop canopies and trees with high accuracy, utilizing techniques like object-based image analysis or structure from motion (SfM) [14, 15]. Grüner *et al.* [14] demonstrated in a recent study that accurate biomass estimation in grasslands can be achieved solely through RGB imaging using SfM processing. Additionally, RGB imagery can be integrated with near-infrared (NIR) and multispectral sensors for a more accurate estimation of crop biomass [16].

Multispectral sensors capture images in a wider range of bands (4 to 12 bands) than RGB sensors. This allows them to detect variations in plant health that may not be visible to the naked eye, accounting for factors such as chlorophyll content, water content, and nutrient levels, and provide more detailed information about crop health and stress levels. Multispectral sensors are more expensive than RGB sensors, but they can provide valuable insights into crop health. Multispectral cameras are categorized into two types: narrowband and broadband. Narrowband cameras possess a smaller spectral range. Generally, narrower spectral ranges provide more precise spectral information, whereas broadband cameras, such as Sequoia with a 40-nm bandwidth, possess wider spectral ranges. These ranges resemble those of satellites, allowing more convenient application of satellite-based sensor algorithms [e.g., for vegetation indices (VIs)] to broadband camera data [17]. However, findings obtained from narrowband and broadband cameras are not always consistent [18].

Hyperspectral sensors capture images in hundreds or even thousands of bands spread across visible, near-infrared, and shortwave infrared bands, providing extremely detailed information about crop health and stress levels [19]. Although hyperspectral sensors are the most expensive type, they provide unparalleled insights into crop health. Nevertheless, the extensive data generated by hyperspectral sensors, encompassing hundreds of bands in high-resolution images, necessitates appropriate analysis and processing methods [20]. Despite many important applications, the adoption of hyperspectral sensors in precision agriculture remains relatively limited compared to other sensor types. This limitation can largely be ascribed to the high cost associated with high-resolution spectroscopy [21]. Also, hyperspectral sensors typically require large drones due to their high payload, resulting in limited time and restricting their application in large field scales. Furthermore, the complexity of data acquisition and analysis contributes to the scarcity of studies in this field, as it requires specialized training. Some researchers anticipate that hyperspectral sensors will primarily serve research purposes. However, ongoing technological advancements are expected to lead to the development of smaller and more affordable devices, allowing hyperspectral-based sensing to become a widely adopted approach in agriculture, forestry, and related fields.

Thermal sensors capture images of the thermal radiation/heat emitted by crops. Thermal drone remote sensing in agriculture has been extensively

Table 16.1 Commonly used drone sensors in agricultural monitoring.

Sensor type	Sensors	Bands	Number of bands[†]
RGB	Parrot Sequoia, DJI Zenmuse XT2, and Sony RX1R II	Red-green-blue (RGB)	3
Multispectral	MicaSense RedEdge, Parrot Sequoia, and DJI Zenmuse Multispectral	RGB, red-edge, near-infrared	4–12
Hyperspectral	Specim, Resonon, and Headwall Photonics	RGB, near-infrared, shortwave-infrared	100–200+
Thermal	FLIR Vue Pro, FLIR One X2, and Seek Thermal Compact XR	Thermal	1

[†]usual number of bands in sensors.

studied, primarily focusing on detecting crop water stress and managing irrigation resources. Other applications include identifying plant pathologies, phenotyping, monitoring, yield estimation, and identifying drainage networks. However, the utilization of thermal sensors in drones has not been as widespread as other cameras due to limitations in geometric resolution and the nature of temperature data. Combining thermal cameras with RGB and multispectral sensors seems to enhance drone capabilities in precision agriculture, especially for early disease detection [22]. Future developments in thermal remote sensing are expected to enhance user-friendliness and automation, simplifying data acquisition and processing.

16.2.3 Satellite Versus Drones

One major difference between drones and satellite imagery is the spatial resolution they offer [23]. Drones generally provide higher spatial resolution images than satellites due to their closer proximity to the ground, resulting in more detailed images of crops, soil, and other features of interest. In contrast, satellite imagery provides a larger field of view and covers much larger areas, which can be useful for monitoring large-scale crop growth patterns over time [24]. Another difference between the two tools is usability. Drones require more skill and effort to operate than satellite imagery, as they require a trained pilot and sophisticated software to capture and analyze images. Satellite imagery is readily available from many providers such as the United States Geological Survey, the European Space Agency, and Indian Space and Research Organization, and it can be easily accessed and downloaded through online platforms. Cost is another factor to consider when comparing the two technologies. Drones require a significant investment in hardware and software, including the UAV itself, cameras, sensors, and analysis software. However, satellite imagery platforms, like Landsat or Sentinel, offer data at no cost to the general public, although some others involve a fee based on the level of resolution and frequency of images needed [24]. Both drones and satellite imagery can produce 2D and 3D maps of agricultural fields, but the accuracy and quality of these maps can vary depending on the tool used. Drones can produce highly accurate and detailed 3D maps, which can be useful for monitoring crop height, biomass, and other metrics. In contrast, satellite imagery generates less detailed 3D maps but can provide large-scale coverage of crop growth patterns over time.

Another benefit of using drone mapping is the flexibility it offers in terms of image capture. Drones can be flown on-demand, allowing farmers to capture images at specific times and in response to specific events, such as sudden weather changes [24]. The advancement of micro-UAVs weighing only a few kilograms has changed the traditional paradigm of remote sensing by providing end-users with greater control over data acquisition features. Satellite imagery, on the other hand, is often limited by cloud cover, and images can only be acquired at specific intervals, which can result in missed opportunities for monitoring and analysis [25]. Certain applications in sustainable agriculture depend heavily on the choice of acquisition dates and times. Since drones can be flown on demand, they can capture images more frequently than satellites, enabling farmers to track crop performance and detect real-time changes. This is particularly useful in precision agriculture, where farmers can make decisions based on near real-time information on crop health status and growth patterns. The close-to-surface data acquisition capability of drones provides the advantage of being quasi-independent or less affected by clouds.

16.2.4 Drone Mapping: A Complementing System

Drones and satellite constellations stand out as two important earth observation systems used for monitoring land surface dynamics. While satellite remote sensing is extensively employed for agricultural monitoring, it comes with some limitations that can be addressed by integrating drone mapping into the monitoring process. For example, satellite remote sensing provides wide area coverage, but the accuracy of the data can be affected by atmospheric conditions and other factors. Drone mapping can help overcome some of these limitations by providing ground-level data for calibration and validation. Drones offer acquisition flexibility and high spatial resolution imagery, while satellites are useful in retrieving time-series data over large geographical areas. For instance, by combining the high spatial resolution of drone imagery with the wide area coverage of satellite imagery, farmers and researchers can obtain a complete and more accurate picture of their agricultural land. The synergy between drone and satellite data is critical in comprehending the dynamics of the Earth's surfaces [26], and techniques such as data calibration, fusion, and comparison can be used to exploit this multi-source synergy [27] while accounting for resolution trade-offs (spatial, spectral, and temporal).

Data calibration: Calibration is an important application of drone data in satellite-based models, both qualitatively and quantitatively. An example of qualitative calibration is the work by Nhamo *et al.* [28], which utilized drone data to train and validate a Landsat-based model for mapping irrigated fields in South Africa. The inclusion of labeled drone data in the model significantly enhanced the accuracy of the irrigated field mapping from 71% to 95%. Likewise, Zhang *et al.* [25] performed a quantitative calibration of Sentinel-2 imagery for monitoring chlorophyll content in winter wheat crops at the reviving stage. They utilized drone observations to perform reflectance correction and found that the inversion model response was improved when the drone-based inversion model was used than the satellite-based model.

Data comparison: Data comparison between drone and satellite images is essential to evaluate the strengths and limitations of each data source, helping determine their suitability for specific agricultural applications. The research conducted by Khaliq *et al.* [24] emphasizes the significance of comparing data from different sources to identify the most suitable data source for specific agricultural objectives. In their research, they compared the effectiveness of Sentinel-2 (spatial resolution, 10 m) and drone imagery (spatial resolution, 5 cm) using NDVI for vineyard variability assessment. They found that the raw spatial resolution of Sentinel-2 imagery data cannot be directly employed for variability assessment due to the inter-row space affecting NDVI computations and leading to biased crop descriptors. Such comparisons are especially needed for row crops or layouts lacking intensity, where the crop canopy does not cover the entire surface, or in case of weed infestation. High-resolution drone images allowed a more detailed assessment of the field and provided clear identification of specific areas affected by various agricultural stresses, whereas the satellite image, although less detailed, provided a broader overview of the field's health status and identified large-scale patterns and variations across the entire field.

Data Fusion: Data fusion is a robust technique that leverages the unique features of multiple data sources to create new, enhanced data, which is significant in remote sensing applications [27], particularly in sustainable agriculture. The most basic approach involves combining data to densify or complete a time series. For example, Gevaert *et al.* [29] used data fusion to monitor crop growth in a potato field by integrating the temporal resolution of Formosat-2 imagery with the high spatial resolution of drone imagery. They used drone-based hyperspectral data with Bayesian theory to address missing spectral information in multispectral satellite data. This fusion allowed for the determination of crop biophysical characteristics such as LAI and chlorophyll content using continuous spectral reflectance at high temporal frequency. Jenerowicz *et al.* [23] developed another remarkable example, in which they devised a pan-sharpening process based on drone-satellite data fusion to combine the spatial information from high-resolution panchromatic imagery with the spectral information from lower-resolution multispectral or hyperspectral imagery, resulting in multispectral or hyperspectral images with enhanced spatial resolution.

16.3 Drone Mapping for Precision Agriculture

16.3.1 Crop Health Monitoring and Assessment

Drones have several advantages when it comes to crop health monitoring and assessment. One of the main advantages of drone mapping is its capacity to generate detailed and timely information about crop health. The availability of multispectral, thermal, and active remote sensing drone sensors has enabled detailed analysis of crop health indicators, such as LAI, VI, and chlorophyll content which allows detection of variations in plant health that are not discernible to the human naked eye. Such information proves helpful in identifying areas of stress or disease, which can then be targeted with precision treatments, reducing the need for broad-spectrum chemical applications. For example, Anthony *et al.* [30] used drones to develop a crop height measurement system that can be used by practitioners to investigate crop development and adjust crop treatments accordingly. Kim *et al.* [31] used VIs, synthesized from drone multispectral sensor imagery, for crop damage assessment of rice (*Oryza sativa*) crops after a chemical exposure.

Passive sensors are predominantly utilized in agricultural remote sensing to acquire electromagnetic wave reflectance information from canopies [32]. The reflectance characteristics from vegetation at various wavelengths of the electromagnetic spectrum are influenced by various factors, including chemical and morphological characteristics of the organ or leaf surface, which can further help extract vegetation properties such as plant type, plant water content, and chlorophyll content. This has resulted in the development of high spectral sensors, leading to the development of various VIs (Table 16.2), to detect vegetation changes utilizing different wavelengths. A few commonly used VIs are explained below with their utilization for precision agriculture:

The normalized difference vegetation index (NDVI) is a widely employed indicator for monitoring crop health. Its calculation involves the normalized ratio between the red and near-infrared regions of the electromagnetic spectrum. NDVI is most efficiently used to characterize crop growth owing to its sensitivity to changes in chlorophyll content and canopy structure. For example, healthy vegetation gives high NDVI values because of more reflectance of near-infrared light and less absorbance of red light. Conversely, stressed or unhealthy vegetation will have a lower NDVI value. Additionally, some studies used drone-based NDVI values to estimate crop growth parameters such as LAI [33, 34].

The Green normalized difference vegetation index (GNDVI) is another frequently employed index for monitoring crop health. It is a modified version of NDVI, calculated as the difference in reflectance between the green and near-infrared portions of the electromagnetic spectrum. It is less affected by canopy cover and especially useful for monitoring crops with high levels of canopy cover. GNDVI is particularly useful for analyzing crops with high chlorophyll content, such as corn (*Zea mays*) and soybean (*Glycine max*). For example, Corti *et al.* [35] found that the GNDVI was the best predictor for the N-related variables in corn, and Wahab *et al.* [36] found GNDVI useful to assess vigor and crop yields at various stages of corn.

Table 16.2 Commonly used vegetation indices for agricultural monitoring.

Index	Calculation formula	Agricultural applications
Normalized difference vegetation index (NDVI)	$\frac{NIR - Red}{NIR + Red}$	Crop health and vigor
Green NDVI (GNDVI)	$\frac{NIR - Green}{NIR + Green}$	Crop health and vigor, especially in areas with high levels of canopy cover
Normalized difference red edge index (NDRE)	$\frac{NIR - Rededge}{NIR + Rededge}$	Crop chlorophyll content, especially in areas with high levels of canopy cover
Canopy chlorophyll content index (CCCI)	$\frac{NDRE - NDRE_{min}}{NDRE_{max} - NDRE_{min}}$	Measure crop chlorophyll content
Enhanced vegetation index (EVI)	$\frac{2.5*(NIR - Red)}{NIR + C_1 * Red - C_2 * Blue + L}$	Crop health and vigor, especially in areas with high levels of atmospheric noise
Soil-adjusted vegetation index (SAVI)	$\frac{(1+L_1)*(NIR - RED)}{(NIR + RED + L_1)}$	Crop health and vigor, especially in areas with low levels of canopy cover
Normalized difference water index (NDWI)	$\frac{Green - NIR}{Green + NIR}$	Monitor water content in water bodies
Crop Water Stress Index (CWSI)	$\frac{NIR - SWIR1}{NIR + SWIR1}$	Monitor crop water stress
Greenness index (GI)	$\frac{2*G - R - B}{2*R + G + B}$	Crop greenness
Optimized soil-adjusted vegetation index (OSAVI)	$\frac{(1+L_2)*(NIR - Red)}{NIR + Red + L_2}$	Crop health and vigor, especially in areas with low levels of canopy cover and high levels of atmospheric noise

Note: C_1 and C_2 are the aerosol scattering correction factors, L is the soil and canopy background adjustment, L_1 is a constant and is related to the slope of the soil line in a feature-space plot, L_2 is an adjustment factor to reduce soil noise (generally 0.16), and K depends on the type of vegetation and the atmospheric conditions.

The canopy chlorophyll content index (CCCI) is another index that is commonly used for crop health monitoring. It is a sensitive index of chlorophyll content and is less affected by canopy cover and soil background than NDVI. It is especially useful for monitoring crops with low levels of canopy cover. CCCI measures the amount of chlorophyll in a plant. Photosynthesis, the process through which plants convert light energy into chemical energy, relies on the presence of chlorophyll. Thus, chlorophyll is an essential component for photosynthesis in plants. A high CCCI value suggests healthy crop vegetation with elevated chlorophyll content, while a low CCCI value indicates stressed or unhealthy vegetation with low chlorophyll content. CCCI has a strong relation to leaf nitrogen content [37, 38]. This index is particularly useful in differentiating nitrogen from water stress [39].

In addition to these indices, several other indices can be used for crop health monitoring and assessment, including the enhanced vegetation index (EVI) [40], the soil-adjusted vegetation index (SAVI) [41], and the normalized difference water index (NDWI) [42].

Several recent studies have demonstrated the effectiveness of drones in crop health monitoring and assessment. One such study by Li *et al.* (2019) used drone imagery and NDVI to analyze the effects of nitrogen fertilizer on wheat. The study found that the use of drone imagery and NDVI could effectively detect the effects of nitrogen fertilizer on wheat and could be used to optimize nitrogen fertilizer application rates to improve wheat yield.

16.3.2 Field Management Zones and Variable Rate Application

Feld management zones are areas within a field that exhibit consistent variation in soil properties, topography, crop properties, and/or yield potential. By dividing a field into management zones and treating each zone according to its unique characteristics, farmers can optimize their inputs and ensure that each plant receives the appropriate amount of water, nutrients, and other inputs while minimizing waste and negative impacts on the environment. This can lead to healthier plants with higher yields and improved quality. For example, areas with high nutrient levels may require less fertilizer than areas with low nutrient levels. Similarly, field areas possessing high water-holding capacity may require less irrigation water than areas with low water-holding capacity.

Mapping the within-field crop status variability is of great importance to guide precision agriculture practices such as variable rate application (VRA) [43]. Sustainable agriculture aims to minimize the negative environmental impacts of agriculture while maintaining or increasing agricultural productivity. Precision agriculture techniques, such as VRA of inputs based on field variability, have emerged as a promising approach to reducing agricultural inputs, such as fertilizers, water, and pesticides, without compromising yields. VRA involves adjusting agriculture input application rate as per field variability, seeking a balance between agricultural inputs and spatial crop demands. One of the most effective ways to optimize VRA is to map field crop variability and divide the field into different management zones.

Traditionally, satellites have been used to map field crop variability using vegetation or other indices and dividing fields into different management zones. Tools, such as the decision support system- CropSAT, have been providing field variability VI maps along with variable rate application prescriptions that can be easily uploaded to field application systems through a USB [44, 45]; however, satellite data dependence on clear and cloud-free limits applications of such tools. As discussed earlier, drones overcome these limitations and provide opportunities for multispectral sensor-based management zone delineation with higher accuracy at a finer spatial resolution, and they are becoming increasingly popular in creating field management zones due to their versatility, affordability, and precision. Drone images can be analyzed to identify patterns and variations in soil moisture, nutrient levels, and plant health, which can further be used to create field management zone maps. Drone-produced crop health maps through vegetation indices combined with field-observed soil, nutrient, and yield maps can help in relating vegetation growth with field properties and later improve field management zone delineation to optimize VRAs. On a similar note, Maresma *et al.* [46] utilized drone multispectral images to compute VIs, SPAD (soil plant analysis development), and crop height to assess whether an additional side-dress application of N fertilizer was required at the flowering stage of corn growth. Similarly, Rasmussen *et al.* [47] used drone multispectral images to divide fields into 1-m^2 patches with two categories, with and without green vegetation, to map weeds and optimize herbicide applications and Sugiura *et al.* [48] used drone RGB images to assess disease severity in a potato field, later to be utilized for optimizing fungicide applications.

16.3.3 Weed, Insect, and Disease Detection and Management

Weed, pest, and disease infestation are major problems in agriculture, and they can cause significant yield losses. Traditional methods of weed/pest/disease control, such as applying herbicides/pesticides/fungicides to the entire field, can be expensive and harmful to the environment. Drone mapping is an innovative solution that can help farmers detect and control such infestations more effectively. Drone-captured high-resolution multispectral images of fields and developed VIs can assist in identifying affected areas, even in the early stages of growth along with their location, size, and density of infestation [49]. Thermal remote sensing can also help identify anomalies in temperature patterns, which can be indicative of underlying issues such as disease infestation or pest presence [50]. Infected plants often exhibit different surface temperatures compared to healthy plants due to physiological changes caused by the pest/disease. Certain pests, such as insects or mites, can cause localized damage to plants by feeding on their tissues. This feeding activity can result in changes in plant physiology and alter their surface temperature. As such, thermal imaging can prove useful in continuous monitoring of plant health and insect or disease progression over time.

Infestation mapping before taking control measures can assist in herbicide/pesticide selection and optimizing spray application parameters based on the type and density of infestation. This targeted approach improves efficiency and reduces the amount of chemicals being used. For the last few years, scientists have been working toward real-time weed detection and targeted weed treatments [51, 52]. Ground-based sensors deployed on sprayers have also been used for targeted chemical applications against infestations; however, these methods are limited with area coverage for a short period. Drones provide the opportunity for infestation mapping over large areas in a short period [53]. Moreover, drone mapping can be used to monitor the effectiveness of chemicals or other control measures, allowing for adjustments to the treatment plan as needed. Therefore, drone mapping provides a safe and effective way to detect and manage infestations, making it a valuable tool for farmers seeking to improve their weed and pest control practices and protect their crops [54], especially farmers opting for organic farming [49]. Traditional methods of weed control, involving herbicide applications, are not an option for organic farmers and the alternate non-chemical weed control practices, such as heat treatment, infrared radiations, steaming, and soil solarizations, require targeted weed control applications, which further enhance the importance of weed mapping using drones.

Certain crops attract specific pests and diseases, and continuous cultivation of the same crop in a field can facilitate the buildup of pests and pathogens. By suggesting crop rotations through drone data analysis, farmers can disrupt pest life cycles and reduce disease incidence. This can result in reduced reliance on chemical pesticides, lower pest control costs, and improved overall plant health. Crop rotation can be an effective strategy for managing weeds as well. Drone analysis can identify weed hotspots or areas prone to weed infestations and be used for suitable crop rotations possible that include weed-suppressive crops or crops with different weed susceptibility. Thus, farmers can reduce weed pressure, minimize herbicide use, and save on weed management expenses.

16.3.4 Yield Forecasting

Yield forecasting is a critical aspect of agricultural management, as it helps farmers and stakeholders estimate the potential crop production for a given season. Drones can generate accurate 3D maps and perform biomass analysis, providing insights into crop growth and yield potential. Generated mapping data can be utilized in conjunction with data-driven models or advanced analytics techniques to estimate and forecast crop yields [55]. Yield predictions help farmers in making informed decisions on plan harvest schedules, and marketing strategies, and optimize resource allocation for harvesting operations. Efficient harvest planning reduces labor costs and ensures timely and effective harvesting.

Drone-captured maps can also be used to understand factors contributing to field spatial yield variability such as relating yield variability with different levels of plant vigor, stress, or disease incidence across the farm. LAI and biomass are two critical indicators of crop health and development that are used for in-season yield estimation [56]. Remote sensing through UAVs can help retrieve these important parameters and generate yield prediction at a large spatial scale. Crop yield estimation from UAV data can be accomplished through crop modeling or statistical relationships [56]. Obtaining spatial maps of crop biophysical parameters or indices at multiple time points during the growing season offers improved estimations of biomass and yield compared to relying on a single snapshot [57]. Moreover, collected time-series data over a growing season can be studied to get insights into changes in crop physiological conditions during the season and fluctuation in yield potentials. This could help in understanding yield trends during different seasonal environmental conditions. Maresma *et al.* [46] used a regression-based methodology to investigate the relationship between corn yield and biomass, as well as spectral indices measured at the V12 stage. Their findings reveal that red-based indices, including NDVI and WDRVI (wide dynamic range vegetation index), exhibited the strongest correlation with grain yields across various fertilizer application rates.

16.4 Economic Perspectives

Drone mapping is a relatively new technology, but it can revolutionize agriculture. As indicated in the above discussions, the use of drone mapping in agriculture could bring significant economic benefits to the agricultural sector, contributing to both profitability and sustainability. While the initial investment in drone technology and training can be high, the long-term benefits are significant. For instance, by providing detailed and timely information on crop health and conditions, drones can assist farmers in efficient planning and maximizing crop yields. Optimizing the use of agricultural inputs using field management zones, VRA, and site-specific targeted applications (weed, pest, and disease control) are other key economic benefits of precision agricultural farming using drones. The targeted chemical applications can reduce input wastage and save costs over expensive inputs. Adopting irrigation management strategies using drone-mapped field zones can lead to water savings and cost reduction. Additionally, real-time information helps the farmers to identify and address issues quickly, reducing downtime and improving farmer's time productivity, leading to significant cost savings in the case of large-scale farming.

Moreover, drone mapping can also help farmers make better decisions about land use and crop rotation and contribute to long-term economic sustainability. By providing detailed information on soil health and fertility, drones can help farmers identify areas that are suitable for different crops or uses, optimizing land use, and reducing the risk of crop failure. Moreover, integrating crop models with drone measurements can provide precise recommendations for suitable crop rotations for the field. Implementing these rotations can enhance soil fertility, reduce weed/pest/disease pressure, and ultimately contribute to increased crop yields and the long-term economic sustainability of the farm.

16.5 Challenges

Despite the many potential benefits of utilizing drone mapping for agriculture, its widespread adoption for agricultural sustainability faces a few challenges. Two major challenges are related to practical applications in farmer fields and the reproducibility and accuracy of drone mapping data.

Practical applications on farmer fields: While drone mapping has demonstrated its value in agricultural research and consultancy settings, its practical application on farmer fields is still limited. Several factors contribute to this challenge, including a lack of technical knowledge and familiarity with drone technology among farmers, compliance with regulatory requirements governing drone operations, and the cost associated with acquiring drones and necessary equipment.

The knowledge gap regarding technical aspects of operating drones and effectively utilizing the data is one of the main limiting factors for drones' practical application. Many farmers lack the necessary training and expertise to integrate drones into their agricultural practices. Additionally, there are other requirement issues such as computational requirements for data processing and interpreting data that need to be addressed and improved. There have been efforts for real-time data processing and generating maps, however not much adopted yet. Educational programs, workshops, and training initiatives can help bridge these knowledge gaps and empower farmers to effectively utilize drones in their farming practices.

Moreover, the regulatory framework surrounding drone mapping imposes certain restrictions and compliance requirements that need to be followed. Farmers must navigate through regulations related to flight altitudes, airspace restrictions, licensing, and privacy concerns. Compliance with these regulations can be challenging and may require additional time, effort, and resources. Simplifying information on regulations and providing clear guidelines tailored to the agricultural sector can facilitate compliance and encourage farmers to embrace drone technology.

Another significant challenge is the cost associated with acquiring drones and the required equipment. Drones and their accompanying sensors, software, and accessories can be costly, making them inaccessible for many farmers, especially those with limited financial resources. The initial investment and ongoing maintenance expenses can be prohibitive, creating a barrier to entry for farmers who wish to adopt drone technology for agricultural sustainability. Government agricultural or environmental agencies should step in here and provide financial incentives or assistance programs to make drones more accessible to farmers.

To overcome these challenges, collaboration between drone manufacturers, agricultural organizations, educational institutes, and government agencies can play a crucial role in promoting the adoption of drones in agriculture by addressing these limitations and creating a conducive environment for their successful implementation in farmer fields.

Reproducibility and accuracy of drone mapping data: Data reproducibility and accuracy are crucial aspects of drone mapping. Ensuring reproducibility in drone mapping involves capturing consistent and comparable data over time and across different areas. Factors such as weather conditions, flight parameters, and sensor settings can affect data collection, leading to variations in results. Other minor factors are sensor quality, image resolution, and data processing techniques. Bansod *et al.* [58] reported reproducibility issues with drone mapping. Rasmussen *et al.* [47] reported data accuracy issues with an effect of cloud covers on VIs while stating no effect of flight altitude and data resolution. However, most of the reported reproducibility and data accuracy issues are from studies with uncalibrated sensors [43].

Addressing these challenges requires a multidimensional approach. As a first step, standardization of data acquisition and processing methodologies can enhance reproducibility, allowing for consistent and comparable results such as collecting data during cloud-free days, near solar noon, and with a consistent flight altitude and overlapping plans. Other than that rigorous sensor calibration, validation using ground truth measurements, and quality control measures can further enhance data accuracy and reproducibility.

16.6 Future Perspectives

The prospects of drone mapping in agricultural sustainability are promising, with several potential advancements and applications on the horizon. To ensure its widespread, all the challenges discussed above need to be addressed. Manufacturers should focus on improving sensors for enhanced data acquisition, reducing costs, and developing user-friendly systems, making drone mapping more accessible to farmers. The significance of drone mapping in agricultural practices is increasingly being acknowledged.

Furthermore, the future scope of drone mapping in agriculture lies in the integration of state-of-the-art techniques such as machine learning, artificial intelligence, and the Internet of Things (IoT). By leveraging these technologies, drone mapping can be utilized for advanced applications in agricultural practices. Machine learning algorithms can analyze large datasets collected by drones to identify patterns and correlations, aiding in decision-making processes. Artificial intelligence can enable real-time data processing, facilitating on-demand monitoring and providing farmers with up-to-date information for timely interventions. Integrating drones with IoT devices, such as weather sensors and crop health sensors, will create a comprehensive network of interconnected devices, enabling smart farming practices.

Lastly, fostering the development of open-access data platforms and promoting data sharing among research institutions globally is essential. Currently, numerous research institutes are engaged in drone-based agricultural research to varying degrees. However, these research endeavors tend to be exclusive, impeding the pace of digital advancement in agriculture compared to other industries. Therefore, it is crucial to prioritize data sharing and collaboration to establish standardized and robust methods for drone operation, data processing, and analytics in precision agriculture applications.

16.7 Conclusions

Drone mapping is poised to play a significant role in shaping the future of agricultural sustainability, and its integration into farming practices can contribute to a more resilient, efficient, and sustainable agricultural sector. By embracing drone mapping and harnessing its potential, we can pave the way for sustainable agriculture, where resources are utilized efficiently, environmental impact is minimized, and agricultural practices are optimized for productivity and long-term viability. Precision agriculture has the potential for the next green revolution and drones will be playing a key component in emerging technologies of precision agriculture while addressing the limitations of satellite remote sensing and ground-based monitoring systems. Drone mapping has demonstrated its effectiveness in various agricultural applications, including crop monitoring, weed and pest detection, monitoring water and nutrient stress, and enabling site-specific agricultural management. However, there are a few limitations associated with drone mapping and hindering its widespread agricultural practices such as high initial cost, complex data processing, and strict rules and regulations. Overall, the future of drone mapping in agriculture holds tremendous opportunities. Real-time data processing, predictive analytics, and precision applications will revolutionize farming practices, enabling farmers to make informed decisions and optimize their operations and ultimately contributing to the enhancement and promotion of agricultural sustainability.

References

1. L. Rosa, Adapting agriculture to climate change via sustainable irrigation: Biophysical potentials and feedbacks, *Environ. Res. Lett.*, vol. 17, no. 6, pp. 063008, 2022.
2. H. Ritchie and M. Roser, "Environmental impacts of food production", Published online at OurWorldInData.org. Retrieved from: 'https://our-worldindata.org/environmental-impacts-of-food' [Online Resource], 2022.
3. W. F. Laurance, J. Sayer and K. G. Cassman, Agricultural expansion and its impacts on tropical nature, *Trends Ecol. Evol.*, vol. 29, no. 2, pp. 107-116, 2014.
4. A. Casson, B. Ortuani, V. Giovenzana, L. Brancadoro, S. Corsi, O. Gharsallah, R. Guidetti and A. Facchi, A multidisciplinary approach to assess environmental and economic impact of conventional and innovative vineyards management systems in Northern Italy, *Sci. Total Environ.*, vol. 838, pp. 156181, 2022.
5. L. Schilirò, C. Robiati, L. Smeraglia, F. Vinci, A. Iannace, M. Parente and S. Tavani, An integrated approach for the reconstruction of rockfall scenarios from UAV and satellite-based data in the Sorrento Peninsula (southern Italy), *Eng. Geol.*, vol. 308, pp. 106795, 2022.
6. J. Su, C. Liu, X. Hu, X. Xu, L. Guo and W.-H. Chen, Spatio-temporal monitoring of wheat yellow rust using UAV multispectral imagery, *Comput. Electron. Agric.*, vol. 167, pp. 105035, 2019.
7. N. Elliott, G. Backoulou, M. Brewer and K. Giles, NDVI to detect sugarcane aphid injury to grain sorghum, *J. Econ. Entomol.*, vol. 108, no. 3, pp. 1452-1455, 2015.
8. S. Liaghat and S. K. Balasundram, A review: The role of remote sensing in precision agriculture, *Am. J. Agric. Biol. Sci.*, vol. 5, no. 1, pp. 50-55, 2010.
9. C. Zhang, A. Marzougui and S. Sankaran, High-resolution satellite imagery applications in crop phenotyping: an overview, *Comput. Electron. Agric.*, vol. 175, pp. 105584, 2020.
10. C. Karakizi, M. Oikonomou and K. Karantzalos, Vineyard detection and vine variety discrimination from very high-resolution satellite data, *Remote Sens.*, vol. 8, no. 3, pp. 235, 2016.
11. C. Wei, J. Huang, L. R. Mansaray, Z. Li, W. Liu and J. Han, Estimation and mapping of winter oilseed rape LAI from high spatial resolution satellite data based on a hybrid method, *Remote Sens.*, vol. 9, no. 5, pp. 488, 2017.
12. M. Sibanda, O. Mutanga, M. Rouget and L. Kumar, Estimating biomass of native grass grown under complex management treatments using world-view-3 spectral derivatives, *Remote Sens.*, vol. 9, no. 1, pp. 55, 2017.
13. A. Dalla Marta, D. Grifoni, M. Mancini, F. Orlando, F. Guasconi and S. Orlandini, Durum wheat in-field monitoring and early-yield prediction: assessment of potential use of high resolution satellite imagery in a hilly area of Tuscany, Central Italy, *J. Agric. Sci.*, vol. 153, no. 1, pp. 68-77, 2015.

14. E. Grüner, T. Astor and M. Wachendorf, Biomass prediction of heterogeneous temperate grasslands using an SfM approach based on UAV imaging, *Agronomy*, vol. 9, no. 2, pp. 54, 2019.
15. J. Torres-Sanchez, A. I. de Castro, J. M. Pena, F. M. Jimenez-Brenes, O. Arquero, M. Lovera and F. Lopez-Granados, Mapping the 3D structure of almond trees using UAV acquired photogrammetric point clouds and object-based image analysis, *Biosyst. Eng.*, vol. 176, pp. 172-184, 2018.
16. L. Roth and B. Streit, Predicting cover crop biomass by lightweight UAS-based RGB and NIR photography: an applied photogrammetric approach, *Precis. Agric.*, vol. 19, pp. 93-114, 2018.
17. J. M. Fernández-Guisuraga, E. Sanz-Ablanedo, S. Suárez-Seoane and L. Calvo, Using unmanned aerial vehicles in postfire vegetation survey campaigns through large and heterogeneous areas: Opportunities and challenges, *Sensors*, vol. 18, no. 2, pp. 586, 2018.
18. D. Zhao, L. Huang, J. Li and J. Qi, A comparative analysis of broadband and narrowband derived vegetation indices in predicting LAI and CCD of a cotton canopy, *ISPRS J. Photogramm. Remote Sens.*, vol. 62, no. 1, pp. 25-33, 2007.
19. G. S. Baath, K. C. Flynn, P. H. Gowda, V. G. Kakani and B. K. Northup, Detecting Biophysical Characteristics and Nitrogen Status of Finger Millet at Hyperspectral and Multispectral Resolutions, *Front. Agron.*, vol. 2, pp. 604598, 2021.
20. T. Adão, J. Hruška, L. Pádua, J. Bessa, E. Peres, R. Morais and J. J. Sousa, Hyperspectral imaging: A review on UAV-based sensors, data processing and applications for Agriculture and Forestry, *Remote Sens.*, vol. 9, no. 11, pp. 1110, 2017.
21. B. Lu, P. D. Dao, J. Liu, Y. He and J. Shang, Recent advances of hyperspectral Imaging Technology and Applications in agriculture, *Remote Sens.*, vol. 12, no. 16, pp. 2659, 2020.
22. J. Zhang, Y. Huang, R. Pu, P. Gonzalez-Moreno, L. Yuan, K. Wu and W. Huang, Monitoring plant diseases and pests through remote sensing technology: A review, *Comput. Electron. Agric.*, vol. 165, pp. 104943, 2019.
23. A. Jenerowicz, K. Siok, M. Woroszkiewicz and A. Orych, The fusion of satellite and UAV data: simulation of high spatial resolution band. XIX, vol. 10421, pp. 364-375, 2017.
24. A. Khaliq, L. Comba, A. Biglia, D. Ricauda Aimonino, M. Chiaberge and P. Gay, Comparison of satellite and UAV-based multispectral imagery for vineyard variability assessment, *Remote Sens.*, vol. 11, no. 4, pp. 436, 2019.
25. S. Zhang, G. Zhao, K. Lang, B. Su, X. Chen, X. Xi and H. Zhang, Integrated satellite, unmanned aerial vehicle (UAV) and ground inversion of the SPAD of winter wheat in the reviving stage, *Sensors*, vol. 19, no. 7, pp. 1485, 2019.
26. C. Kuenzer, S. Dech and W. Wagner, Remote sensing time series revealing land surface dynamics: Status quo and the pathway ahead, In: Kuenzer, C.,

Dech, S., Wagner, W. (eds) *Remote Sensing Time Series. Remote Sensing and Digital Image Processing*, vol 22, Springer International Publishing, 2015.
27. E. Alvarez-Vanhard, T. Corpetti and T. Houet, UAV & satellite synergies for optical remote sensing applications: A literature review, *Sci. Remote Sens.*, vol. 3, pp. 100019, 2021.
28. L. Nhamo, R. Van Dijk, J. Magidi, D. Wiberg and K. Tshikolomo, Improving the accuracy of remotely sensed irrigated areas using post-classification enhancement through UAV capability, *Remote Sens.*, vol. 10, no. 5, pp. 712, 2018.
29. C. M. Gevaert, J. Tang, F. J. García-Haro, J. Suomalainen and L. Kooistra, Combining hyperspectral UAV and multispectral Formosat-2 imagery for precision agriculture applications. In *2014 6th Workshop on Hyperspectral Image and Signal Processing: Evolution in Remote Sensing (WHISPERS)*, pp. 1-4, IEEE, 2014.
30. D. Anthony, S. Elbaum, A. Lorenz and C. Detweiler, On crop height estimation with UAVs. In *2014 IEEE/RSJ International Conference on Intelligent Robots and Systems*, pp. 4805-4812. IEEE, 2014.
31. H. Kim, W. Kim and S. D. Kim, Damage assessment of Rice crop after toluene exposure based on the vegetation index (VI) and UAV multispectral imagery, *Remote Sens.*, vol. 13, no. 1, pp. 25, 2020.
32. J. Xue and B. Su, Significant remote sensing vegetation indices: A review of developments and applications, *J. Sens.*, vol. 2017, 2017.
33. S. Fuentes, C. Poblete-Echeverría, S. Ortega-Farias, S. Tyerman and R. De Bei, Automated estimation of leaf area index from grapevine canopies using cover photography, video and computational analysis methods, *Aust. J. Grape Wine Res.*, vol. 20, no. 3, pp. 465-473, 2014.
34. J. Berni, P. Zarco-Tejada, G. Sepulcre-Cantó, E. Fereres and F. Villalobos, Mapping canopy conductance and CWSI in olive orchards using high resolution thermal remote sensing imagery, *Remote Sens. Environ.*, vol. 113, no. 11, pp. 2380-2388, 2009.
35. M. Corti, D. Cavalli, G. Cabassi, A. Vigoni, L. Degano and P. Marino Gallina, Application of a low-cost camera on a UAV to estimate maize nitrogen-related variables, *Precis. Agric.*, vol. 20, pp. 675-696, 2019.
36. I. Wahab, O. Hall and M. Jirström, Remote sensing of yields: Application of UAV imagery-derived NDVI for estimating maize vigor and yields in complex farming systems in sub-Saharan Africa, *Drones*, vol. 2, no. 3, pp. 28, 2018.
37. E. Barnes, T. Clarke, S. Richards, P. Colaizzi, J. Haberland, M. Kostrzewski, P. Waller, C. Choi, E. Riley and T. Thompson, Coincident detection of crop water stress, nitrogen status and canopy density using ground-based multispectral data. In *Proceedings of the Fifth International Conference on Precision Agriculture*, vol. 1619, p. 6. 2000.
38. D. M. El-Shikha, E. M. Barnes, T. R. Clarke, D. J. Hunsaker, J. A. Haberland, P. Pinter Jr., P. M. Waller and T. L. Thompson, Remote sensing of cotton

nitrogen status using the canopy chlorophyll content index (CCCI), *Trans. ASABE*, vol. 51, no. 1, pp. 73-82, 2008.
39. G. Fitzgerald, D. Rodriguez and G. O'Leary, Measuring and predicting canopy nitrogen nutrition in wheat using a spectral index—The canopy chlorophyll content index (CCCI), *Field Crops Res.*, vol. 116, no. 3, pp. 318-324, 2010.
40. P. Pipatsitee, R. Tisarum, K. Taota, T. Samphumphuang, A. Eiumnoh, H. P. Singh and S. Cha-Um, Effectiveness of vegetation indices and UAV-multispectral imageries in assessing the response of hybrid maize (Zea mays L.) to water deficit stress under field environment, *Environ. Monit. Assess.*, vol. 195, no. 1, pp. 128, 2023.
41. S. Candiago, F. Remondino, M. De Giglio, M. Dubbini and M. Gattelli, Evaluating multispectral images and vegetation indices for precision farming applications from UAV images, *Remote Sens.*, vol. 7, no. 4, pp. 4026-4047, 2015.
42. R. Latif, A. Saddik and A. Elouardi, Evaluation of agricultural precision algorithms on UAV images. In *2019 International Conference of Computer Science and Renewable Energies (ICCSRE)*, pp. 1-4. IEEE, 2019.
43. J. Rasmussen, S. Azim, S. K. Boldsen, T. Nitschke, S. M. Jensen, J. Nielsen and S. Christensen, The challenge of reproducing remote sensing data from satellites and unmanned aerial vehicles (UAVs) in the context of management zones and precision agriculture, *Precis. Agric.*, vol. 22, pp. 834-851, 2021.
44. M. Söderström, K. Piikki, M. Stenberg, H. Stadig and J. Martinsson, Producing nitrogen (N) uptake maps in winter wheat by combining proximal crop measurements with Sentinel-2 and DMC satellite images in a decision support system for farmers, *Acta Agric. Scand. Sect B*, vol. 67, no. 7, pp. 637-650, 2017.
45. M. Söderström, H. Stadig, J. Martinsson, K. Piikki and M. Stenberg, CropSAT—a public satellite-based decision support system for variable-rate nitrogen fertilization in Scandinavia. In *Proceedings of the 13th International Conference on Precision Agriculture* (unpaginated, online). Monticello, IL: International Society of Precision Agriculture. 2016.
46. Á. Maresma, M. Ariza, E. Martínez, J. Lloveras and J. A. Martínez-Casasnovas, Analysis of vegetation indices to determine nitrogen application and yield prediction in maize (Zea mays L.) from a standard UAV service, *Remote Sens.*, vol. 8, no. 12, pp. 973, 2016.
47. J. Rasmussen, J. Nielsen, J. Streibig, J. Jensen, K. Pedersen and S. Olsen, Pre-harvest weed mapping of Cirsium arvense in wheat and barley with off-the-shelf UAVs, *Precis. Agric.*, vol. 20, pp. 983-999, 2019.
48. R. Sugiura, S. Tsuda, S. Tamiya, A. Itoh, K. Nishiwaki, N. Murakami, Y. Shibuya, M. Hirafuji and S. Nuske, Field phenotyping system for the assessment of potato late blight resistance using RGB imagery from an unmanned aerial vehicle, *Biosyst. Eng.*, vol. 148, pp. 1-10, 2016.

49. O. H. Y. Lam, M. Dogotari, M. Prüm, H. N. Vithlani, C. Roers, B. Melville, F. Zimmer and R. Becker, An open source workflow for weed mapping in native grassland using unmanned aerial vehicle: Using Rumex obtusifolius as a case study, *Eur. J. Remote Sens.*, vol. 54, no. sup1, pp. 71-88, 2021.
50. K. Subramanian, S. Pazhanivelan, G. Srinivasan, R. Santhi and N. Sathiah, Drones in insect pest management, *Front. Agron.*, vol. 3, pp. 640885, 2021.
51. L. Li, Y. Fan, X. Huang and L. Tian, Real-time UAV weed scout for selective weed control by adaptive robust control and machine learning algorithm. In *2016 ASABE Annual International Meeting*, p. 1, 2016.
52. S. Khan, M. Tufail, M. T. Khan, Z. A. Khan, J. Iqbal and A. Wasim, Real-time recognition of spraying area for UAV sprayers using a deep learning approach, *PLoS One*, vol. 16, no. 4, pp. e0249436, 2021.
53. D. L. Shaner and H. J. Beckie, The future for weed control and technology, *Pest. Manage. Sci.*, vol. 70, no. 9, pp. 1329-1339, 2014.
54. F. López-Granados, Weed detection for site-specific weed management: mapping and real-time approaches, *Weed Res.*, vol. 51, no. 1, pp. 1-11, 2011.
55. A. Bawa, S. Samanta, S. K. Himanshu, J. Singh, J. Kim, T. Zhang, A. Chang, J. Jung, P. DeLaune and J. Bordovsky, A support vector machine and image processing based approach for counting open cotton bolls and estimating lint yield from UAV imagery, *Smart Agric. Technol.*, vol. 3, pp. 100140, 2023.
56. Y. Peng, Y. Li, C. Dai, S. Fang, Y. Gong, X. Wu, R. Zhu and K. Liu, Remote prediction of yield based on LAI estimation in oilseed rape under different planting methods and nitrogen fertilizer applications, *Agric. For. Meteorol.*, vol. 271, pp. 116-125, 2019.
57. I. Campos, L. González-Gómez, J. Villodre, M. Calera, J. Campoy, N. Jiménez, C. Plaza, S. Sánchez-Prieto and A. Calera, Mapping within-field variability in wheat yield and biomass using remote sensing vegetation indices, *Precis. Agric.*, vol. 20, pp. 214-236, 2019.
58. B. Bansod, R. Singh, R. Thakur and G. Singhal, A comparison between satellite-based and drone-based remote sensing technology to achieve sustainable development: A review, *J. Agric. Environ. Int. Dev.*, vol. 111, no. 2, pp. 383-407, 2017.
59. Hassler, S. C., & Baysal-Gurel, F. Unmanned aircraft system (UAS) technology and applications in agriculture. *Agronomy*, 9, 10, 618, 2019.

17

Advanced Use of Drones in Irrigation and Water Management

Swati Sharma* **and Pragya Singh**

Amity Institute of Geoinformatics and Remote Sensing, Amity University, Noida, Uttar Pradesh, India

Abstract

A growing number of problems in irrigation management and agriculture are being solved by drones. This study presents the use of a drone for mapping a command area irrigation project in Devariya Adampur of the Balrampur region of Uttar Pradesh, India. The drone survey was carried out on about 796.657-ha area. The objective of the drone survey was to identify the crops and cost-effective methods for irrigation and alignment for the proposed irrigation canal in the study area. The outcome of the drone survey resulted in an accurate estimation of the area to be irrigated and identifying the best network of canals to be constructed according to gravity flow and accurate identification of crops. This work demonstrates the application of unmanned aerial vehicles (UAV) as an effective remote sensing tool in gathering rapid-resolution spatial data at low cost and time. L-Sections of different earthen gullies were made by using AutoCAD software which is a versatile and useful element in construction. The Earthen gullies were an important parameter to consider in the design and construction of the drainage system in the proposed irrigation canal.

Keywords: Drone survey, UAV, irrigation canal planning, Balrampur region

17.1 Introduction

Around the world, achieving sustainable development and effective water management is a difficult task. Water fulfills home and industrial

**Corresponding author*: swati.9238@gmail.com

Disha Thakur, Sanjay Kumar, Har Amrit Singh Sandhu and Chander Prakash (eds.) Sustainable Development Using Geospatial Techniques, (437–454) © 2024 Scrivener Publishing LLC

requirements, feeds cattle, sustains organic life, and provides safety from hunger. To address challenges faced in agriculture and the irrigation sector, technologies like drones are now frequently used in water management. Drones, also known as unmanned aerial vehicles (UAVs), are remotely piloted or autonomous flying devices that can be used for a variety of purposes. Planned development of a structure with the help of remote sensing technology can reorganize a country's economy [1]. In the planning stage of any project using remote sensing, the spatial components from the base maps [2] are studied for the execution of the project.

When the LiDAR (light detection and ranging) and GPS methods were introduced for land surveying, they faced some level of doubt [3]. Similarly, the UAV has been put under doubt regarding its accuracy and the capabilities of UAV-based products [3, 4]. A topographical survey involves data collection about the boundaries of physical features. In recent years, UAVs have been frequently used because of features like low cost, time effectiveness, and low-altitude remote sensing techniques and that is why steadily it is being applied to large-scale topographic mapping [5, 6]. The data acquired by drone/UAV mapping can be utilized for land and resource management practices. One of the key advantages of using UAVs for topographical surveys is the speed and efficiency with which data can be collected [6, 7]. Additionally, the accuracy of the data collected by drones is often higher than traditional surveying methods [8]. UAVs can capture data with much higher precision [9] and detail than human surveyors, and the use of LiDAR can create extremely accurate terrain models with high resolution. Overall, the use of UAVs for topographical surveys has revolutionized the field of surveying, making the process faster, more accurate, and more cost-effective in the fields of crop monitoring, security, map making, logistics in times of disasters, and wildlife monitoring [8–15]. The main objective of this study was to use UAV-captured data in computer-based tools to facilitate the planning of an irrigation plan.

17.2 Study Area

Devariya Adampur is a small village in which six different small villages fall in the Balrampur District of Uttar Pradesh. The latitude and longitude of the area are 27°5'9.05"N and 82°20'45.60"E and the total study area of the village is about 796.657 ha (Figure 17.1). As it was a vast area, the area was subdivided into different names such as k1, k2, k3, k4, k5… to k20 units. Table 17.1 shows the nomenclature of various subdivided villages from serial numbers 1 to 20.

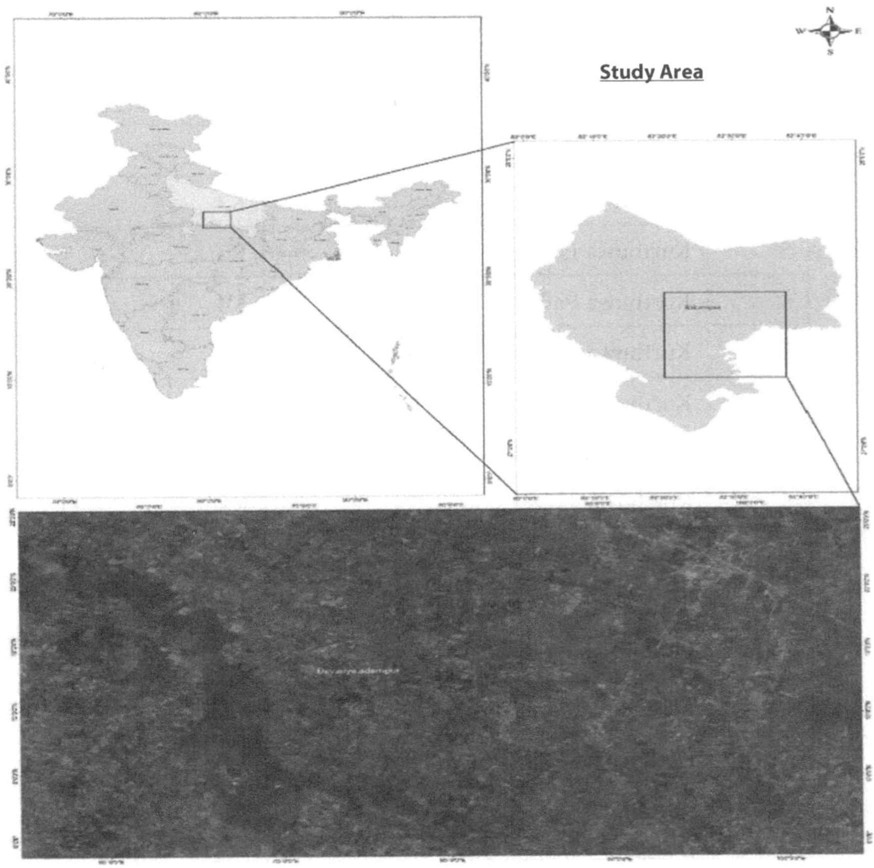

Figure 17.1 Study area.

17.3 Methodology

Figure 17.2 shows the workflow for planning irrigation systems using drones in the study area.

17.3.1 Flight Planning

Important flight planning criteria include drone speed, flying height, intended overlaps for photos, camera specifications, and flight path. The region to be scanned in a single trip is determined by the drone's range and battery life. Flight planning in the irrigation field typically refers to the use

Table 17.1 Names of the villages and their nomenclature in the study area.

S.no.	Village names that come under AOI	Nomenclature
1.	Kurthuwa Part 1	K1
2.	Kurthuwa Part 2	K2
3.	Kurthuwa Part 3	K3
4.	Kurthuwa Part 4	K4
5.	Kurthuwa, Jigni	K5
6.	Kurthuwa	K6
7.	Maddhoghat, Jigni	K7
8.	Kurthuwa	K8
9.	Maddhoghat, Jigni	K9
10.	Maddhoghat	K10
11.	Maddhoghat	K11
12.	Madhoghat, Kartaha	K12
13.	Kartaha	K13
14.	Kartaha	K14
15.	Kurthuwa, Rehmatpur	K15
16.	Kathara, Awajpur	K16
17.	Rehmatpur, Kathara, Awajpur	K17
18.	Rehmatpur	K18
19.	Rehmatpur	K19
20.	Rehmatpur	K20

of drones or UAVs to collect data about crop health, water usage, and other factors that can help farmers optimize their irrigation practices.

The following steps were followed for planning the drone flight:

Determined the area to be surveyed: The first step was to determine the area of the field that needed to be surveyed.

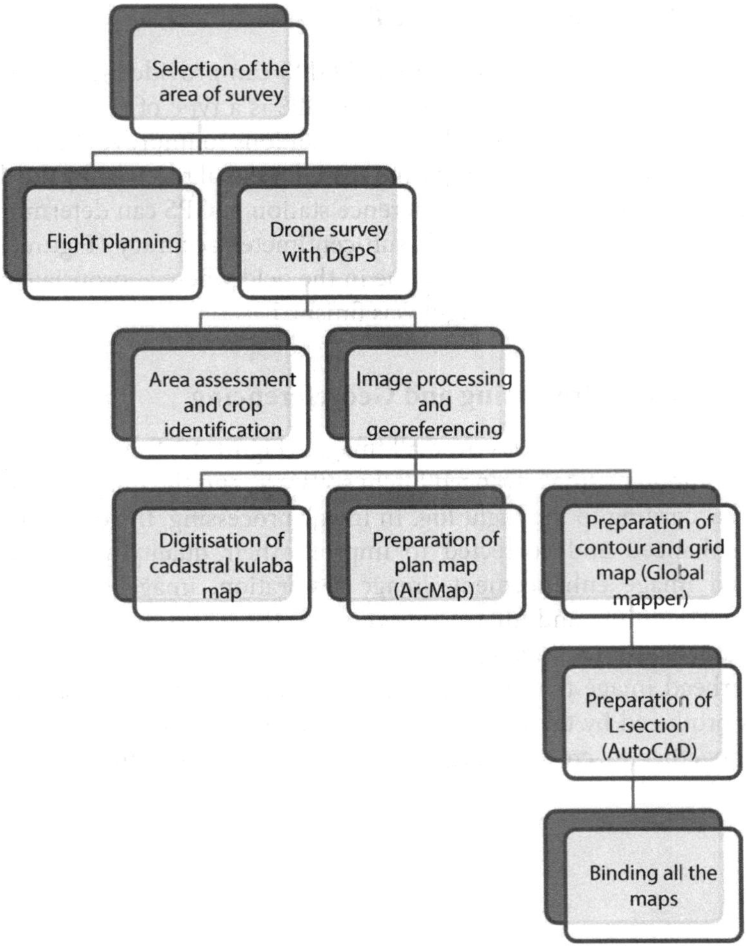

Figure 17.2 Methodology flowchart.

Planned the flight path: In this step, the altitude, speed, and direction of the drone were decided.
Drone Sensor setting: This involved sensor calibration.
Flight Operation: Finally, the flight was conducted to map the entire study area.

The entire flight planning steps were done by using Litchi and global Mapper software and Google Earth (for DJI Phantom Drone).

17.3.2 Drone Survey with DGPS

Drone surveys with DGPS (differential GPS) can provide highly accurate aerial data for a range of applications. DGPS is a type of GPS that uses a network of fixed ground-based reference stations to improve the accuracy of GPS measurements. By comparing the GPS signal received by the drone with the signal received by the reference station, DGPS can determine the exact position of the drone with sub-centimeter accuracy (Figure 17.3). Therefore, drone surveying was done in the field where appropriate photo overlaps were kept, and the flight was finished.

17.3.3 Image Processing and Georeferencing

Once the drone survey was over, all images (Figure 17.4) were transferred to the computer using advanced Litchi software, and the images were geo-tagged according to the flight log. In image processing, the digital images were enhanced and corrected to improve their quality which mainly involved image enhancement, image restoration, image segmentation, image recognition, and image compression. The ortho-mosaic image was prepared (Figure 17.5) using photogrammetry software which represented an overhead image that gave the Earth's surface view for the study area. It was produced by taking a series of overlapping aerial images, stitching them together to create a larger composite image, and then applying a

Figure 17.3 Field measurements with DGPS.

Figure 17.4 Drone images of the study area.

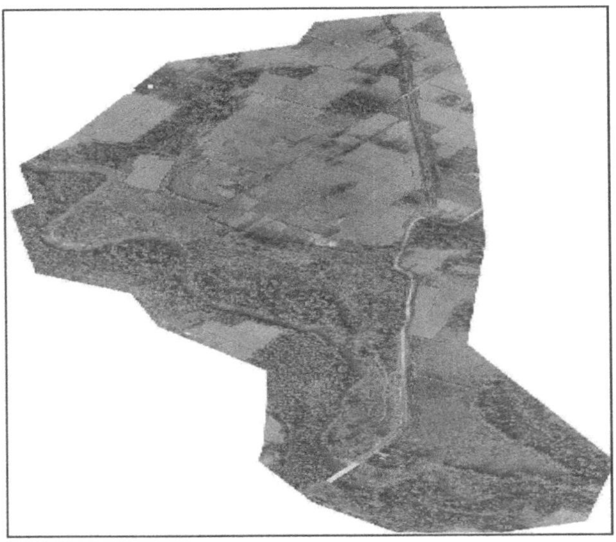

Figure 17.5 Ortho-image of the surveyed area.

process called orthorectification to correct for distortions caused by differences in terrain elevation and camera tilt.

Ortho-mosaic imageries are of great use for agriculture and land management.

17.3.4 Area Assessment and Crop Identification

In the area assessment process, superimposing of the map was done with the study area outline/plan map (Figure 17.6) which was a hand-made map procured from the Bhunaksha site of the state of Uttar Pradesh, India, having all important information about the agriculture fields and to whom the field belongs. Furthermore, using the ortho-mosaic image, the crops were identified. The visual interpretation was done through the drone's images. The image (Figure 17.5) shows the ortho-mosaic image of the k20 area of Balarampur and the field crop.

17.3.5 Digitization of Plan Map

Digitization was done using the ArcMap involving the process of converting analog maps, and diagrams into digital form (Figure 17.7). This was

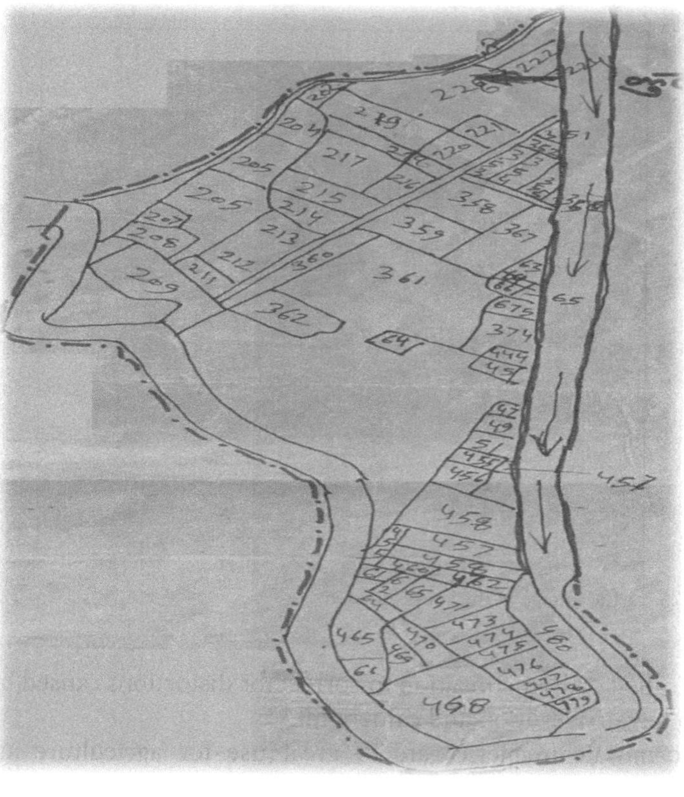

Figure 17.6 Hand-made plan map of the study area.

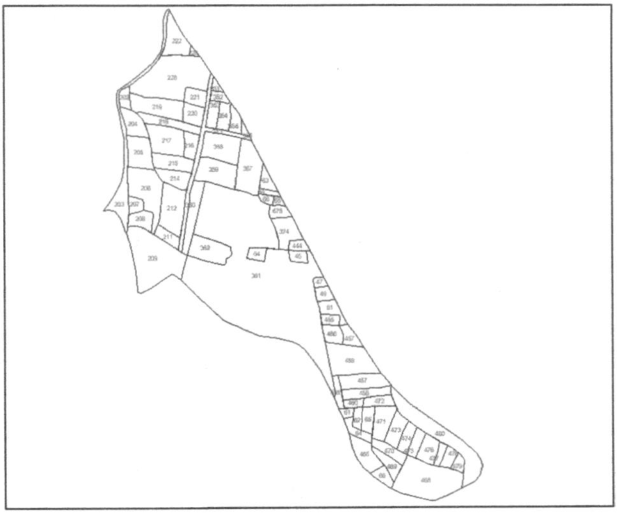

Figure 17.7 Digitized map from ortho-mosaic of drone imageries.

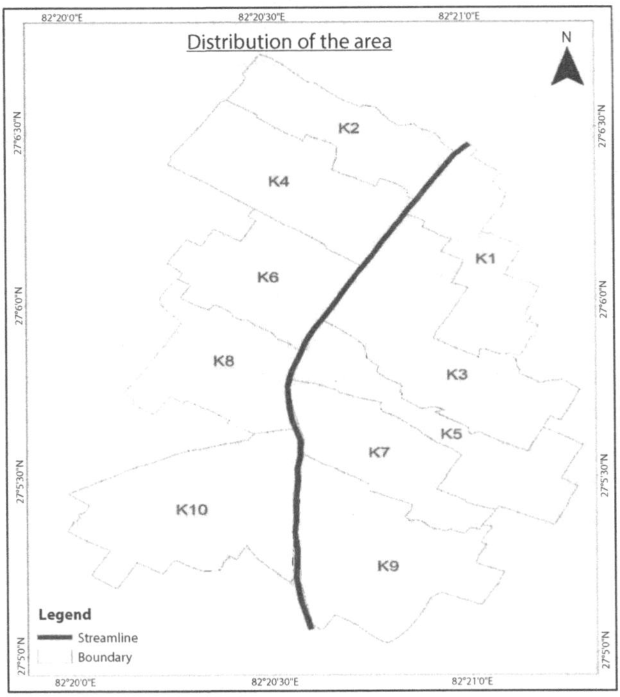

Figure 17.8 Distribution of the area from K1 to K10 with the mainstream in the study area.

done by tracing the features by using image processing techniques to automatically convert the scanned version of the plan map into vector data. Using the digitized version of the study area map, natural features like drainage were inserted in the map to view its preparation of contour and grid map in Global Mapper (Figure 17.8).

17.3.6 Preparation of L-Section in AutoCAD

After using the drone imageries to produce an ortho mosaic of the study area, the plan map of the area for the irrigation project was developed as shown in Figures 17.6, 17.7, 17.8, and 17.11. Now, the target was to make L-sections for planning earthen gullies that could connect the crop fields with the main water canal or reservoir. The L-section refers to the view of a cross-section of a 3D object which is extracted by slicing the object in the vertical plane. In an L-section drawing, the existing ground level is typically represented by a line or series of lines that indicate the elevation of the ground at various points along the section line. To create an existing ground line in an L-section drawing, the following steps were used: draw a line along the L-section, use elevation data to determine the existing ground level, draw a series of horizontal lines at the elevation of the existing ground level, connecting the horizontal lines to the section line to create a continuous line that represents the existing ground level.

AutoCAD commands such as "line", "offset", and "trim" were used to create the existing ground line in the L-section drawing. The concept here was to draw L-sections for planning earthen gullies to connect with the main water stream (canal) of the study area which can continuously supply water to all the crop fields.

F.S.L. (full supply level) refers to the level of water in an irrigation canal or reservoir required to meet the irrigation needs of crops. In areas where irrigation is necessary for crop production, F.S.L. is an important consideration for farmers and water managers. It is used to determine the amount of water required for crop fields throughout the growing season. The F.S.L. mainly depends on how much water is needed in crop fields, what is the crop type, and what is the capacity of the canal or reservoir to supply water in the earthen gullies that connect to the crop fields which is an important factor in maintaining the quality of crops.

17.3.6.1 Slope

In the case of slope in the L-section, the slope refers to the angle of inclination of the surface that forms the vertical leg of the L. This is usually expressed as a ratio or percentage of the rise (vertical distance) to the run (horizontal distance) based on which the slope is determined. The rise is the vertical distance between the two surfaces forming the L, and the run is the horizontal distance along the surface that was used. The slope was then calculated by dividing the rise by the run and expressing it as a ratio or percentage. Understanding the slope in the L-section is important in civil engineering construction, as it helps to determine the stability and drainage of the terrain, and can inform decisions about site preparation, excavation, and grading.

17.4 Result and Discussion

17.4.1 Contour Map

The contour map represents the shape and features of the terrain in a two-dimensional representation. Figure 17.9 shows the contour map prepared for the study area.

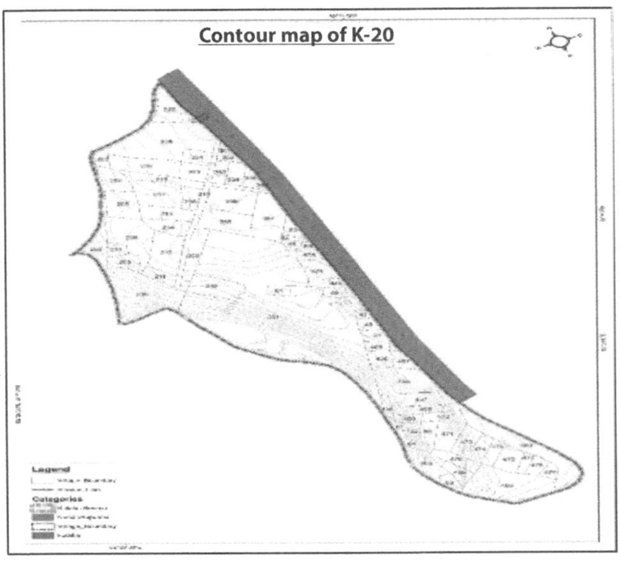

Figure 17.9 Contour map of the study area.

17.4.2 Grid Map

The grid map represented the elevation changes every 30 m or 100 feet in the study area, depending on the scale of the map. It helped to identify the high and low points in the terrain, as well as slopes and ridges (Figure 17.10).

17.4.3 Final Plan Map

In this map structures, village, and field boundaries, unlined earthen gullies and lined earthen gullies along the streamline were made according to the area, its slope, and the crop types. This map reflects where gullies and structures for irrigation purposes can be made according to the need. Diversion boxes were suggested where two gullies (small earthen stream channels) join to divert the water in the appropriate crop field. The tail box was put where unlined and lined earthen gullies were meeting. The green line shows the unlined gullies and the red line shows the lined gullies (Figure 17.11).

17.4.4 L-Sections

L-Sections of different guls were made by using AutoCAD software which is a versatile and useful element in construction that can serve a variety of purposes. In this study, an L-section of an earthen gully refers to

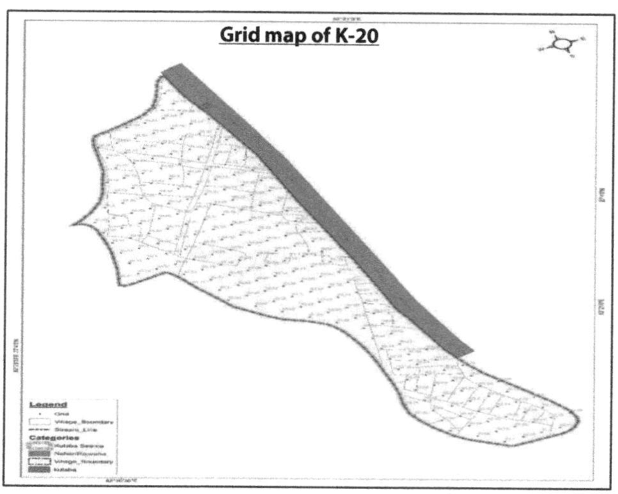

Figure 17.10 Grid map.

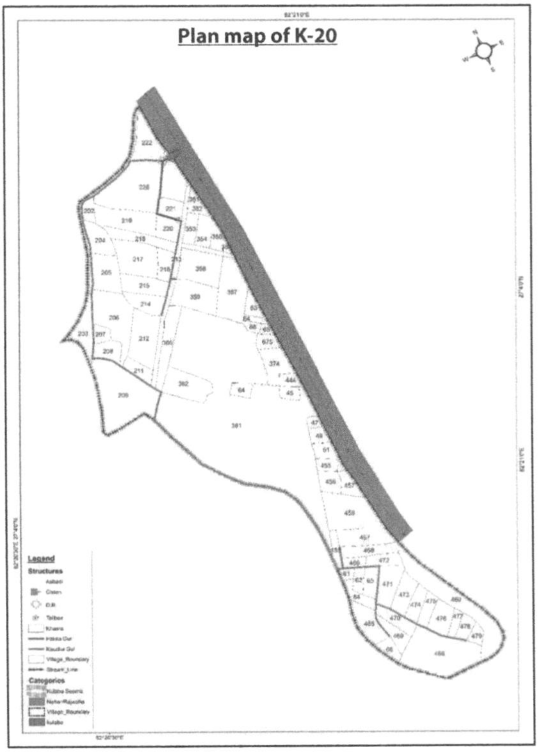

Figure 17.11 Plan map.

a cross-sectional view of a gully that has an L-shape. Earthen gullies are natural or human-made channels that are formed by the erosive action of water or wind on the soil. They can be found in a variety of environments, from agricultural fields to desert environments. The L-section of an earthen gully was the most effective part of this study to plan the water and land management which provided important information about the shape, size, and depth of the gullies to be used, as well as their potential for erosion. In an L-section of an earthen gully, the vertical leg of the L-shape represented the steep side of the gully, while the horizontal leg represented the flat surface. The slope of the vertical leg was an important parameter in designing erosion control measures. The width and depth of the horizontal leg of the L-section were important factors in planning the water storage structures.

Therefore, understanding the L-section can help in making informed decisions about how to manage and save the crop fields from erosion, and

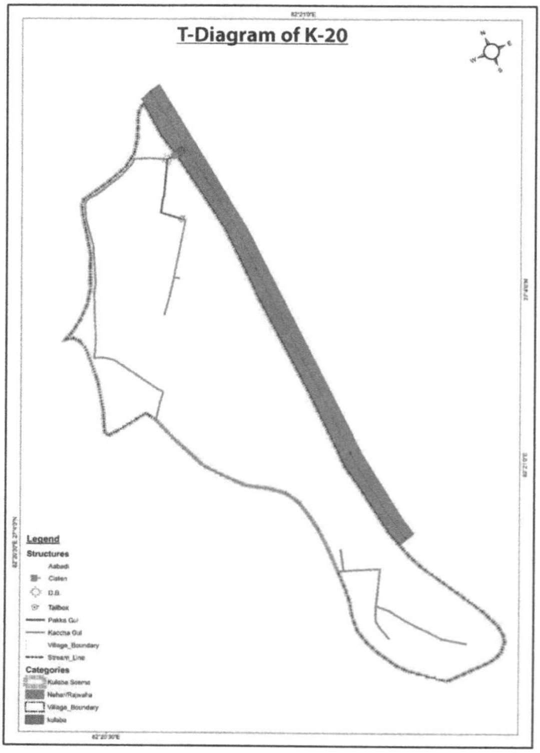

Figure 17.12 Diagram of the irrigation plan map in the study area.

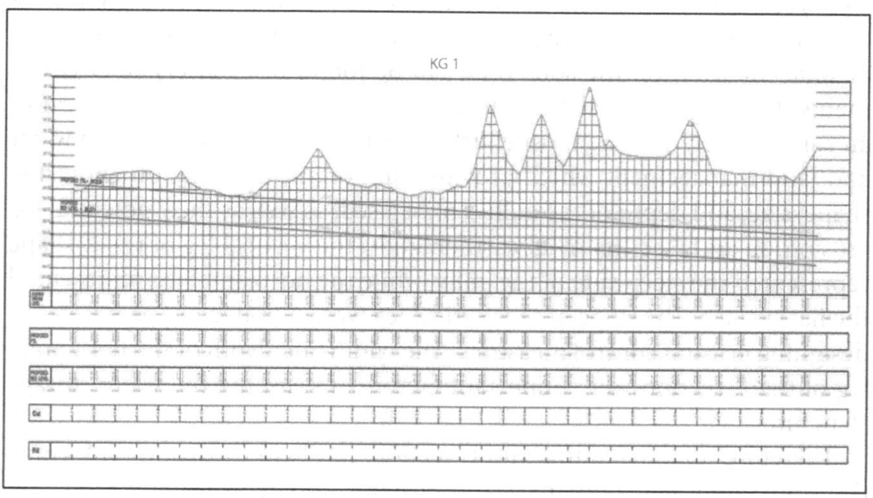

Figure 17.13 L-Section of KG1 (kaccha/unlined gully).

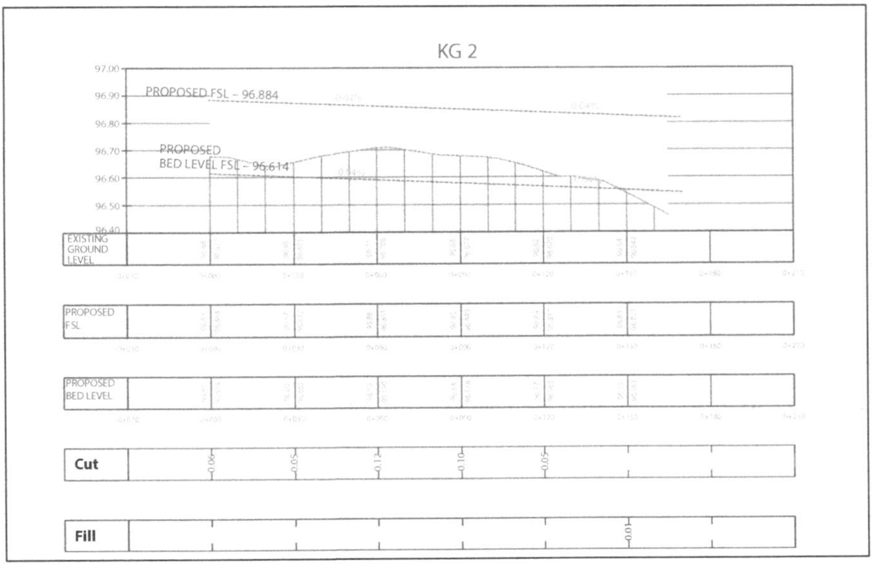

Figure 17.14 L-Section of KG2 (unlined/kaccha gully).

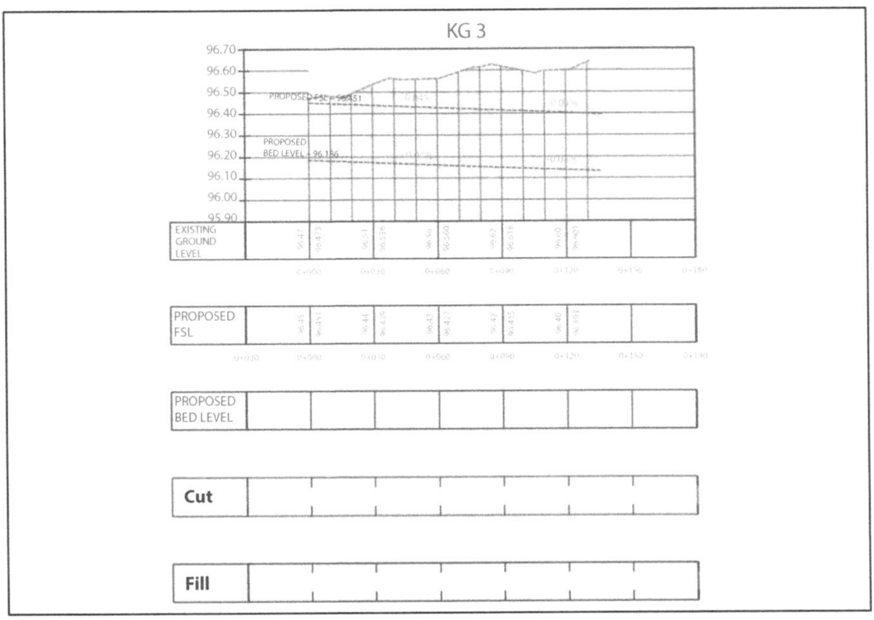

Figure 17.15 L-Section of KG3 (unlined/kaccha gully).

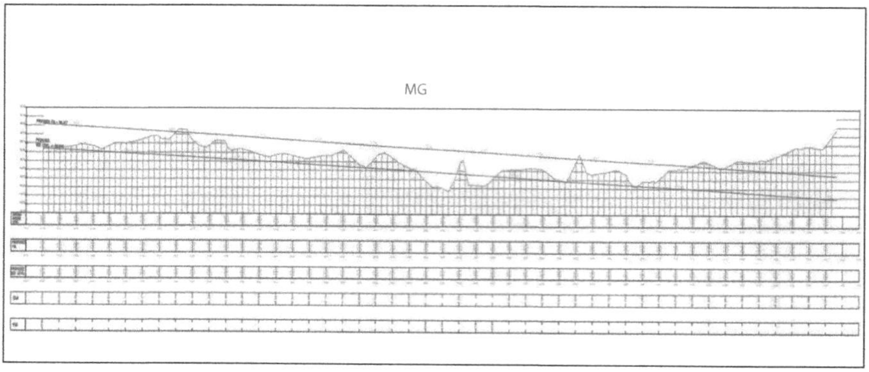

Figure 17.16 L-Section of MG (main gully).

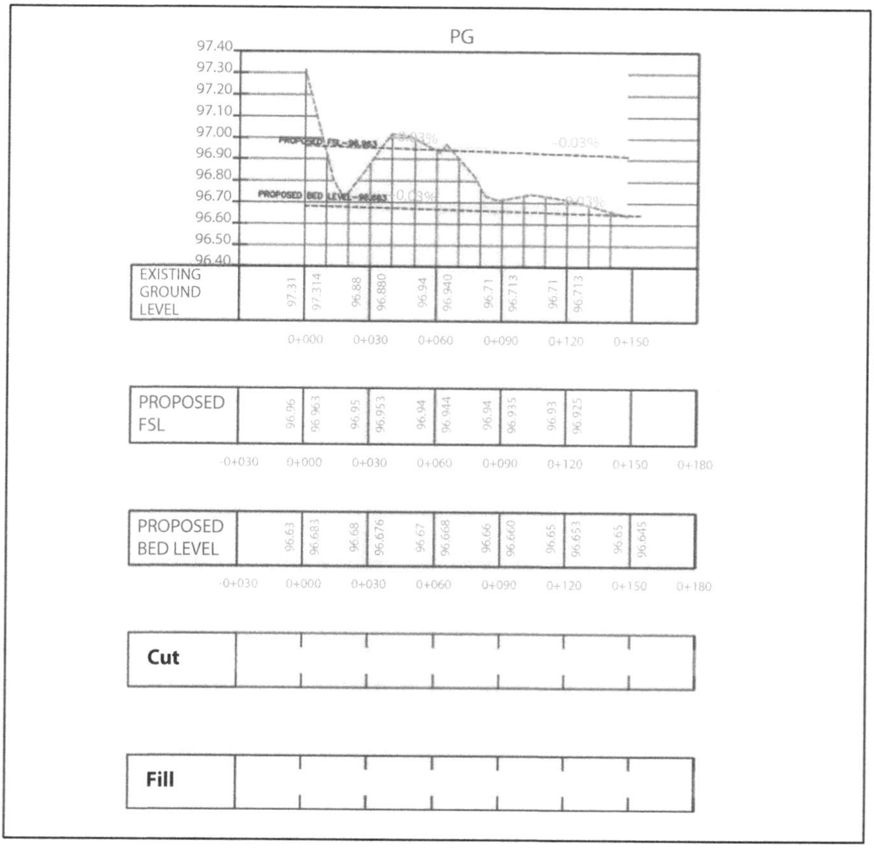

Figure 17.17 L-Section of PG (lined/pakka gully).

other kinds of hazards such as flooding etc. A pre-requisite can help to design the project to avoid hazards by adopting practices like planting trees and installing check dams. The L sections resulting from the study area are shown in Figures 17.13, 17.14, 17.15, 17.16, and 17.17.

17.5 Conclusion

The main objective of this study was to plan an irrigation project with the help of time-efficient drone mapping. Firstly, the drone imageries helped to visualize the overall study area through the ortho-mosaic which was overlapped with the hand-made cadastral map of the study area for checking the correctness. Once the ortho-mosaic was prepared using Litchi software, the plan map with grid and contours was generated using the Global Mapper using the DEM and DTM after which the earthen lined (pukka gullies) and unlined (kaccha gullies) man-made water channels a.k.a gully were planned. A map with lined and unlined Gully was prepared (Figure 17.12) so that the water can reach all the fields and crops in each estimate by connecting them to the main canal or water reservoir in the study area. The slope rise and fall, direction of the earthen lined and unlined gullies was planned based on the L-sections with vertical and horizontal extents where the vertical leg represented the slope of gullies, whereas the horizontal leg represented the horizontal surface suitable for water storage. Overall, the L-sections played an important role in planning the directions, slope rise and fall, and water storage structure planning in the study area.

This study using drone technology was not only time-efficient but also helped to avoid human error in mapping by even accessing the areas which clearly showed the extent of the main water channel in the village of Balrampur, Uttar Pradesh. This study helped to plan the irrigation and land management strategy by using the old cadastral map and by developing a new plan map of the study area which not only will help in irrigation planning but also updated the old records of land and its extent.

Acknowledgments

The authors would like to thank the office of Amity Institute of Geoinformatics and Remote Sensing, Amity University, Noida for all types of research and logistic support during the work.

References

1. Anna Wellenstein., & Wael Zakout. (2017). https://blogs.worldbank.org/sustainablecities/reversing-geospatial-digital-divide-one-stepor-leap-time.
2. Stock, K., & Guesgen, H. (2016). Geospatial reasoning with open data. In *Automating open-source intelligence* (pp. 171-204). Syngress.
3. Fitzpatrick, B.P., 2016. Unmanned Aerial Systems for Surveying and Mapping: Cost Comparison of UAS versus Traditional Methods of Data Acquisition (Doctoral dissertation, University of Southern California).
4. Gao, Z. & Song, Y. & Li, C. & Zeng, F. & Wang, F. (2017). Research on the Application of Rapid Surveying and Mapping for Large Scare Topographic Maps by UAV Aerial Photography System. *ISPRS - Int. Arch. Photogramm. Remote Sens. Spat. Inf. Sci.* XLII-2/W6. 121-125. 10.5194/ isprs-archives-XLII-2-W6-121-2017.
5. Chen Jie, Tong Xiaohua, Liu Xiangfeng, etc. Photogrammetric Processing of Unmanned Aerial Vehicle Imagery in Heihe Middle Reaches [J]. *Geomatics World,* 2014, 21(1) : 63-67.
6. Toutin T., Chenier R. GCP Requirement for High resolution Satellite Mapping [J]. *Int. Arch. Photogramm. Remote Sens. Spat. Inf. Sci.*, 2004, 35(3):836-839.
7. Yang Ailing, Yu. Hongwei, Zheng Canhui. Quality Discussion on the Aerial Images of Light UAV [J]. *Geomatics Spat. Inf. Technol.*, 2011, 34(2) : 185-187.
8. Shakhatreh, H., Sawalmeh, A.H., Al-Fuqaha, A., Dou, Z., Almaita, E., Khalil, I., Othman, N.S., Khreishah, A. and Guizani, M., 2019. Unmanned aerial vehicles (UAVs): A survey on civil applications and key research challenges. *IEEE Access*, 7, pp. 48572-48634.
9. Turner, I.L., Harley, M.D. and Drummond, C.D., 2016. UAVs for coastal surveying. *Coastal Eng.*, 114, pp. 19-24.
10. del Cerro, J., Cruz Ulloa, C., Barrientos, A. and de León Rivas, J., 2021. Unmanned aerial vehicles in agriculture: A survey. *Agronomy*, 11(2), p.203.
11. Barbedo, J.G.A., 2019. A review on the use of unmanned aerial vehicles and imaging sensors for monitoring and assessing plant stresses. *Drones*, 3(2), p. 40.
12. Uysal, M., Toprak, A.S. and Polat, N., 2015. DEM generation with UAV Photogrammetry and accuracy analysis in Sahitler Hill. *Measurement*, 73, pp. 539-543.
13. Fiori, L., Doshi, A., Martinez, E., Orams, M.B. and Bollard-Breen, B., 2017. The use of unmanned aerial systems in marine mammal research. *Remote Sens.*, 9(6), p. 543.
14. Tsouros, D.C., Bibi, S. and Sarigiannidis, P.G., 2019. A review on UAV-based applications for precision agriculture. *Information*, 10(11), p. 349.
15. Torresan, C., Berton, A., Carotenuto, F., Di Gennaro, S.F., Gioli, B., Matese, A., Miglietta, F., Vagnoli, C., Zaldei, A. and Wallace, L., 2017. Forestry applications of UAVs in Europe: A review. *Int. J. Remote Sens.*, 38(8-10), pp. 2427-2447.

Index

Also of Interest

Check out these other related titles from Scrivener Publishing

Sustainable Production Innovations: Bioremediation and Other Biotechnologies, Edited by Alok Patel and Amit Kumar Sharma, ISBN: 9781119791904. Presenting the latest technologies and practices in this ever-changing field, this groundbreaking new volume covers the gambit for providing solutions and practical applications of smart and efficient energy systems.

Sustainable Management of Electronic Waste, Edited by Abhishek Kumar, Pramod Singh Rathore, Ashutosh Kumar Dubey, Arun Lal Srivastav, Vishal Dutt, and T. Ananthkumar, ISBN: 9781394166176. Written and edited by a group of industry professionals, this new volume provides cutting-edge insights into how the sustainability of managing electronic waste can be achieved, for engineers, scientists, and students.

Disaster Management and Environmental Sustainability, Edited by Sanjay Kumar, Suraj Kumar Singh, Shruti Kanga, Gowhar Meraj, Majid Farooq, and M.S. Nathawat, ISBN: 9781394166350. Learn how environmental degradation is driving disaster risk and discover strategies to minimize it through environment-sensitive development in this insightful book.

DIGITAL AGRICULTURAL ECOSYSTEM: Revolutionary Advancements in Agriculture, Edited by Kuldeep Singh and Prasanna Kolar, ISBN: 9781394242931. The book comprehensively explores the dynamic synergy between modern technology and agriculture, showcasing how advancements such as artificial intelligence, data analytics, and smart farming practices are reshaping the landscape to ensure food security in the era of climate change, as well as bridging the gap between cutting-edge research and practical implementation.

THE DIGITAL AGRICULTURAL REVOLUTION: Innovations and Challenges in Agriculture through Technology Disruptions, Edited by Roheet Bhatnagar, Nitin Kumar Tripathi, Nitu Bhatnagar and Chandan Kumar Panda, ISBN: 9781119823339. The book integrates computational intelligence, applied artificial intelligence, and modern agricultural practices and will appeal to scientists, agriculturists, and those in plant and crop science and management.

HANDBOOK OF AGRICULTURAL BIOTECHNOLOGY: Volume I Nanopesticides, Edited by Charles Oluwaseun Adetunji and Julius Kola Oloke, ISBN: 9781119836148. The book provides detailed information on the application of nanopesticides for the management of numerous agricultural pests and pathogens, as well as to foster an innovative pathway toward future sustainable agriculture and food systems.

HANDBOOK OF AGRICULTURAL BIOTECHNOLOGY: Volume II Nanobioherbicides, Edited by Charles Oluwaseun Adetunji, Julius Kola Oloke, ISBN: 9781119836155. The book provides a detailed examination of the application of nanobioherbicides that come from plants including information on the different metabolites derived from numerous plants that could become bioherbicides.

HANDBOOK OF AGRICULTURAL BIOTECHNOLOGY: Volume III Nanofungicides, Edited by Charles Oluwaseun Adetunji and Julius Kola Oloke, ISBN: 9781119836162. This book provides state-of-the-art information about recent advancements in the application of nanobiofungicides for effective management of post-harvest diseases and agricultural fungal diseases, including blights, mildews, molds, and rusts.

HANDBOOK OF AGRICULTURAL BIOTECHNOLOGY: Volume IV Nanoinsecticides, Edited by Charles Oluwaseun Adetunji and Julius Kola Oloke, ISBN: 9781119836179. The book provides detailed information about the application of repellent products that contain plant-based ingredients known as nanobioinsecticides, including the pesticide evaluation scheme guidelines for repellent testing, relevant information about the procedures to evaluate several repellent compounds, the development of new products that offer high repellency, and guidelines for consumer safety.

ECORESTORATION FOR SUSTAINABILITY, Edited by Arnab Banerjee, Manoj Kumar Jhariya, Surendra Singh Bargali, and Debnath Palit, ISBN: 9781119879718. With climate change and land degradation advancing at an alarming rate, the state of the global ecosystem and how it can be restored in our march toward sustainability is incredibly important, and this new, groundbreaking volume tackles these important issues head on.

FUNCTIONAL POLYMERS IN FOOD SCIENCE, Part 1: Food Packaging, Edited by Giuseppe Cirillo, Umile Gianfranco Spizzirri, and Francesca Iemma, ISBN: 9781118594896. Provides an extensive overview of the link between polymer properties and food quality. *NOW AVAILABLE!*

FUNCTIONAL POLYMERS IN FOOD SCIENCE, Part 2: Food Processing, Edited by Giuseppe Cirillo, Umile Gianfranco Spizzirri, and Francesca Iemma, ISBN: 9781118595183. Provides an extensive overview of the link between polymer properties and food quality. *NOW AVAILABLE!*

ADVANCES IN FOOD SCIENCE AND TECHNOLOGY Volume 1, Edited by Visakh, P.M., Sabu Thomas, Laura B. Iturriaga, and Pablo Daniel Ribotta, ISBN: 9781118121023. Written in a systematic and comprehensive manner the book reports recent advances in the development of food science and food technology areas. *NOW AVAILABLE!*

ADVANCES IN FOOD SCIENCE AND NUTRITION Volume 2, Edited by Visakh P.M., Laura B. Iturriaga, and Pablo Daniel Ribotta, ISBN: 9781118137093. This important book comprehensively reviews research on new developments in all areas of food chemistry/science and nutrition. *NOW AVAILABLE!*

Applied Water Science Volume 1: Fundamentals and Applications, Edited by Inamuddin, Mohd Imran Ahamed, Rajender Boddula and Tauseef Ahmad Rangreez, ISBN 9781119724766. Edited by one of the most well-respected and prolific engineers in the world and his team, this is the first volume in a two-volume set that is the most thorough, up-to-date, and comprehensive volume on applied water science available today. *COMING IN SUMMER 2021*

Printed in the USA/Agawam, MA
September 16, 2026

910661.153